Contents

Preface

This residential handbook is geared for use by anyone interested in residential design, construction, mechanical systems, and estimating.

Design (Chapters 1 through 6) covers the fundamentals of residential design, including site selection, basic plan shapes, room sizes and layouts, and other considerations.

Construction Assemblies and Materials (Chapters 7 through 17) presents the basic construction techniques and the materials used in residential construction, including the procedure for sizing wood framing members.

Mechanical Systems (Chapters 18 through 20) discusses plumbing, heating, air conditioning, and electrical systems. Topics include terminology, materials, fittings, and the types of systems used.

Estimating (Chapters 21 through 33) covers the basics required to estimate the amounts of materials required to build the building.

This material should be supplemented with exposure to the construction industry by visiting residences under construction, reading current magazines, and establishing a personal library on topics of interest. Many groups (such as the American Plywood Association) offer a variety of technical material at low cost. Most important in any business, or area of interest, is to watch and listen. There is no single ''correct'' or ''best'' way to do most things. Pay attention to what others are doing and why. Keep your mind open to ideas that are new to you, and to the approaches others are using. A closed mind is like a closed trap, nothing can get in and nothing can get out.

Part I

Residential Design

Chapter 1

Residential Design

1.1
PRELIMINARY CONSIDERATIONS

There is probably no more interesting architectural design than residential. It is both interesting and frustrating. The number of factors to be considered at one time seems staggering when the land, building orientation, building codes, zoning regulations, architectural committees, and restrictive covenants must be checked while the designer determines the approximate number of rooms, size of rooms, and style of house that the clients desire—and all in trying to keep within a budget. The designer should spend enough time with the clients to understand how they live, entertain, work, and sleep; and their hobbies, likes, and dislikes. Every client is different, and when dealing with husbands and wives, the differences may affect the design and conflicts may arise. It is important to know the clients well or it will be impossible to design for *them*. Note that a designer designs for the *clients,* and on no other basis. Conflicts between clients are usually worked out between them, not with the designer. The designer should discuss with them the areas in question only in relation to the advantages and disadvantages of each idea.

The various aspects of the design stage are included in this section of the book. All these items need to be considered early in the design. Nothing can be left until later, or it could turn out that not enough space was allowed for the bathroom or that the furniture will not fit in the bedroom.

1.2
CODES AND STANDARDS

Generally all housing being designed must meet the building code, zoning, and any other code or property standards in force in a given locale. These requirements vary considerably in different areas, and the designer must become familiar with all applicable codes and follow the most stringent requirements. Also to be considered when designing a house is how it is to be financed. If government-backed financing is involved or contemplated, it is necessary to have available the standards for such financing, to make certain that they are met.

Standards such as Housing and Urban Development's Minimum Property Standards for One- and Two-Family Dwellings generally include requirements regarding site

3

design, building design, materials, construction methods, and testing. All applicable codes and standards should be a part of the designer's library.

1.3
COST

Cost is a factor in almost every design and should be considered fully and carefully by the clients and designer at the outset. The designer must determine exactly what the clients' budget will be and what is included. Basically, the budget should be broken down into figures for land, house (with built-in appliances), furnishings, and landscaping. An amount must be figured for each item, and each may vary considerably, depending on the clients. Some clients may already own the land, others will move the furnishings they have and use them, and some will use some of their furnishings and purchase some new ones. A certain amount of landscaping will be done on all new homes, even if only the planting of grass. Landscaping costs may vary from a few hundred to several thousand dollars, depending on the complexity of the landscape design. Appliances that will be included inside the home must also be carefully discussed and noted. It is important that even the brand names preferred be noted when there is a preference. Clients who want the biggest and best of appliances generally will want the finest of everything in the house. This type of approach by the clients generally means that the cost of the house will be 10 to 25 percent higher than that for clients who select middle-range appliances and built-ins.

The designer needs to know the budget in order to help the clients stay within it. Through experience and frequent contact with building contractors and current prices of materials and labor, the designer should be able to estimate the approximate cost on a square-foot basis. By taking the square footage of the house times the approximate cost per square foot, the designer will know if it is close to the budget. Any extra items that the clients will want must also be considered.

It is part of the designer's job to let the clients know if the budget is being exceeded. Ideally, this should first be done face to face or on the telephone so that it may be thoroughly discussed; then it should be put into a letter and sent to the clients. In this manner the clients will not feel later that they were not kept completely up to date. For all these reasons it is important that the designer be given accurate figures by the clients. Many clients are reluctant to give the full amount that they are prepared to spend because they feel that the designer will not make attempts to come under that amount. Then, if actual prices are over the budget, the clients may not be able to afford the house as designed. The designer must be aware of this feeling and be ready to assure the client that he can and will work within the budget. Clients can often be encouraged to give an ideal budget figure *and* an amount they refuse to exceed. In this case early in the design stage the designer can tell them whether they are within their budget. As an example, in a recent design the clients' house budget allowed for a maximum house size of 2,000 square feet, but the number and arrangements of rooms that the client wanted would not even fit into an area of 2,200 square feet. The clients were informed immediately and had to make the basic decision to either increase the budget by about 10 percent or to revise the design and space requirements. There is no sense wishing and hoping. Make the client aware of any possible cost increases. If the designer is firm and honest with clients, they will understand and respect him. Clients should never be kept in doubt as to costs because rarely will they be pleasantly surprised when they do hear. Whenever there is a question as to whether a budget can be met, the clients should be fully and completely informed.

1.4
SQUARE FOOTAGE

Whenever two people discuss the size of a building in terms of square footage, there will be differences in what is included in the figure. The square footage of the house is considered to be that portion of the house that is heated, or enclosed by exterior walls. This typically does not include garages, screened porches, exterior decks, and balconies, plus any unheated spaces such as storage areas or basements. The square footage is taken as the outside dimensions of the building. For example, a house with total overall dimensions of 60 by 24 feet, no garage, has 1,440 square feet. Actual usable space, this gross square footage minus the area that exterior and interior walls take up, amounts to considerably less than 1,440 square feet. When zoning or deed restrictions have minimum building requirements, they should be checked to determine if they specify how the square footage must be calculated and what may be included in the figures. Where there is no definition it may be reasonably assumed that heated square footage, the gross area, is what is meant.

Any garages, porches, decks, and basements should be listed separately (Figure 1-1), and in calculating budget cost varying

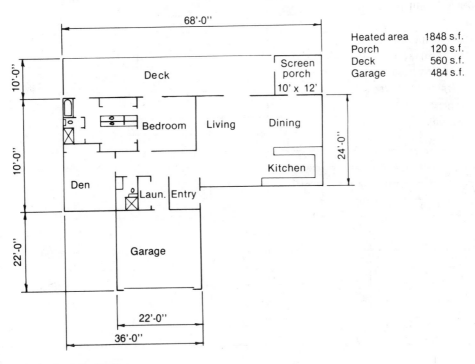

Figure 1-1. Square footage.

weights will be given to each item. Each designer must use his own cost values based on experience. When working up budget figures some designers will take garages, porch, and basement areas and calculate that these will cost about one third to one half of the heated square-footage cost. Decks and overhangs may be figured from one sixth to one fourth of the house cost. Since there are so many variables, a preferred method is to calculate a budget figure for the heated area and then add to that the amount the garage, porches, decks, and fireplaces will cost. Each figure should be listed separately so that it may be easily used for reference, revised as required, or deleted if requested. The price for each item will vary according to size, shape, and materials used to construct it. A typical form used for budget estimates is shown in Figure 1-2.

PRELIMINARY BUDGET						
ITEM	S F	COST/S.F.	LUMP SUM	MATERIALS	LABOR	SUBTOTALS
LAND						
HEATED SPACE						
BASEMENT						
PORCHES						
DECKS						
KITCHEN APPLIANCES						
LAUNDRY APPLIANCES						
FIREPLACE						
BRICK EXTERIOR						
AIR CONDITIONING						
LANDSCAPING						
PAVED DRIVEWAY						
WELL						
SEPTIC TANK						
SURVEY						
BUILT-IN CABINETS						
GARAGE						
ARCHITECTS FEE						
					TOTAL	

Figure 1-2. Budget form.

Figure 1–3a shows a home with a recessed portion along the front. Note that by figuring the house square footage of the plan, there is a total of 2,350 square feet. Yet it will cost virtually the same as the plan in Figure 1–3b, which is the same house without the recess.

Note that the square footage totals 2,775. If the house will cost $95,000 to build, it will cost $95,000 ÷ 2,611, or $36.38 per square foot, using the recessed plan. By comparison, the rectangular plan would cost $95,000 ÷ 2,775 or $34.23 per square foot.

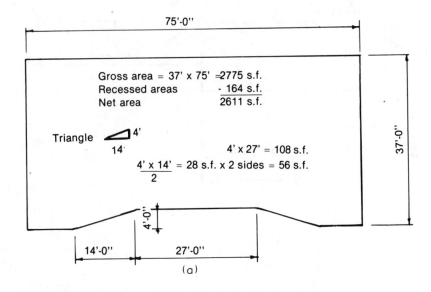

(a)

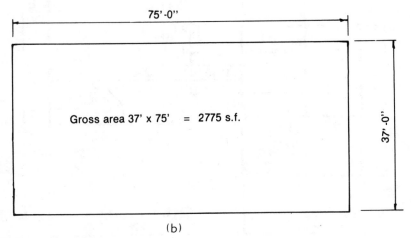

(b)

Figure 1-3. Building areas.

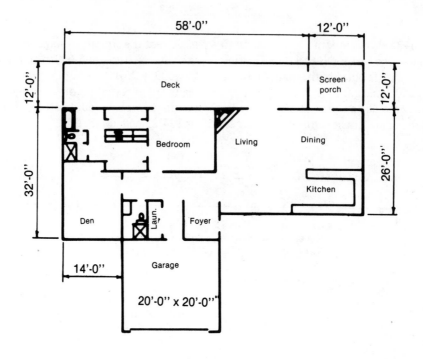

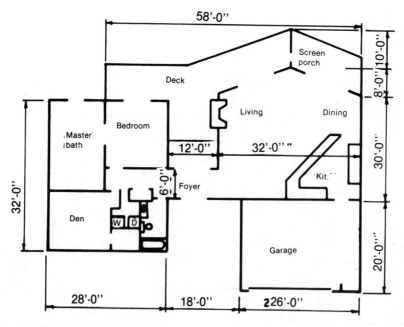

Figure 1-4. Calculation of square footage.

1.5
ARCHITECTURAL COMMITTEES

Many communities and neighborhoods have committees that have been set up to review and approve the size, style, materials, and all exterior features of houses and lots. These committees are set up to "protect" the existing properties and property values of the area by regulating what can and cannot be built. Most frequently, it is the more exclusive neighborhoods that have relied on this form of control, but an increasing number of vacation areas and middle-income neighborhoods are being set up with such committees.

Committee requirements vary considerably, but most require floor plans and exterior elevations, and many require an exterior perspective and landscape plans. Similarly, the guidelines for approval vary considerably; some groups meet weekly, some monthly, and some state simply that they will meet within a certain period of time after they receive a set of drawings.

The meeting schedule and drawing requirements of all committees need to be known by the designer. Since working with such committees is a part of normal procedures in many areas, the designer should know who is on the various committees, how the drawings are presented for consideration, and in general what the committee wants to see as it reviews the drawings. The general approach by the designer should be one of complete cooperation, because it is in the interest of the client to obtain fast approval.

Generally, the committees contain, at best, one person who is familiar with drawings and how to read them. This is usually a local architect. The other members, typically, have no actual construction or design experience; they are simply members of the community. The drawings given to them should be as complete as possible. The elevations should show materials as clearly as possible, because many committee members have trouble "picturing in their minds" things such as vertical siding and shingles; and when they are not shown, it gives these members reason to ask for more information before approving the plans.

When possible, it is good to become friendly with the people on such committees so that any questions which arise will more likely be raised in a phone call than in a letter in which approval is delayed. Unfortunately, some committees spend a great deal of their time involved in "local politics," and they mix this in lightly with approval of drawings.

Whenever a designer feels that a committee may object to a design, it may be wise to discuss it ahead of time with a committee member and get a reaction to it. This will give the designer an indication ahead of time of the committee's feelings, and then revisions to the design can possibly be made.

Chapter 2

Site Selection

2.1
SITE SELECTION

It is impossible to design a home for someone without first selecting a site or, if a site has already been purchased, without a thorough review of the site. The vast majority of people who want a home designed have already purchased the land. The rest have narrowed the site selection to a few choices and will then ask for advice. It is unfortunate that more people do not ask for site-selection advice from anyone other than the real estate agent who is selling the land.

Too often, people purchase a lot that will not be suitable for the type or style of house they would like to live in. Each site has certain characteristics which will influence the type of house that will look best on it and be the most economical.

Before even beginning to look at land, certain things must be found out about the clients: how they live, what preferences in house styles they have, what their total budget for land and house is, and how the money should be divided between land and house. Other information, such as family and individual interests, whether the adults work or are retired, how much land they would like, and the importance of closeness to shopping, schools, and places of worship, is required. It is important to understand the clients or it will be impossible to assist them.

The designer who undertakes to assist in land selection must be aware of all the community areas available, and the facilities, lot sizes, restrictions, price ranges, and types of people in the community. The types of communities should be discussed with the clients to determine what they seem to prefer.

The type of information the designer will need is found in part by observation and by studying the community, and in part through the people who sell real estate. Real estate agents can keep a designer informed of the sites that are available and the prices. They will provide updating on available sites and changing prices.

The various factors to be considered in site selection are discussed in this chapter. Some things are obvious to the eye and some require research.

2.2
ZONING

A zoning board is a department of the local authorities that has jurisdiction in a given locale. This department makes zoning regulations, which are local laws that regulate the use of land and buildings in a given area. For example, the zoning may allow only one use, such as single-family homes or agriculture, or the zoning may allow multiple use, such as agriculture, shopping centers, and single-family homes all in one area. So this is the first item to be checked by the designer, and the clients' needs must be kept in mind. The clients may intend to have an office in the home, but it may not be allowed in an area zoned for single-family residences.

Zoning affects any proposed construction in many ways, because it usually contains requirements pertaining to the building setback from streets and property lines (limiting how close to these lines a building can be built), amount of parking required, types of construction, exterior finishes allowed, minimum building size, control of the appearance of the building by requiring the use of particular materials, requiring exterior design type, landscaping, and many other items.

The effects of the zoning on the proposed project should be considered before the property is purchased. Zoning regulations can be amended, but it is important to determine whether they are amended by a zoning board or by a vote of the residents in the zone affected. Their reaction to any amendments can be determined by reviewing their past actions. If they tend to turn down most requests for amendments, it is a sign they prefer the status quo and will probably turn down future amendments. As a further check the designer should find out who seems to be actively involved on the side that consistently wins and get their opinion on proposed zoning amendments.

Property should never be purchased based on an assumption that the zoning will be amended. If it is not amended, the owner may possess a piece of property that may be useless to him. Instead, purchase property contin-

gent on the zoning amendment being passed. Then if it is not passed, the contract is no longer in effect. In this manner the seller of the property and the real estate agent will do whatever they can to assist in getting the zoning amended.

Minimum building size should be checked carefully. It may be that the minimum is larger than the client wants or can afford, and this would eliminate a particular area from consideration. Many good sites remain unsold because the minimum building size required is too high. At the same time, if the minimum building size in the area is 1,200 square feet, it is not likely that clients who want a 2,000-square-foot house will build in that area. They would have an expensive house in an area of less-expensive homes, and the house might be difficult for them to sell. It is not easy to get zoning amendments that reduce building size.

Setbacks from streets and property lines are equally important. Small setback requirements may mean that houses will be placed very close together. A sideyard setback (Figure 2–1) of less than 15 feet is absolutely unacceptable for suburban lots 100 feet wide or more, yet in the city, where lots are often 40 to 80 feet wide, it may be more than could be allowed. Setbacks should be clearly understood, because it is difficult to get zoning amendments on setbacks.

Although zoning often imposes restrictions that seem unreasonable, a buyer should be twice as wary of locales with no zoning. Unzoned land means that anything may be done with it, and after the house is built, a steel company or oil refinery might decide that the surrounding land is ideal for a new plant. Zoning restricts, but it also protects. When it is agreed that it is unreasonably restrictive, it can be amended. Unzoned land should not be purchased for residential use.

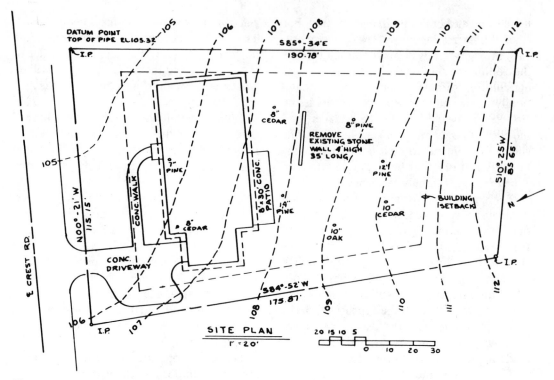

Figure 2-1. Building setbacks.

2.3
CODES

The requirements of the building codes in the locale should be checked. Codes generally cover the general construction, and plumbing, heating, and electrical construction. In addition, they contain requirements concerning types of construction allowable, fire resistance, ventilation, and light requirements. Codes that are highly restrictive may greatly increase the cost of the building. A talk with the building inspector will determine what regulations he is most critical in checking, costs of permits, what inspections are required, and any costs. A building code offers the protection of adequate building-con-

struction practices. Realistically, however, it is the inspector and his ability and interest in enforcing the code that are most important.

2.4
NEIGHBORHOOD

Make a careful study of the neighborhood by driving through all the surrounding streets. Keep an eye out for things such as railroads, dumps, swamps, busy intersections and roads, dogs, play areas, and motorcycle trails. Is the area well kept, with houses and yards well maintained, or is it on the downgrade? Look carefully, because these things will affect the pleasure that will be found in the new house. Nothing is more frustrating than to move into an area where you are chased by dogs as you walk down the street (a leash law is usually of little help) or are almost run over by a motorcycle.

The style of the other houses in the neighborhood should also be checked. If all the homes are contemporary, the clients should be planning something that will blend with the neighborhood and not be completely different, such as a Williamsburg colonial. This does not mean that the house should be identical to the others, only that it should not be totally different. Some of the nicest neighborhoods are a blend of all styles of homes, all mixed in on reasonably large lots with room for landscaping.

2.5
UTILITIES

Utilities (water, sewer, gas, electricity, and telephone) in a prospective area should be carefully checked. Bringing in any utility can be expensive. If public water is not available, it will be necessary to have a well put in with a pump and tank, and water quality will have to be continually checked. The cost of tapping public water mains should also be checked. A well should be put in deep enough to provide an adequate supply of water even during hot, dry spells when the water table drops. If a sewer is not available, a private septic system must be provided, and this requires soil tests to make certain that the soil can absorb the waste. Electricity can be brought to almost any site, but if there is none now, a check should be made on possible costs to the owner and how long it would take to get it in. A similar check should be made on the telephone service and natural gas.

2.6
DEED RESTRICTIONS

The deed must be examined for restrictive covenants (Figure 2-2), easements, and any other building restrictions. These covenants read much like zoning regulations. It may be easier or more difficult to have them amended, depending on what process has been set up for amendments. In one development it may be only one person who determines whether a deed restriction may be waived. In other areas, the restrictions were put in the deeds long ago, and no one is left to amend (or enforce) them. Any restrictions that are illegal, such as restrictions to race or national origin, cannot be enforced, even though they are in the deed.

Typically, the restrictions cover the same items as might be found in zoning. Restrictions on land use, type and style of buildings, minimum and maximum building size, setbacks, materials, and rights-of-way may all be included and must be checked carefully. Deeds often make reference to restrictions listed elsewhere; the designer should take the time to find and read them.

Where zoning regulations and deed restrictions set different limits on something, the more stringent restriction is followed. For example, if the zoning calls for a minimum building size of 1,500 square feet and the deed restriction calls for 1,800 square feet, 1,800 square feet will be used. The same is true of all other restrictions and regulations.

Easements give the utility companies permission to make use of a portion of the land for electric, water, and gas lines, and to come on the land at any time to inspect, service, or repair them. The size and locations of the easements should be carefully noted to determine if they will affect any proposed plans for the property.

BUILDING USE AND ARCHITECTURAL CONTROL

a. No building, septic tank, seepage pit, well, sign, dock, pier, incinerator, trash or garbage receptacle, fence, wall or other structure shall be commenced, erected or maintained upon the Properties, nor shall any exterior addition to or change or alteration therein be made until plans and specifications showing the nature, kind, shape, heights, materials and location of the same shall have been submitted to and approved in writing as to harmony of external design and location in relation to the surrounding structures and topography by the Developer or an architectural committee composed of three (3) or more Owners appointed by the Developer. In the event the developer or its designated committee fail to approve or disapprove such design or location within sixty (60) days after said plans and specifications have been submitted to it, said plans and specifications shall be deemed approved.

b. The Properties shall be used for residential purposes only and for no other purpose whatsoever and no business, commercial or manufacturing enterprise, shall be conducted on said premises. No building shall be erected, altered, placed or permitted to remain on any lot other than one single family dwelling. One private garage or boathouse, or combination garage and boathouse for family automobiles and boats.

c. The outside finishing of all buildings must be completed within one (1) year after construction has started, and no asphalt shingles, imitation brick, building paper, insulation board or sheathing or similar non-exterior materials shall be used for the exterior finish of any such building; exterior finish shall be wood or asbestos shingles or siding, logs, brick, stone or concrete.

d. Every dwelling house shall have not less than 750 square feet of enclosed living space exclusive of porches, breezeways, carports, patios, pool areas, garages and other accessory uses.

e. No trailer, mobile home or similar type structure, basement, tent, shack, garage, barn or other out-building shall at any time be used as a residence, temporarily or permanently, nor shall any structure of a temporary character or any building in the process of construction, be used as a residence. No signs of any nature not previously specifically approved in writing by the Developer shall be permitted on The Properties.

f. No animals or birds of any kind shall be raised, bred or kept on any lot, except that not more than two dogs and two cats may be kept provided that they are not kept, bred or maintained for any commercial purposes. Noxious or poisonous weeds shall not be permitted to grow on The Properties.

Figure 2-2. Typical restrictive covenants.

2.7
SITE TOPOGRAPHY

Sites of interest should be checked carefully. Although a flat lot is easiest to build on, it is not ideal for someone who wants a split-level home. Sloping lots can be economical for certain types of homes, especially if the slope is sufficient to open up part of the lower level (Figure 2–3) to the exterior, but it is often a little more difficult to design and construct the house. The sloping lot may allow for decks off

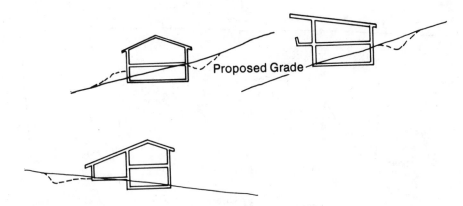

Proposed Grade

Figure 2-3. Sloping lots.

the back with the most beautiful views (provided that there is something to look at), yet expensive retaining walls may be required around the house. Each site will have its advantages and disadvantages.

A sketch of the site should be made for later site comparison. The sketch (Figure 2-4) should show any large rocks, hills, valleys, generally level areas, slope of land, large trees, groups of trees, existing walls or buildings, north, and what the view is in each direction. Also, the locations of surrounding buildings should be noted. All advantages and disadvantages of each lot should be noted. If this is done for each site, it will be much easier later to discuss the various sites.

The soil conditions in each locale must also be known, because they may affect the style and construction of the house to be built. If the water table is 4 feet below the surface, it will not be economical to put in a full basement below grade, but a raised ranch (Figure 2-5) or split-level house might be built. Sites that have been recently filled with soil may settle under the weight of a house, causing cracking of the walls.

Once the site has been selected, the buyer should be certain to make only a minimum down payment and be sure to get a sales agree-

ment. This is the time to make any purchase contingencies, such as zoning amendments, a part of the contract. Even before the down payment is made, a lawyer should be hired. Not all lawyers do this type of work, and not all are good at it. The lawyer must be familiar with all the legal restrictions and be able to offer advice on the best way to handle the clients' best interests. The designer should keep in touch with, and recommend, one of several lawyers who do exceptional work in this field. A good lawyer saves a client many times the cost of paying him (unfortunately, a poor lawyer will charge the client a fee and still not protect him). All this to say: select your lawyer carefully.

After the down payment is made, the lawyer will make a search of the title to the property and determine that the property does belong to the seller. The lawyer will check to determine if anyone has legal claims (liens) against the property, and, if so, will advise the client as to how to proceed.

A client may also purchase title insurance, which guarantees that the title to the property is clear. Then the insurance company will pay any liens against the land if such claims were overlooked. A good lawyer makes this unnecessary.

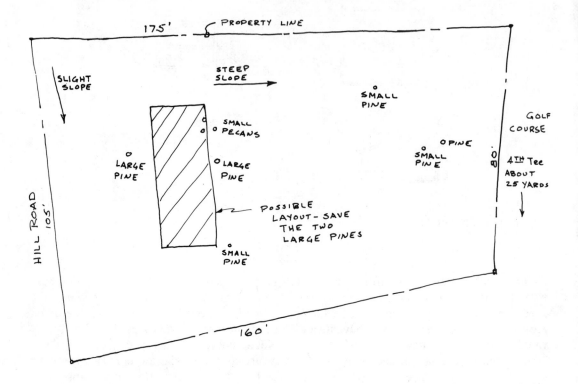

Figure 2-4. Site sketches.

_____ 2.8 _____
LOT SIZE

The size of lot that the client feels to be adequate should be determined. No particular size is average. Lot sizes may range from 40 by 120 feet to several acres in size. The lot size will depend on the clients' needs and interests, the number of people in the family, and what the family members do. An acre of land may be quickly filled with a play area consisting of gym set, playhouse, digging area for the children, vegetable and flower garden, out-

side entertainment area, and an area for trees and shrubs. The use of the lot, as well as the lot size, will determine how much yardwork will be involved. It is possible, by using pine needles, pine bark, and stone, to keep exterior work to a minimum, with perhaps just a strip of grass around the house. Entire yards do not need grass to be attractive.

Lots are usually considered large if they are 100 to 150 feet wide and 130 to 180 feet deep. A ranch house will stretch out and need wide lots, and many ranches are 100 feet long, including garages. Many colonial designs also

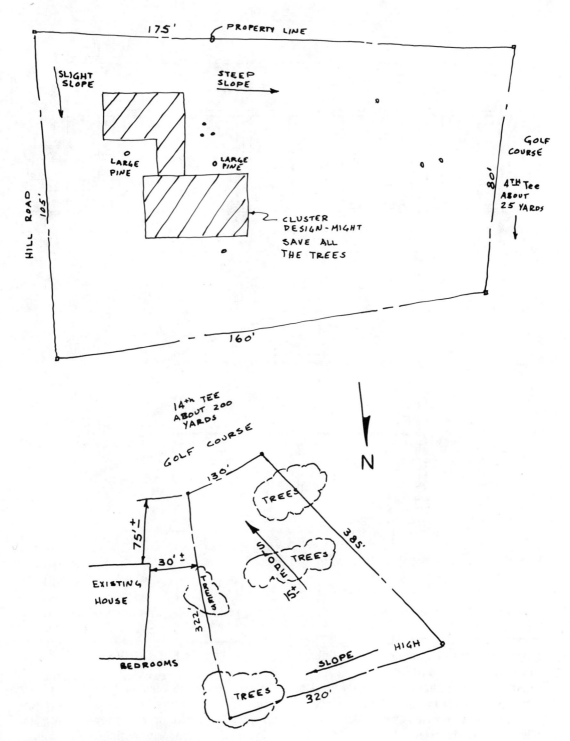

Figure 2-4. Site sketches (continued).

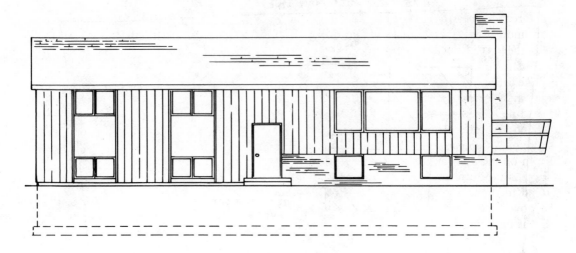

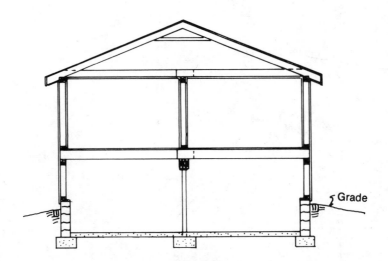

Figure 2-5. Raised ranch.

emphasize the length of the house. These types of homes often fill the lot from setback to setback.

An increasing number of clients want at least $1/2$ to 1 acre. The needs, and budget, of the client must be seriously considered in the selection of the lot size.

2.9
SITE SURVEY

After the site has been purchased, the owner should have a plan of the site made by a surveyor. This site (plot) plan shows the size and shape of the property by showing the length and direction (bearing) of the property lines, surrounding streets, any sidewalks and curbs, trees, the contours (showing the slope), direction of north, and it would also show any other existing features, such as walls, buildings, large rocks, and the like. The designer should make certain that the surveyor is instructed to locate trees, or he might not do it. All setbacks and easements should also be shown on the plan by the surveyor. Note in Figure 2–1 that the property owned does not extend to the street. That land is owned by the governing municipality, and it may be used as the location of water or sewer pipes, as a sidewalk, or as part of a road-widening project.

Preliminary designs may begin before the survey, but final plans should not be made. Without a survey it is difficult to exactly locate the trees to be saved and the slope of the land. It is extremely difficult to determine the slope of a piece of property by eye; the information from the survey is necessary.

2.10
LOCATION OF BUILDING ON SITE

A site analysis should be done taking into consideration trees, views, setbacks, types of home desired, prevailing winds, septic tank and water placement, and the slope of the land. Figure 2–1 shows a typical site plan from the surveyor. It has a large dropoff toward the golf course running along the east line of the site. The building setbacks are shown and the road located along the west property line. The client has indicated a desire to look out over the golf course, which runs north-south. The first plan (Figure 2–6a) would require the cutting of only two large trees on the site. The notes from the site indicate that both trees are in good health and should be saved. Typically, no tree closer than 4 feet to the edge of the building can be saved. At this point, the designer decided that a real attempt must be made to save both trees, and the client agreed. The plans in Figure 2–6b were designed in an

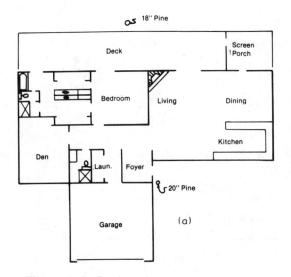

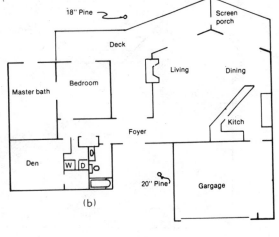

Figure 2-6. Design plans.

attempt to give the client all the rooms and
sizes he wanted, save the trees, and stay within
the budget.

Some sites require much planning, some
less. House orientation usually ends up one of
two ways: the house entry is at the front, and
the living room faces the front or the back.
Occasionally, designs are used that place the
house at an angle on the site (Figure 2–7), but
most clients have an initial reaction against
them because such placement is not as common as houses that are straight and have their
long dimension facing the front.

Modest-sized lots are best taken advantage
of and made to seem much larger by using as
much of the lot as possible. This is usually accomplished by the use of landscaping and
fences, outdoor living, planting, walk, and
play areas. Best results are often achieved if
the house is designed to open out onto the
yard by use of large windows and glass doors.

The use of large windows and glass doors

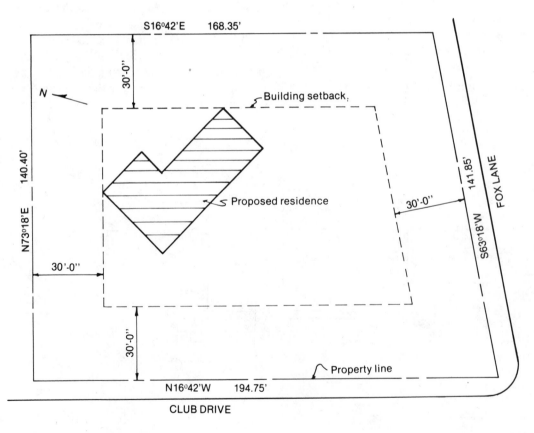

Figure 2–7. Site orientation.

also makes small rooms seem much larger, and of course large rooms seem larger yet. This is true whenever the interior of a room blends in with the exterior. This effective design approach is used on multi-story designs by the use of balconies.

When possible, the living areas of the house are oriented toward the south side, but no guideline is ever completely true because many clients in warmer climates wish to avoid the warmer southern exposure. In colder climates the few months of summer are cherished, and a southern exposure is often preferred.

Chapter 3

Basic Plans

3.1
NUMBER OF STORIES

Many clients have a definite preference for either one-story, two-story, or split-level homes. A one-story house has the advantage of having everything on one level, which means that there are no steps. Steps can be difficult for older people to use and may be a safety hazard for small children as well. Generally, one story is easiest to build and easier for the homeowner to maintain. Long one-story houses are often called ranch homes.

Two-story houses are generally less expensive than one-story houses of the same square footage and comparable features. This is because there is less footing, foundation, and exterior wall, and because by putting a second floor over the first, in effect one half of the roof construction is saved. But the same basic amount of floor construction is used. There is a resulting savings of material and labor for these items. Additional costs of a multi-story house are the cost of stairs and the extra space required by the stairs. At the same time, by using two stories it may be possible to reduce hallways sufficiently to offset the added stairway area. Each particular situation is different and must be considered separately. For example, a fireplace would have to have the chimney built about 10 feet higher on a two-

story building, adding slightly to the cost. Also to be considered is whether an additional half- or full bathroom is required by going to two stories. The two-story home provides a natural design breakdown of living on the first floor, with kitchen, living, dining, family, play, and any other rooms, and the second floor for sleeping with bedrooms and bathrooms. Advantages include a rather natural zoning of activities, generally lower cost, and use of a minimum of the land area, leaving a maximum of the area for enjoyment. Disadvantages tend to center around the desire not to have to climb stairs. Even many people who have lived in two-story houses claim that they are up and down the stairs a considerable amount, even when they try to plan their day so that excess step climbing is avoided.

Split-level homes (Figure 3–1) are an attempt to compromise between one- and two-story houses by reducing the steps from level to level (although the total number of steps may be the same), and to take advantage of the cost savings of two-story construction in at least part of the house. They are particularly fitting and economical on sloping sites (Figure 3–2).

The basic decision as to the number of floors desired should be left to the owner. If the owner wants a single-floor layout on a sloping site, the possibility of a multi-story layout should be discussed. If a single-level

22

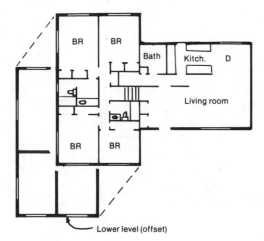

Figure 3-1. Split-level.

Figure 3-2. Split-level on site.

layout is still the most desirable to them, it is possible that they would consider a design that would allow future use of the lower level as a recreation room or storage, or for other use by future owners. This prospect should be carefully and completely discussed with the clients. Simply because the designer prefers or feels that a particular solution is best does not mean that it will be best; the clients' best interest comes first and is far more important. Even so, there is nothing wrong with presenting various ideas and plans to the clients, but the ideas should not be pushed. After a full, fair presentation of all ideas and plans, it is the clients who should make the final selection,

because they are the ones who will be paying for and living in the house after it is completed.

3.2
PLAN SHAPES

Commonly used plan shapes are shown in Figures 3-3 and 3-5. The shape of the plan will affect the cost of the house as well as the exterior styling. The plan should be such that it will create interest, but this can be done with plans of any shape if the designer uses a measure of creativity in the exterior design. A rectangular or square plan (Figure 3-3) can be designed with plain or fancy exteriors (Figure 3-4), depending on the designer's ability and the clients' approval.

More-irregular plans are often used to add interest to the exterior elevations as well as to get additional natural light into each room. Since the more irregular the plan, the more the house will cost, eliminate any extra corners when economy becomes important. Irregular plans are used a great deal on those lots where saving trees is important, as shown in Figure 3-5. In this case the clients know that saving the trees will cost them a little extra and are willing to pay it. The ell, multiple ell, H, U, and cluster plans are extremely useful when trees are to be saved.

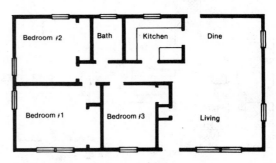

Figure 3-3. Rectangular plan.

Figure 3-4. Elevations.

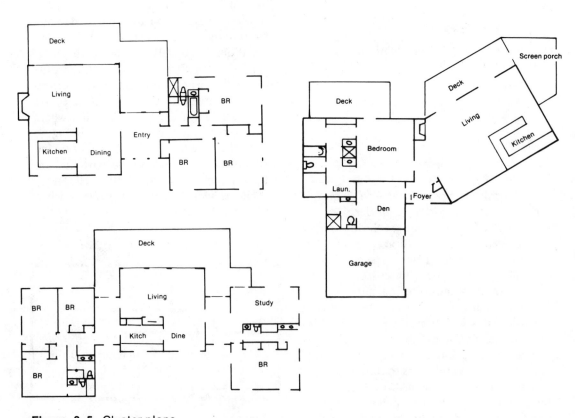

Figure 3-5. Cluster plans.

Chapter 4

Room Sizes and Layouts

4.1
LIVING ROOM

The living room is the center of activity for many families and typically serves at different times (and sometimes all at once) as a television room, family conversation room, playroom, and guest entertainment room. Many other families use the living room in a more formal sense, primarily for the entertainment of friends and for business entertaining, and use a family room for more relaxed entertaining of the family and close friends. Some residences even have separate rooms as playrooms for the children.

Living room size may range from a minimum size of about 160 square feet, with the average size closer to 300 square feet (15 x 20 feet), to very, very large living rooms. Many people put as much space as they can spare into the living areas of the house, because these are where most of the hours that a person is awake and at home are spent.

The design of the living room space should include furniture grouping and circulation. The designer does not have to be an interior decorator to check these factors, just use good sense. The living room is usually located just off the front entry, because visitors usually enter the house and go directly into the living room. Where space is limited, the front door opens directly into the living room, but generally homes with more than about 1,400 square feet have the front entry open in a foyer, which is simply a small space to greet guests, have them come in, remove coats and boots, straighten their clothes, and then enter the appropriate room. Many homes have a half-bath at or near the front entrance to be used by guests if they need to "freshen up" when they arrive as well as throughout their stay. This half-bath is commonly referred to as a powder room.

The front door into the foyer or living room is usually a single 3-foot-wide door, or double doors that total 5 or 6 feet in width. Wide openings are usually designed between the living room and adjoining spaces, such as dining rooms and foyer. Often, no door is used on these openings or the door may be decorative, so that the area can be closed off until later in the evening. For example, when the dining room adjoins the living room, an 8-foot opening would often be appropriate. A set of folding doors with decorative hardware in this opening would allow the dining room area to be closed off throughout a portion of the evening of entertaining and then opened at the appropriate time for snacks, meal, drinks, and so on, with at least some element of surprise. The flow of people will be greatly enhanced with the wide openings.

Window size and location will greatly affect the furniture arrangement in the space. Large picture windows, floor-to-ceiling windows, and groups of sliding glass doors are often used so that a view may be enjoyed and to bring in a feeling of the outside. The plan in Figure 4–1 has 20 feet of sliding glass doors opening out to a deck that overlooks a golf course and ponds. When the curtains are opened, the room will feel even larger than it is, and the sliding glass doors offer a good flow of people when entertaining. The furniture arrangement is often greatly affected by the use of a large glass area because it may be necessary to arrange the furniture out in the room rather than having it backed up to a wall. In this case a larger living room is required to achieve the elegant feeling of open space.

The living room is usually designed around a central focal point, such as a fireplace or a window or group of windows; in many homes the focal point is the television set. Typically, the furniture is then arranged in groups around this point. Large living rooms have a central focal point and also have smaller groups of furniture for other purposes, such as a game table, piano, or other activity. All living rooms can have objects that serve as focal points, such as a painting, a piece of sculpture, or something made by the owner of the house. No matter what it is that is made the focal point, it should be something that the owner is proud of.

Instead of always using walls between the living room and other spaces, a wall or divider may be formed by use of cabinets and bookshelves, or wine-bottle holders, a fireplace, or a bar. The use of decorative wall finishes should also be considered as highlights in the living room. Such finishes as a wall of cork, a

piece of natural travertine marble (holes and all—not filled and polished), siding from an old barn, and articles of interest to the owner such as a fisherman's net, a lobster trap, old photographs and cameras, and similar items can be used to highlight a space.

Considerable built-in furniture may be used in the living room for storage, records, television, bookcases, and desks. Once again, the client will have to decide what and how much is actually needed. Many clients prefer ready made units, as they can be easily moved and rearranged as the need arises.

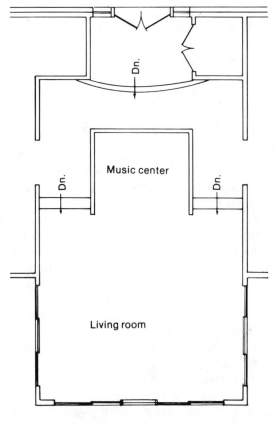

Figure 4-1. Glass wall.

The traffic flow pattern of people from the front door, through the living room, and into adjoining rooms must be checked and re-checked with the furniture laid out in sketches. When possible the living room should be located so that the daily traffic of the house will not go through the living room. When space permits, the living room should be located where it will not be used as a hall. The design, shape, and size of the foyer (Figures 4–1 and 4–2) will help control where people will walk, and it will generally take a little extra square footage. In some designs the living room area may be lowered or raised from the surrounding floor levels to discourage through-traffic.

Designing the Living Room

1. The designer should determine how the clients will use the room. Will it be for everyday use or will it be formal? Just how often will it be used, and by how many people? What furniture pieces will be placed in the room? Will the furniture be existing furniture or new? If new, an interior designer should be included in the discussions from the beginning.

2. The designer should determine what an average-sized living room is for your client. It will vary considerably.

3. In talking with the clients, the designer should decide what the focal point of the room will be.

4. Next, the designer should sketch out several layouts of furniture arrangements to be certain that the furniture will fit into the room easily and comfortably.

5. Once these ideas are tentatively approved, and the basic size and shape of the room determined, the preliminary design portion is over. The room may vary somewhat in final form, but the basic work is done.

Interior Elevations

Any wall in the living room with built-in cabinets, fireplace, or other features, such as special wall finishes or treatments, should be shown in an interior elevation. Details of construction and installation are required also. Fireplace elevations, enlarged plans, and details (Figure 4–3) are often placed adjoining the interior elevation or grouped together on a separate sheet.

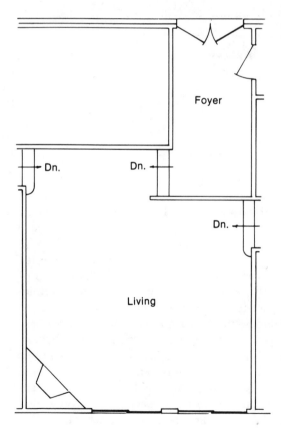

Figure 4–2. Foyer.

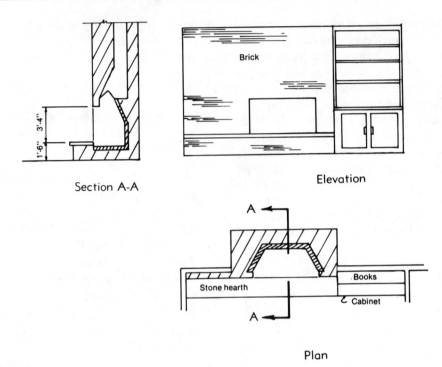

Section A-A

Elevation

Plan

Figure 4-3. Interior elevations.

--- 4.2 ---

DINING ROOM

When economizing is important, a separate dining room is commonly omitted and the space better used for other purposes. Yet a separate dining room is an extremely important room to a great many clients. The basic purpose of the dining room is to provide a special space or room for eating. Alternative approaches include a dining area set up at one end of a family or living room, separate from the kitchen but planned as a part of the kitchen. Another approach is to have a "working kitchen," a kitchen without table and chairs but perhaps an eating bar and stools, and in the same room an area set aside for dining. This area might be more formal in appearance than the usual kitchen by the use of wall paneling, carpeted floors, and special lighting fixtures.

The clients' life style will determine the type of dining area to be designed. The size of the family, the type and amount of entertaining that they do, and the furniture owned and to be purchased must be considered also. Dining rooms and areas used for formal entertaining or group entertaining will need to be closely related to the kitchen and living room (Figure 4-4a) and, if large groups are anticipated, to the kitchen, living room, and family room (Figure 4-4b). Separate dining rooms should have a large opening from the living room, which may or may not have doors. The dining room may be emphasized or dramatized by raising or lowering the floor level. Dining rooms that are part of a living or family room may be visually separated by use of built-in cabinets, bookshelves, fireplace, screens, or other dividers, stationary or movable.

Dining rooms vary greatly in size depending on the number of people being planned for and the furniture. A dining room 10 x 12 feet will seat 4 to 6 people around a rectangular table, with a buffet against one wall. A dining room 12 x 15 feet will seat 6 to 8 people and a buffet, and a room 14 x 18 feet will serve 10 to 12 people and accommodate several pieces of furniture. The designer must determine whether the client intends to use the dining room for all meals, all family meals, or for formal gatherings. This will help determine whether to plan it primarily as functional for family use, or decorative for formal use.

The dining room should be located next to the kitchen to provide for easy serving. A pass-through is often put in the wall between the dining room and kitchen to reduce the number of steps needed when serving a meal. The pass-through usually ranges from 2 to 5 feet wide and needs to have ample counter room on the kitchen side. The pass-through commonly has folding or sliding doors so that they can be closed.

The door from the kitchen to the dining room should be as large as possible preferably 3 feet wide, to make passage with food, dishes, and trays easier. Pocket sliding doors are often used in this doorway so that the door can simply be pushed out of the way and will not interfere with the layout of furniture.

Window size and arrangement will affect the arrangement of furniture, although the table is usually placed in the center of the room. Dining rooms with views should have large windows so the view may be enjoyed.

Often these are sliding glass doors opening onto a porch or patio. If the patio connects to a living room that has sliding glass doors, a good flow of people will be possible.

Designing the Dining Room

1. The designer should determine how many people the room should be designed to seat and what shape and size of furniture they will use.
2. The designer should sketch furniture, window, and door layouts to arrive at an approximate dining room size and shape.
3. The designer must check the layouts with the clients. Once approved, the layouts serve as the guide to the sizes used in the design sketches of the house.

4.3
KITCHEN

Kitchen planning receives a great deal of attention in the design of a house because the kitchen is often the most important room in the house. The kitchen must be cheerful, efficient, and of the proper size. The primary use of a kitchen is food preparation, but it is also used for dining, sewing, laundry, and storage. It is necessary for the designer to determine how the clients will use the kitchen and exactly what activities are planned for it.

The design of the food preparation area is the most important portion of the kitchen layout. The location of the sink, refrigerator, and cooking unit will control most of the activities in the kitchen, and these points are commonly connected to form a work triangle. The general efficiency of the kitchen is measured by the length around the triangle. The smaller the triangle, the more efficient the

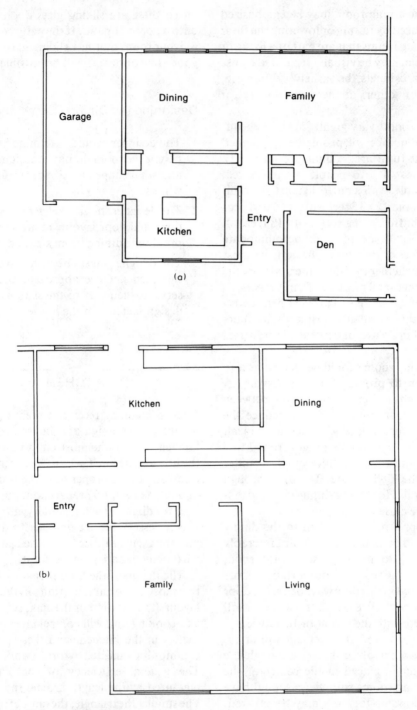

Figure 4-4. Dining room layout.

kitchen. That does not mean that everyone wants the most efficient kitchen possible—it is but one thing that should be checked during the design.

The clients should begin by selecting a basic kitchen layout. Commonly used layouts are shown in Figures 4–5 through 4–10 and include the corridor (also referred to as pullman and two-wall), L-shaped, U-shaped, one-wall, the peninsula, and the island.

The *corridor kitchen* (Figure 4–5) (also referred to as *pullman* and *two-wall*) is simply a corridor at least 4 feet wide with cabinets and appliances on both sides. Basically there is room for only one person to work, but it provides efficiency and saves steps. Doors are often placed at one or both ends so that it can be closed off. If is often used in smaller homes and apartments where the eating space has been put into a dining room instead of a kitchen. An example is the floor plan of a small house designed for low-cost housing.

The *L-shaped plan* (Figure 4–6) has cabinets and appliances on two adjacent walls. This widely used plan has a relatively large work triangle, but the design is such that there will be few people passing through the work area. The additional space in a square or rectangular room can be used as an eating area with table and chairs.

The *U-shaped plan* (Figure 4–7) is almost as efficient as the corridor plan, and with proper door placement there is no cross traffic through the work area. This three-wall layout usually has a sink in the middle, a stove on one side, and a refrigerator on the other. Space for eating could be allowed along the fourth wall.

The *one-wall* (straight line) *kitchen* (Figure 4–8) is used for cottages, apartments, and designs that allow the use of only one wall for cabinets and appliances. This design is used only in special design situations because of its limited space and long work area.

Peninsula kitchens (Figure 4–9) are kitchens with a counter or eating bar that projects into the room. They are frequently used in kitchens that are combination kitchen-dining rooms (Figure 4–9), where informal meals are eaten at the counter and formal meals are eaten in the dining room. In other arrangements, the peninsula may be used as a cooking or food-preparation center.

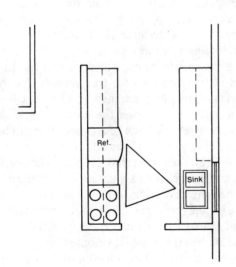

Figure 4-5. Corridor kitchen.

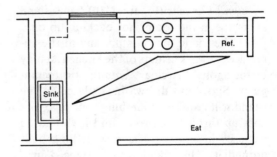

Figure 4-6. L-shaped kitchen.

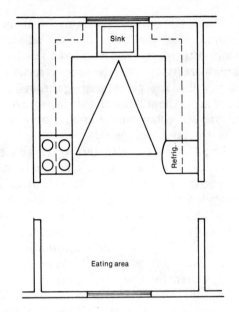

Figure 4-7. U-shaped kitchen.

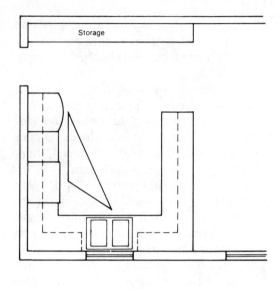

Figure 4-9. Peninsula kitchen.

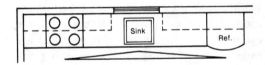

Figure 4-8. One-wall kitchen.

Island kitchens (Figure 4–10) have island areas out in the room, not backing up on a wall. The island may contain any of the appliances or work centers of the kitchen. It may be octagonal, circular, square, or rectangular. Sinks and dishwashers placed in the island will require a plumbing vent pipe that goes up through the roof, and it should be enclosed in a wall or have something built around it. The island may be placed in a kitchen of any shape.

In any design the food-storage and cooking utensils should be placed near the area in which they will be used. Extra food is often kept in storage areas, but the food currently being used should be near the food-preparation area. Many kitchens have storage units for food designed right into the kitchen. Increasingly, all space is being used. Cabinets over the refrigerator, which may be hard to get at, may be used for tray storage (Figure 4–11). Spices may be lined up in a rack just under the counter over the food-preparation center. Even the back of a closet door can be utilized (Figure 4–12) by adding shelves.

The kitchen is commonly located adjacent to the garage, for easy access with groceries. When the kitchen has a utility area or room, the garage entry is usually into the utility area, which also acts as a mudroom, and then into the kitchen (Figure 4–13). Some clients prefer easy access from the front door to the kitchen, especially when there is no garage or carport or when the kitchen is located at the front of

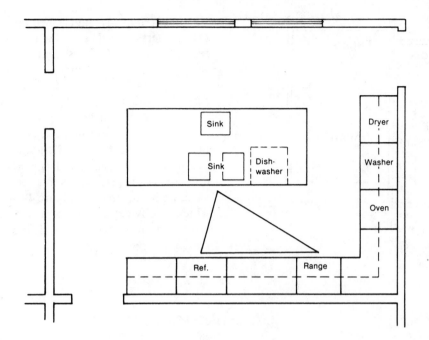

Figure 4-10. Island kitchen.

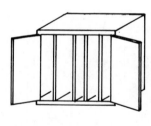

Vertical dividers in cabinet-storage for trays, slicing boards, platters.

Figure 4-11. Tray storage.

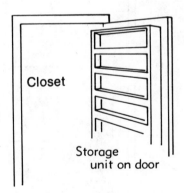

Figure 4-12. Can storage.

the house. When possible, the front door–kitchen access should not be through the dining, living, or other rooms. This may require a hall, which adds square footage to the house.

Windows are necessary to make the kitchen bright and cheerful. A window is commonly placed over the sink. Originally this window provided a view for the woman as

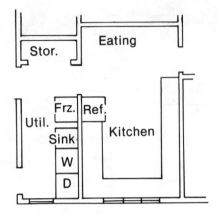

Figure 4-13. Utility area.

she prepared meals and cleaned up; but with dishwashers and other appliances, as well as readily prepared food, less time is spent in the kitchen. Today, the window is still placed over the sink; most clients want a window there, and homes sell better with a window there. Most definitely, if there is no other location for a window in the kitchen, an ideal place for it is over the sink. Large kitchens and eat-in kitchens often have large picture windows or sliding glass doors that look out to a yard, patio, or deck. In Figure 4-14 the bow window provides a spot for a small kitchen table and gives the client a view out to the garden.

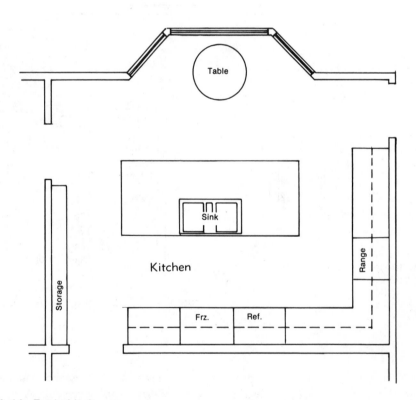

Figure 4-14. Eat-in kitchen.

Appliances

Kitchen appliances include refrigerators, freezers, ranges, range-hood, sink, food disposals, dishwashers, waste compactors, and built-in small appliance centers that do things such as mixing and blending foods. The wide range of styles, sizes, colors, and features for each appliance makes it impossible to list them all. Refrigerators are available as under-the-counter units; combination refrigerator-freezers with the freezer at the top, bottom, or side, with ice, crushed ice, and/or ice water; and recently one that has a cassette tape deck built into the outside of the refrigerator door. Refrigerators range from 9 to 21 cubic feet in capacity. If the freezer is separate from the refrigerator, it may be undercounter, medium or full height, with front-opening doors, or a chest type, which opens at the top. Freezers may be located in the kitchen but are often put in a utility area located convenient to the kitchen.

No matter how the units are purchased or installed, a house will need a unit for baking (an oven) and another for boiling and frying (a cooktop). When these are combined into a single unit, it is called a range. Separate units are often preferred, with the cooktop and the oven side by side. Cooking units may be operated by gas or electricity, and it is important to determine which the client prefers. Cooktop units are available in three- and four-burner models, and the newer models have porcelain tops. Ovens are available that are self-cleaning, continuous cleaning, or that you clean yourself. Oven doors may be solid, have a small window or be all window, which can be seen through when the inside oven light is on. The list seems endless; new models are available with features such as radios and televisions. Many clients want two stoves, for cooking at different temperatures at the same time. Microwave ovens are also increasingly popular.

Range hoods are desirable for taking smoke and odors out of the kitchen. The hood may be a simple exhaust over the range or a specially designed unit over the entire cooking area. Ideally, they are vented to the exterior.

Sinks are available in single, double, and triple compartments in enamel or stainless steel. A garbage disposal is often used, especially where central sewer systems are used. Dishwashers are also popular and may be fitted directly into the standard kitchen counter.

Waste compactors are becoming increasingly important. They compress garbage into a much smaller volume, saving fuss, mess, and space. It would be a great help if everyone had one, but it is still considered a luxury item for the average homemaker.

Designing the Kitchen

1. First, the designer should determine the shape of kitchen desired by the clients and any special requirements they might have.

2. The designer should list all appliances that must have space allowed for them. The size or capacity desired for each appliance and all brand-name preferences should be noted. Sizes should then be determined for each of the appliances selected from the manufacturers' information, and then used as the kitchen is laid out and designed.

3. Next, the amount of storage and cabinet space required by the clients must be determined.

4. Starting with the shape of kitchen selected by the clients the designer should sketch possible layouts, locating doors, windows, eating, and laundry or utility areas.

Interior Elevations

Kitchen interior elevations are a "must" for most residences. Elevations (Figure 4–15) are shown for all walls with cabinets, appliances, or built-in items such as desks and sewing or work centers. Details of construction and installation of built-ins should be shown also.

LAUNDRY

Currently, there seems to be two basic approaches to locating the washer and dryer. The first and more traditional approach is to locate the laundry in or adjacent to the

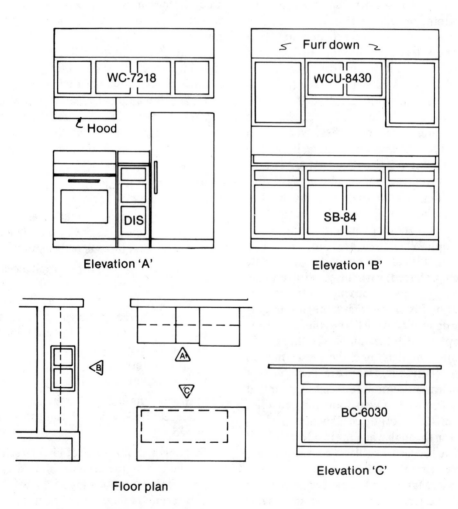

Elevation 'A'

Elevation 'B'

Floor plan

Elevation 'C'

Figure 4–15. Kitchen elevations.

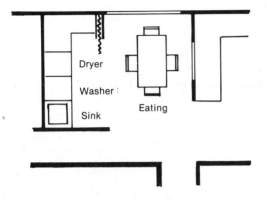

Figure 4-16. Laundry areas.

Designing the Laundry

1. The designer should determine what features the clients want in the laundry area and the preferred location.
2. Next, the designer makes tentative layouts to determine the size required.

_____ 4.5 _____

BEDROOM

The minimum area allowed in most codes and standards is 120 square feet for a master bedroom and 80 square feet for single bedrooms. These sizes are used primarily in low-cost and government-sponsored housing. Most residential designs have much larger bedrooms. When designing small bedrooms the designer has an opportunity to be truly creative, since it will probably be necessary to design built-in furniture, cabinets, and storage walls so that the most effective use can be made of space.

The number of bedrooms generally ranges from two to four (more for large families, of course). A two-bedroom house is not desirable in some areas, because it is harder to sell, and most clients who want a two-bedroom house include a den-study (Figure 4-17), which can serve as a third bedroom if required. Typically, a three-bedroom home has greatest sales potential. Generally, the bedrooms are grouped together in one section of the house, generally referred to as a "quiet zone" or "sleeping area." Grouped with the bedrooms are bathrooms, dressing rooms, and storage.

When laying out the bedrooms, each room should have direct access to the hall and a closet that is accessible from the bedroom. A bathroom should be convenient to the bedroom group. The master bedroom often has

kitchen so that the laundry can be done together with the general cleaning and cooking. Often there will be a laundry center where washer, dryer, ironing, storage, and tub (in which to soak things) are grouped (Figures 4-10 and 4-16).

An alternative approach is to locate a washer and dryer in the bedroom section of the house, sometimes putting them in the bathroom. Some people feel that the vast majority of items to be laundered come from the bedroom section, so they might as well be laundered in the same area.

Laundry facilities required may include a washer and dryer alone or items such as a collection point (a hamper or cabinet), sink for presoaking, table for folding, ironing table, storage for soap and supplies, and even a sewing center.

The washer and dryer may be side by side (a minimum of 5 feet 2 inches is allowed for) or stacked one over the other. Stacked units save floor space and may simply be two separate units on a metal rack or a specially designed stacked unit.

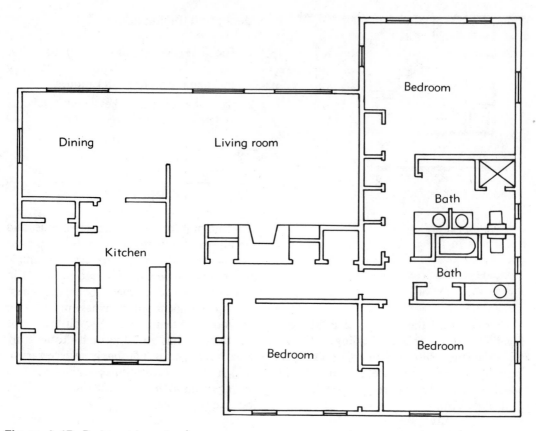

Figure 4-17. Bedroom layout.

its own bathroom. There is an increasing trend for each bedroom to have its own bathroom.

The location of doors will influence the layout of furniture in the bedroom. The hall-to-bedroom door should swing in; any bedroom-bathroom door would swing into the bathroom; and closet doors may be sliding, folding, or swing into the bedroom. The swing of each door should be shown on the plan to be certain it will not interfere with furniture placement. Bedroom doors are usually 2'-6" but a 2'-8" or 2'-10" door provides easier furniture movement. Door sizes into

bathrooms are usually 2'-0" minimum; and 2'-4" or 2'-6" are also used. Pocket sliding doors are also used in bathrooms adjoining bedrooms and eliminate concern regarding the swing of the door. A door located near the corner of the room usually allows more flexibility in the room design because it wastes less space.

Glass sliding doors are often used on first-floor bedrooms that open out to private patios. This tends to make the room seem larger when the curtains or drapes are opened. Occasionally, the sliding doors will open onto patios and decks which also serve the living room, family room, or other portions of the house. On multi-story housing, the sliding glass doors would open onto a balcony or deck. Keep in mind that they take up a great deal of wall space.

Window locations are also important because their placement will influence the location of furniture. Ideally, furniture arrangement is made first, because items such as beds, dressers, and chests of drawers cannot be placed in front of windows. For this reason a certain amount of plain wall space is required, and so the window and door placement must be planned carefully. Homes have been built in which window placement was planned so poorly that there was no wall space wide enough for a double bed, let alone a queen- or king-sized bed.

Another way to handle the placement of windows is to use short windows placed high in the room so that furniture may be situated against the wall below them. Ideally, windows will be located on two walls for cross ventilation when the windows are opened, but they should be located so that any draft would not blow across the bed. Of course, room layouts often prevent placement of windows in two bedroom walls. Also, the ventilation aspect has become less important with the increased use of totally air-conditioned homes, which provide heating, cooling, ventilation, and even humidity control.

Designing the Bedroom

1. The designer should determine from the clients how many people will be living in the house and what sleeping arrangements are contemplated. Typically, a family puts no more than two children in a room, and the trend is to put one child per room. In addition, many clients want an extra bedroom to serve as a combination guest bedroom-den-study-sewing-playroom when it is anticipated that they may need an additional room in the future. It is much less expensive to have the extra room included in the original plan. Separate but adjoining bedrooms may be desired for the husband and wife. The point is this: get to know your clients and their living habits (and their sleeping habits in this case).

2. Next, the type, size, and amount of furniture to be put in each bedroom must be determined by the designer. Nothing should be assumed, because clients vary so much that each room must be carefully checked. The following bedroom checklist contains the information that must be checked for size and the number that will be required:

bed	night stand
dresser	telephone stand
chest	closet (and type)
chairs	dressing area
desk	bathroom (and fixtures)

3. Once a list for *each* bedroom has been compiled, typical room arrangements and sizes may be determined. The common sizes of bedroom furniture are shown in Figure 4–18. The bed sizes shown are all readily available, but there may be more variation in other furniture sizes. Room layouts are quickly and easily drawn by use of furniture templates. A variety of layouts may be quickly and easily drawn by their use. Minimum spacing between pieces of bedroom furniture is necessary for good room design. Typical clearances are shown in Figure 4–19. Some typical bedroom layouts, from small through large, are shown in Figure 4–20.

4. Now, with some basic bedroom sizes roughed out, the designer will move into surrounding rooms in a continuation of the work on these individual pieces of the

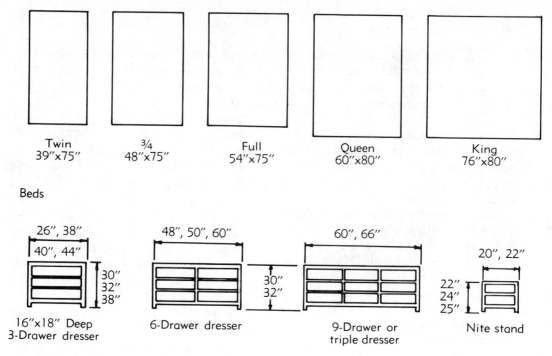

Twin
39"x75"

¾
48"x75"

Full
54"x75"

Queen
60"x80"

King
76"x80"

Beds

26", 38"
40", 44"
30"
32"
38"

16"x18" Deep
3-Drawer dresser

48", 50", 60"
30"
32"

6-Drawer dresser

60", 66"

9-Drawer or
triple dresser

20", 22"
22"
24"
25"

Nite stand

Figure 4-18. Bedroom furniture.

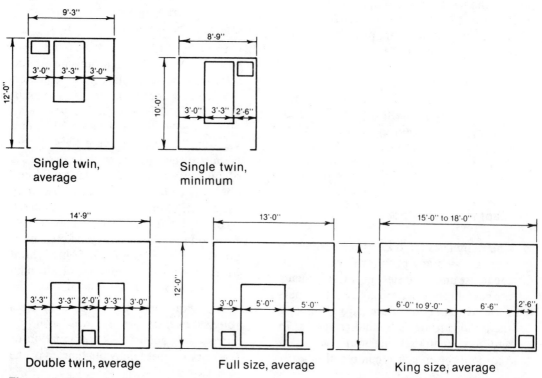

9'-3"
12'-0"
3'-0" 3'-3" 3'-0"

Single twin,
average

8'-9"
10'-0"
3'-0" 3'-3" 2'-6"

Single twin,
minimum

14'-9"
12'-0"
3'-3" 3'-3" 2'-0" 3'-3" 3'-0"

Double twin, average

13'-0"
3'-0" 5'-0" 5'-0"

Full size, average

15'-0" to 18'-0"
6'-0" to 9'-0" 6'-6" 2'-6"

King size, average

Figure 4-19. Typical bedroom clearances.

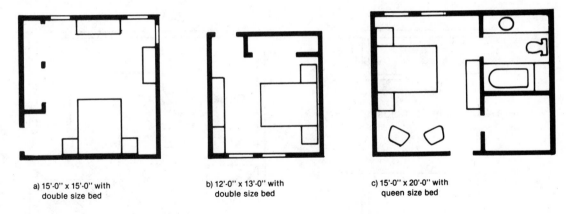

a) 15'-0" x 15'-0" with
double size bed

b) 12'-0" x 13'-0" with
double size bed

c) 15'-0" x 20'-0" with
queen size bed

Figure 4-20. Typical bedroom layouts.

jigsaw. The designer should draw up a list of typical room sizes so that they will be ready to trace when they are needed.

4.6
CLOSETS AND STORAGE

Bedroom Closets

Closets and storage space are important in any bedroom layout. Properly designed and planned, they can provide adequate space to hang and store clothes and may be used as sound buffers against any noise from the rest of the house.

Closets must have a minimum clear hanging space of 2'-0" (with 2'-4" preferred) and a minimum hanging length of 5'-0" for double occupancy and 3'-0" for single occupancy. These minimum lengths of hanging space are just that, minimum, and they are used only where economy is most important. Generally, no closet should be less than 5'-0", because only small children will require less—and then only until they are pre-teenagers. The master bedroom will generally require at least 4'-0"

of hanging space for men and 6'-0" for women. One whole wall will often be lined with closets in a master bedroom. Full access to the hanging space is desirable also. A 5-foot closet with a 2'-6" door (Figure 4-21) will have some space at each end of the closet that is difficult to use. The same closet with accordian, folding, or sliding doors (Figure 4-22) that are as large as possible will provide better access and a more usable hanging space. Walk-in closets need to be a minimum of 5'-6" wide to allow a passageway between the clothes hanging on both sides (Figure 4-23). Other arrangements are possible also, such as closets that are part of a compartmented bathroom (Figure 4-24) and deep, narrow closets with clothes on a sliding rod (Figure 4-25). Closets with storage units along each side and the bottom (Figure 4-26) are often built in on the job, and these types of units may also be used in a compartmented bathroom that adjoins a bedroom.

Each closet requires a clothes-hanging rod placed 5'-8" above the floor. The rod may be of metal or wood and should not be over 1 5/16 inches in diameter or clothes hangers will not fit on it. A shelf is usually provided above the

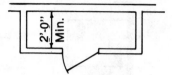

Figure 4-21. Single-door closet.

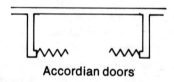

Sliding doors

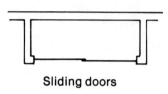

Accordian doors

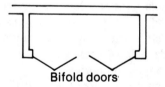

Bifold doors

Figure 4-22. Closet doors.

Figure 4-23. Walk-in closet.

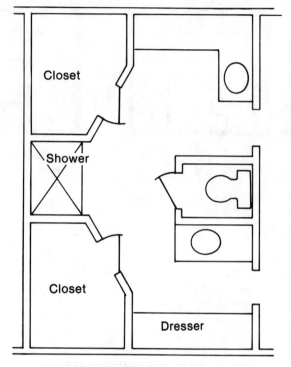

Figure 4-24. Closets in the bathroom.

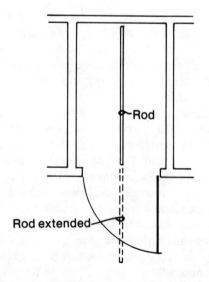

Figure 4-25. Closet with sliding rod.

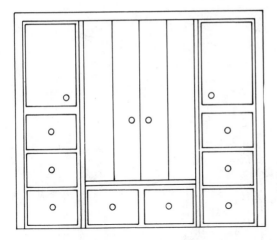

Figure 4-26. Closet with storage.

hanging rod unless storage units are built in. The shelf should be at least 12 inches wide to provide some extra storage space.

Lights are commonly put in closets and may be operated by a pull chain or switch. Although lights are not necessary in every closet, they are a must in long, narrow closets and in walk-in closets.

Linen Closets

Linen closets are often located in the bathroom or hall space adjoining the bathroom. Linen closets vary in depth from 12 to 18 inches and in width from 2 to 4 feet with a shelf-to-shelf distance of 12 inches.

Guest Closets

A closet located at or near the front entrance for the hanging of coats, storage of boots, and so on, by both family and guests, is always desirable. A minimum depth and width of 2 feet is required.

Storage Space

Most residences will require additional storage space. Space in the basement and attic should be utilized when possible, but the dampness of basements and the temperature extremes of attics preclude storage there of certain types of articles. Storage space must then be supplied on the living levels, and many homes will line an entire wall with convenient, easy-to-use closets. The spaces beneath stairways generally make ideal closet locations. Also, built-in cabinets that provide storage and closet space are used to provide decorative, well-organized units (Figure 4–27).

Interior Elevations

All closet and storage areas should be shown in interior elevations that indicate unit size, shape, and features. Construction and installation details are required for all built-in units.

4.7
BATHROOM

Although small homes may require only one bathroom, for the middle-class house the bathroom seems to have become a status symbol. Modest, 1,200-square-foot homes have 1½ bathrooms, and the increasing trend is toward each bedroom having its own complete bathroom, plus a half-bath (one with a water closet (toilet) and a lavatory) for guests. The increased cost that results from this proliferation of fixtures and plumbing is astonishing.

Typical designs for three- and four-bedroom homes generally include a minimum of two full bathrooms, one off the master bedroom and the other conveniently located

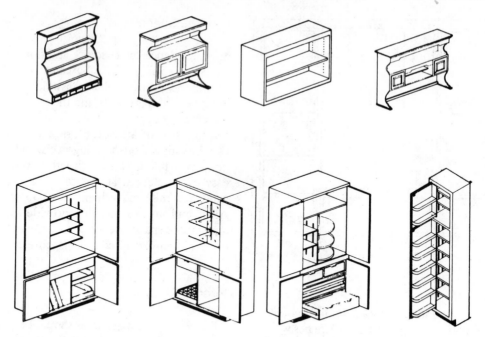

Figure 4-27. Built-in storage.

for the use of the other bedrooms. An optional half-bath is then considered for the entertainment area, near the living room, family room, or off the kitchen. For a split-level or multi-story house, a half-bath is almost always included.

The size, layout, and efficient planning for the bathroom space is necessary early in the planning and schematic phase. The number and type of fixtures required in each bathroom must be known before any of the bathrooms are planned. Too often a house is planned and the bathrooms designed for the leftover space.

The smallest three-fixture—tub, lavatory, and water closet—bathroom size is 5 x 8 feet (Figure 4-28). Although it might be possible to fit the fixtures into a slightly smaller space, it becomes cramped and there will only be the absolute minimum clearances between the fixtures. A variety of typical bathroom layouts and sizes is shown in Figure 4-29, and minimum recommended clearances are noted.

Bathtubs are available in enameled cast iron, enameled formed steel, and fiberglass. Bathtubs may be corner style, full length to 7 feet, set up in the air or sunken in the floor.

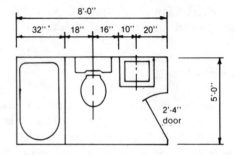

Figure 4-28. Single bathroom.

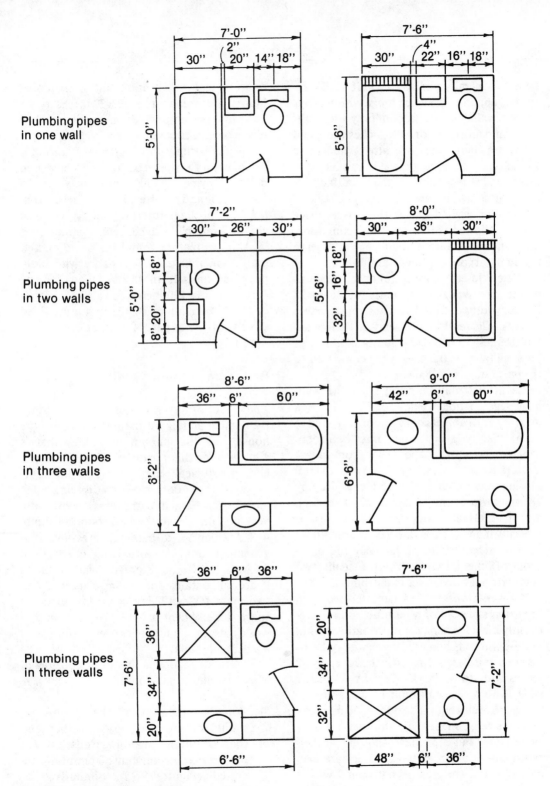

Figure 4-29. Typical bathroom layouts.

Lengths generally available include 42, 44, 48, 54, 60, 66, and 72 inches long, with various widths available. Most commonly used tubs are 60 and 66 inches long, 30 inches wide, and 14 inches high. They are also available in 16-inch heights. Fiberglass tubs are 33 or 36 inches wide, depending on the manufacturer and style selected. The type, size, and style of tub must be decided upon so that proper space may be allowed for it. Bathtubs commonly have shower-fixture attachments and serve the dual purpose of shower and tub.

Water closets, more commonly known as toilets, are available in depths of 26 to 28 inches, widths of 21 to 22 inches, and variable heights. Currently, the trend is to low-silhouette design, with the water tank as low as possible (about 20 inches high) compared to the more standard 28 inches.

Lavatories may be hung from the wall, have legs to hold them up, be part of a lavatory and cabinet combination (referred to as a "vanity") or be set into bathroom cabinets. Lavatory shapes range from square to rectangular to oval to round, with overall outside sizes ranging from 15 inches deep by 18 inches wide, to molded units 20 inches deep and up to 30 inches wide, before they must be specially made. Two lavatories are sometimes used in a bathroom, particularly the bathroom off the master bedroom (usually two countertop lavatories set in a cabinet).

Showers are often used in master-bedroom bathrooms instead of a bathtub and are frequently included in one or more bathrooms in multi-bathroom homes. Showers may be fiberglass units sized from 36 inches deep by 32 inches wide to 37 inches deep by 48 inches wide, with a seat unit built in. Ceramic-tile showers are custom-made on the job and may be any size or shape. Tile showers are used extensively, but fiberglass showers are increasingly popular. Metal showers are also available and are used primarily in low-cost

housing and in vacation and recreational homes. Sizes range from 32 x 32 inches to 36 x 36 inches, and corner units are available when it is necessary to save space.

An increasingly popular item in bathrooms is a bidet, a European approach to increased personal hygiene. The fixture is about 15 inches wide and requires about 28 inches from the front of the unit to the wall. Most bidets are not available with concealed piping.

Another feature found in, or adjoining, bathrooms are saunas. Saunas are available in packaged units from 48 x 48 inches for one person to 96 x 120 inches for a group. Typically a 60 x 96 unit or 48 x 72 unit is provided directly off, or as part of, the bathroom.

Bathroom Layouts

The layout of any bathroom will depend on the number and size of fixtures, the size of the house, and the amount of money that is allocated for this area of the house. Shown here are typical, and not so typical, arrangements of doors, closets, adjoining halls, and the like. Elegant bathrooms are currently very popular. Note the bathroom in Figure 4-24, which is a bathing, dressing, and grooming area also providing closets and drawers for clothes storage. The adjoining bedroom contains only a king-sized bed and two night tables. This elegant layout has a space requirement of 13 x 19 feet and a washer-dryer; in fact, it has everything except a bidet and sauna.

The following items should be considered when planning a bathroom:

1. Whenever convenient the bathrooms should be grouped close together and share common "plumbing walls," which will reduce the amount of plumbing required (Figure 4-30). Similarly, for

economy, single-bathroom homes have the bathroom and kitchen on a common wall.

Once the design includes multiple bathrooms, the prime consideration is to locate the bathroom in the most convenient location, with the fixtures located for space utilization and convenience.

2. Where economy is the key, locate all second-floor plumbing over first-floor plumbing, thus cutting plumbing costs.

3. Bathroom use is increased when the lavatory is placed in a separate compartment.

4. Bathrooms entered from two rooms are inconvenient, because there is always the question of whether someone is in the bath or whether someone forgot to unlock the door.

5. Bathroom windows should not be located above a lavatory or bathtub and are only less objectionable, but not desirable, if placed over a water closet. If no windows are used, the bathroom is ventilated by an exhaust fan.

6. Whenever possible, the water closet should be located so that it will not be seen as the bathroom door is opened.

7. No electrical switches should be located where they could be reached from the tub.

8. Bathroom doors are usually 2'-0", but 2'-4" and 2'-6" wide are used also. Widespread use has been made of sliding pocket doors as a means of compartmenting the interior of the bathroom (Figure 4-31). Such sliding doors are not usually used as the entrance door into the bathroom unless it is a private bathroom entered from a bedroom, because they do not lock securely.

9. The swing of the door into the bathroom should be planned so that it will not interfere with the use of the bathroom fixtures.

10. Plan for mirrors over lavatories and medicine cabinets. Additionally, space must be supplied for washcloths, towels, and other toilet articles. Space below the lavatory in a cabinet and space over the tub may easily and economically be used

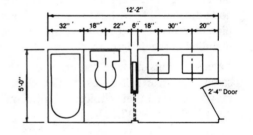

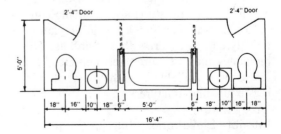

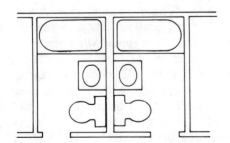

Figure 4-30. Back-to back plumbing.

Figure 4-31. Compartmented bathrooms.

for storage. Proposed bathroom plans should be roughed out to the scale that will be used to design the building. The designer will then have them ready for quick and ready reference. Firms that specialize in home design often have a variety of bathroom layouts drawn up for their reference.

Designing the Bathroom

1. The key point in the design is to determine what will be required in the bathroom. When custom-designing a residence the designer must carefully question the clients on this point, because an increasing number of clients desire compartmented bathrooms.

2. Once the number and size of fixtures is determined, an average size required for the bathroom may be determined by laying out the fixtures, with proper clearance between fixtures and a door tentatively placed and the swing shown on sketch paper. The design and layout are refined and adjusted to fit with the rest of the floor plan in terms of space required and may be included in the design.

3. Later revisions may be required, but the same basic size will be used. The clients may be offered their choice of bathroom layouts (Figure 4–32).

Symbols. The typical symbols used to lay out the bathroom fixtures are shown in Figure 4–33. The draftsman uses a plastic template to quickly and easily draw these symbols. The templates may be purchased commercially and are also available from the major plumbing-fixture manufacturers.

Interior Elevations. Interior elevations are sometimes drawn of the bathroom walls, showing cabinets, mirror sizes and locations,

storage units, glass and soap holders, towel holders, outlets, switches, lights, wall finishes, and any other features. Draw elevations only of the walls with special features. The elevations of the plumbing fixtures are included on most templates and show all required dimensions. Custom-built cabinets will require additional details on construction and installation. The interior elevations serve the purpose of showing exactly what is required, so that there is no question on the part of the owner, designer, or contractor.

_____ 4.8 _____
GARAGE OR CARPORT

The storage facility that clients will want for their automobile(s) may be a garage, completely enclosed with roof, walls, and door; or a carport, a roof with no door and one or more sides or ends left open. In cold climates, where snow and freezing rain is a part of the weather, a garage is desirable. In warmer climates a carport will provide the minimal protection needed for automobile and storage of outside tools and furniture.

Basically, the clients will decide which they prefer. The attached carport is only slightly less expensive than a garage, and many clients feel that, for the small extra cost, an enclosed area is preferred. The space required for garage or carport is the same. For one car a minimum space of 10 x 18 feet is required but will allow for almost no storage. A space of 14 x 22 feet will allow storage on the side and at one end of the space. Minimum space for two cars is 18 x 18 feet, but 20 x 20 feet is preferred. If the clients want storage space, the best design approach is to lay out the space required for the vehicles and then add the desired storage space (Figure 4–34). This same design approach should be taken if a laundry, workshop, or other area is to be a part of the space.

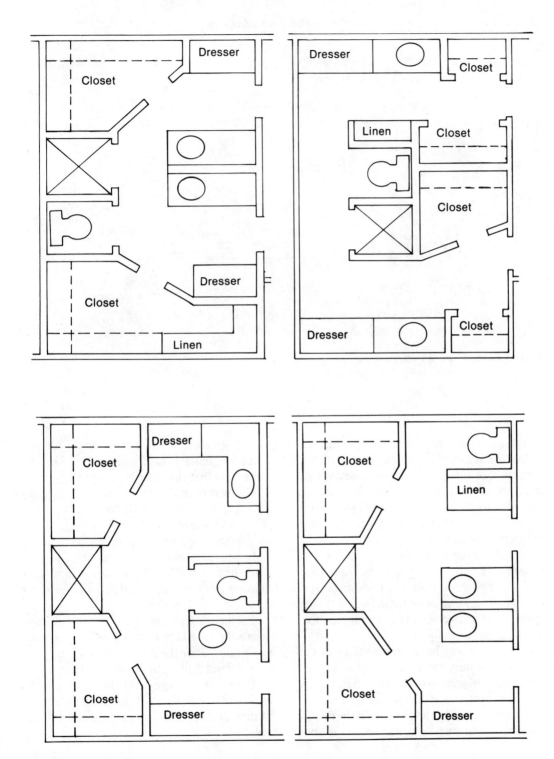

Figure 4-32. Alternate bath layouts.

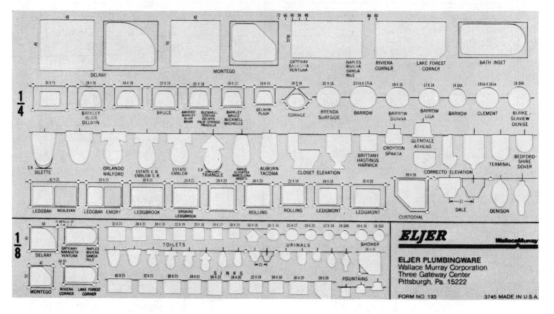

Figure 4-33. Plumbing template.

The garage or carport may be an integral part of the design of the house and attached to the house, or it may be a separate freestanding structure. If it is free-standing, it would be desirable to have a covered walkway to the house. Most garages and carports enter into the kitchen or service area of the house.

Single garage doors are generally 8 feet wide and 7 feet high; double doors are 16 to 18 feet wide and 7 feet high. Double-car garages may have one large or two small doors. Many clients prefer electric-opening doors, especially on the large doors.

The garage may be located on the front or back of the house, but if placed on the front, it should be designed so it blends with the house and does not detract from its appearance. Best results can often be obtained if the garage is entered from the side so that the garage door does not face the street. The driveway and space for guests' cars should be planned at the same time that the garage is planned.

Garages should have a standard, exterior-swinging door 3 feet wide leading from the garage to the exterior so that the garage can be used without opening the large garage door. This provides easy access to storage and workshop areas. Windows to provide natural light are most desirable so that artificial lights do not always have to be used.

Locate the garage floor several inches above the exterior finished grade to keep out any water. Slope the floor toward the door so that any water that might get in will drain out.

Doors into storage areas should be as wide as possible to facilitate use of the whole area. Sliding and folding doors are the most desirable.

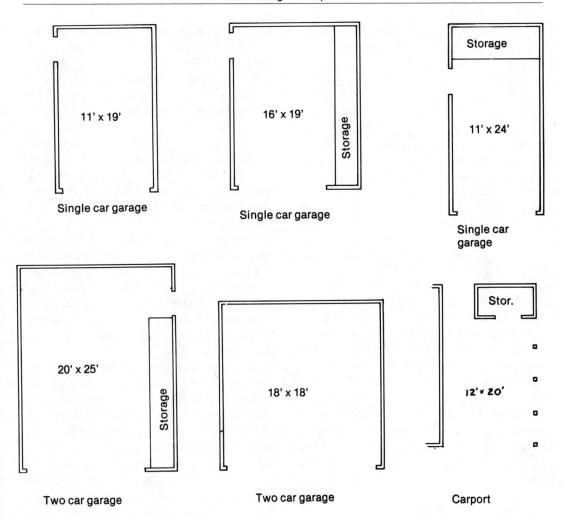

Figure 4-34. Garage and carport layouts.

Designing the Garage

1. The designer should determine from the client if a garage or carport is desirable and how many vehicles it should be designed for.

2. Next, the designer asks what storage facilities are wanted for yard equipment, lawn chairs, bicycles, and other equipment, and determines the width and depth that is most desirable and how wide a door is desired.

3. The best driveway approach and garage location on the site is determined next.

4. The garage size and shape are used to arrive at the basic size and location to be used in the design.

NOTE: Some codes, zoning, and deed restrictions require garages or carports, whether the clients desire one or not. The location of the garage or carport may also be restricted.

4.9
UTILITY ROOM

Somewhere in the residence certain portions of the mechanical equipment, such as a furnace, water heater, electrical box, and so on, must be located. The amount of space and its location vary considerably, depending on the types of mechanical systems that are being used and the type of house that is being built.

Homes with basements commonly have the furnace and hot-water heater located there. They may also be located under the house in homes with crawl spaces and an access door under the house, although they are a little harder to service when repairs are needed. Homes built on grade may have the furnace in a separate room, in a small space inside the house, in the attic (although it would be harder to service unless good access was available to the attic). The water heater may be located near the furnace, but smaller units are often placed under a cabinet top and two units used in the house. Electrical boxes are most commonly located in the garage or utility room and should be convenient.

Gas- and oil-fired furnaces and hot-water heaters must be located near a chimney to take away the fumes. Electric furnaces, baseboard heat, and water heaters do not require chimneys. For economy, homes with fireplaces often have the furnace located nearby so that the furnace chimney can tie into the fireplace chimney.

Air-conditioning units usually require outside condensers. They should be carefully located so that they will be at least partially concealed by a fence, bushes, or shrubs and where any noise from the unit will not be heard inside the house.

Chapter 5

Sound Control in Design

5.1
SOUND CONTROL

Sound control should be considered during the design stage, with emphasis on the use of the structure, possible noise sources from within and without the structure, building and interior space orientation with regard to external noise sources, and acoustic requirements for each area in terms of sound transmission.

Sound may be transmitted through the air (air-borne) or through the structure (structure-borne). Good planning is required to reduce the transmission of sound from exterior to interior and room to room.

Sound Transmission

In each space the goal is to keep the necessary sounds in and the unnecessary (undesirable) sounds out. Sound is not selective; it passes (transmits) through most building materials and any openings, no matter how small. As sound passes through building materials (walls, floors, ceilings, etc.) its intensity is reduced, the amount of reduction depending on the type of material and construction. Sound passes more readily through lighter, more porous materials than through heavy, dense, massive materials and assemblies of materials. Sound passing through openings, even small openings, is affected very little. Sound transmission through walls, ceilings, or floors is not effectively reduced when a sound-absorbent material is used.

Noise may be reduced by the thickness and type of construction used and the elimination of paths for the sound to travel through. The methods to accomplish this are sound, vibration, and impact sound isolation.

Decibels

A decibel (dB) is a standard unit of measure of the intensity of sound levels. It is used to measure the intensity of sound at the frequency ranges that we hear. A chart showing relative sound levels (in dB) is given in Figure 5-1. From this chart, observe that a level of 25 to 40 dB is reasonably quiet, whereas levels of 60 dB and up become increasingly noticeable. The object is to reduce the noise to a reasonable level, not to eliminate the noise.

The important point to remember in dealing with decibels is that the apparent change in loudness varies greatly with the change in sound level. A change of 3 dB is just barely noticeable to the listener, a change of 5 dB is

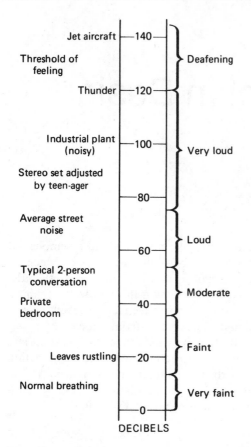

Figure 5-1. Decibels.

easily noticeable, and a change of 10 dB makes a significant difference. A change from 50 to 60 dB seems twice as loud, while a change from 50 to 40 dB seems half as loud. The following table relates sound-level changes to the change in apparent loudness:

SOUND-LEVEL CHANGE (dB)	APPARENT LOUDNESS CHANGE
3	Barely noticeable
5	Easily noticeable
10	Half as loud (or twice as loud)

5.2
ARCHITECTURAL ELEMENTS

Certain architectural elements must be considered in controlling sound. In the design stage the elements are usually considered in the following order:

1. Surrounding environment
2. Arrangement and layout of rooms
3. Room shape

Surrounding Environment

The location chosen for a building should be made with the use of the building in mind. A nursing home or school should not be placed under the approach pattern to an airport. The zoning should be carefully checked to determine what types of businesses (noise sources) may move into the surrounding area. There appears to be a tendency in some cities to include the maximum-allowable noise level, in decibels, in the zoning codes. The type of sound, whether it is radiated at night, and whether it is continuous or intermittent, all have an effect on the allowable level. As increased attention is placed on the environment, it is hoped that more zoning regulations will consider controlling the pollutant "excessive noise." Indeed, jet aircraft are already under sound controls in some areas, and further controls are being discussed.

Exterior noise problems may be reduced by placing as much distance as possible between the noise source and the proposed building, by orienting the building on the site to reduce direct sound transmission and reflection from surrounding buildings (Figure 5-2), and by shielding the buildings from major noise sources by the use of other buildings, barrier walls, and natural topography. Any barrier

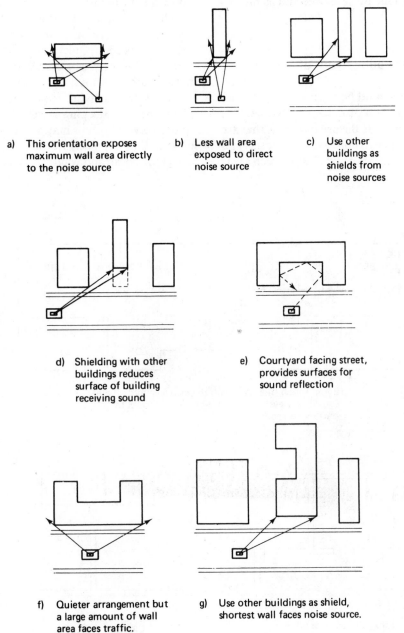

a) This orientation exposes maximum wall area directly to the noise source

b) Less wall area exposed to direct noise source

c) Use other buildings as shields from noise sources

d) Shielding with other buildings reduces surface of building receiving sound

e) Courtyard facing street, provides surfaces for sound reflection

f) Quieter arrangement but a large amount of wall area faces traffic.

g) Use other buildings as shield, shortest wall faces noise source.

Figure 5-2. Building orientation.

used, such as walls or other buildings, should be as close to the noise source and as high as possible.

Arrangement and Layout of Rooms

Noisy rooms should be separated from quiet areas by as great a distance as possible. The plan should include the use of rooms that are not as susceptible to noise, such as closets and corridors, as buffers or baffles between areas that contain noise sources and those which require quiet (Figure 5-3). The rooms from which the noise will originate should be located wherever noise from exterior sources may be expected. Quiet areas should be located as remotely as possible from exterior noise sources. An example of separation of rooms is that a living room should not be placed next to sleeping areas. The travel of sound between rooms may also be controlled by avoiding air paths in the placement and design of doors and windows (Figure 5-4).

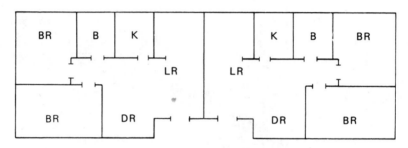

a)   By mirroring (flopping over) the plan an approximately equal noise level on each side of the wall for both apartments is obtained.

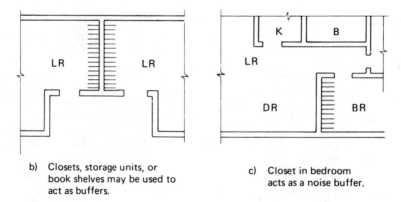

b)   Closets, storage units, or book shelves may be used to act as buffers.

c)   Closet in bedroom acts as a noise buffer.

Figure 5-3. Noise buffers.

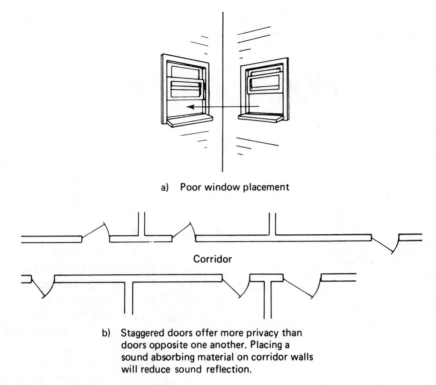

a)   Poor window placement

Corridor

b)   Staggered doors offer more privacy than
doors opposite one another. Placing a
sound absorbing material on corridor walls
will reduce sound reflection.

Figure 5-4. Doors and windows.

Room Shape

The room proportions used will affect the sound reflection in the space. Shapes to be avoided are: long and narrow, corridors with high ceilings, and rooms nearly cubical in dimensions. Each of these conditions will cause excessive reverberation (excessive sound reflection).

Domed or vaulted ceilings (any concave surfaces) tend to focus sound, causing it to be distorted. In large auditoriums with low ceilings it is difficult for some of the audience to hear. Concave surfaces and large flat surfaces may be broken with splayed areas used to diffuse the reflection of sound and direct the sound as desired.

Chapter 6

Roof Shapes

6.1
ROOF SHAPES

The roof shape used on a house greatly affects its overall appearance. The clients should be asked their preference in roof shapes. The roof-material preference of the clients should be noted also.

When the home is being designed a variety of roof shapes and designs should be tried. Figure 6–1 shows the most common roof shapes. Figure 6–2 shows how by changing the roof shape and windows, a variety of exterior designs were achieved for the same basic floor plan.

Gable Roof

The popular *gable* roof is economical and easy to build, readily sheds snow and rain, and is economically ventilated by placing louvers in the peak at the gable end of the attic. It is relatively plain, used on all types of home designs, and varying the pitch of the roof will somewhat alter its appearance. In Figure 6–2 (a, b) a straight gable is shown and also intersecting gables on an L-shaped house. Gables with the ridge slightly extended (Figure 6–2c) are called *flying gables*.

Hip Roof

Although slightly more expensive to build, owing to increased labor to frame the roof, the *hip* roof is a popular adaptation of the gable. With a low pitch it gives a low "ground-hugging" appearance to the house, and with a steeper pitch (say 6 in 12) the house has a rather elegant high roof line (Figure 6–3). One advantage of the hip is that it eliminates the maintenance that would be required on the gable end. Ventilation must be through vents placed under the soffits (under the roof overhang) or by roof-mounted ventilators.

Figure 6–1 shows a straight hip roof. Added interest is achieved when intersecting hips are included (Figure 6–3). The *dutch hip,* a popular adaptation of the hip roof, is shown in Figure 6–1.

Flat Roof

Flat roofs are used extensively on some contemporary designs which tend to deemphasize the roof. They are less expensive to build but add little to the design of most homes. The flat roof requires specialists to install. It consists of built-up layers of tar or asphalt and felt and is often topped with gravel. Special care is

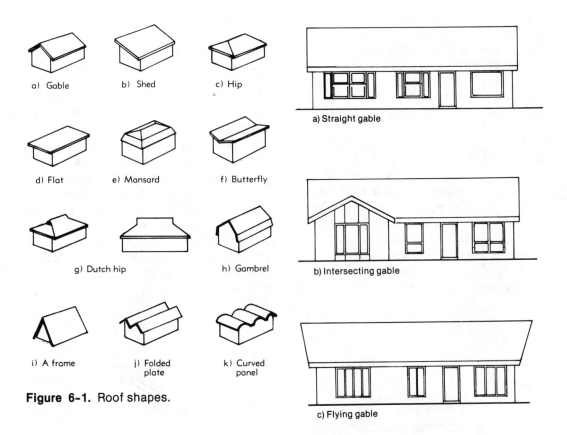

a) Gable

b) Shed

c) Hip

d) Flat

e) Mansard

f) Butterfly

g) Dutch hip

h) Gambrel

i) A frame

j) Folded plate

k) Curved panel

Figure 6-1. Roof shapes.

a) Straight gable

b) Intersecting gable

c) Flying gable

Figure 6-2. Gable roofs.

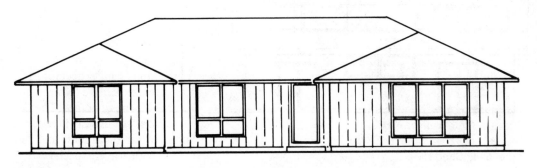

Figure 6-3. Hip roof.

required to reduce any possible leaks. Even a flat roof should have a slight pitch so that any water on the roof will definitely drain off. Pitches of ⅛ to ½ inch per foot are commonly used. In areas of heavy snow the roof will have to be designed to withstand these loads without excessive bending.

Shed Roof

A *shed* roof (Figure 6-4) is a roof with a definite pitch, to permit water to drain. They are not usually used on new housing except in combination with other sheds, when it is sometimes referred to as a *monitor* roof. It is particularly popular for additions and in combination with some of the other roof styles.

Mansard Roof

A *mansard* roof is basically a French design. Its popularity varies considerably through the years. Sometimes it is very popular, and at other times it is difficult to sell homes with this type of roof. Generally the client should be encouraged to stay with roof shapes that tend to be popular year after year, so that the house will not be unnecessarily difficult to sell. At the same time, if a client is definite in his decision to have a mansard roof, it falls to the designer to make the house attractive enough that it will sell anytime. The mansard is generally a little more expensive than the hip roof. Used primarily on two-story residences, it allows extra space upstairs compared to the gable and hip roofs.

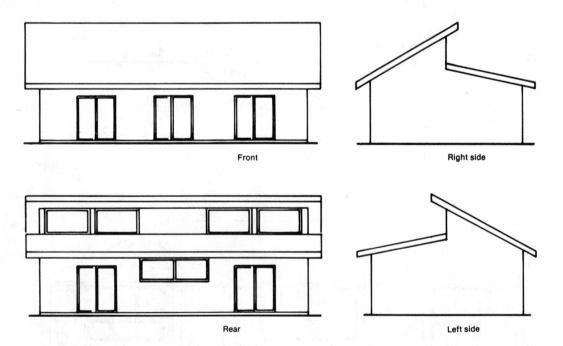

Front

Right side

Rear

Left side

Figure 6-4. Shed roof.

Gambrel Roof

The *gambrel* roof style is also referred to as a *barn* roof and a *Dutch colonial*. Its design provides more headroom than the gable or hip. Many of the more traditional homes use this roof style, and it has a definite appeal to many people. It is used primarily on two-story houses, where its advantages are most readily appreciated in terms of larger upstairs rooms. It costs more than a hip or gable roof but no more than a mansard. It tends to be consistently more popular than the mansard.

Butterfly Roof

The *butterfly* roof is used in some contemporary designs. It is more expensive than the flat or shed roof and slightly less than a hip. Since the water from the roof is directed to the center of the house rather than off it, extra care is required in the flashing at the low point to guard against leaks. Also, either a roof drain will be required to get the water off (added cost involved), or the water will "gush" off the two ends of this valley. This type of roof should not be recommended. Its disadvantages, possible water leaks and poorer resale, outweight the cost savings. If the clients still feel that they want it, it must be designed carefully. Carry the valley flashing well up the slope and put in roof drains or a method of handling the water at the valley ends.

A-frame Roof

In the *A-frame* roof, the framing for the roof makes the walls of the building as well. Once used primarily for small cottages, it is now used for all types of buildings, including residences and churches. If the building is wide enough, there is sufficient space inside, but there is a tendency for usable space in upstairs rooms to be limited (Figure 6-5). Modified A-frames often combine short side walls with the roof structure to minimize odd-shaped rooms on the first floor (Figure 6-6). Generally, this style roof is associated with a rustic approach to design, with exposed beams and wood roof decking. With conventional finishes it can be a very economical system, but with exposed beams and wood decking the cost goes up.

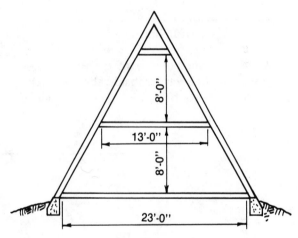

Figure 6-5. A-frame.

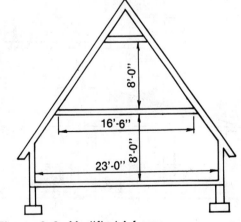

Figure 6-6. Modified A-frame.

Contemporary Roof

Contemporary roof styles, such as the *folded plate, curved panel,* and *paraboloid,* are sometimes used on contemporary and modern homes. Their costs vary and tend to be more than the other roof shapes discussed. Clients who prefer these designs will tend to want more personalized homes, and the neighborhoods should be carefully selected, especially when resale is considered.

----------- 6.2 -----------
ROOF PITCH

The appearance of any of the roof shapes discussed will vary considerably with the angle of the roof (Figure 6–7). This angle is referred to as pitch or slope. The greater the pitch or slope, the higher the roof will be. Although both pitch and slope refer to the roof angle, there is considerable confusion between the two terms and they are often incorrectly interchanged in conversation.

Pitch is equal to the vertical rise of the roof over the span of the roof. Referring to Figure 6–8, if the rise is 10'-0" and the span is 40'-0", the pitch is $^{10}/_{40}$ or ¼ and is referred to as a ¼ pitch.

Slope is equal to the vertical rise of the roof over the span of the roof. For roof slopes it is expressed as a ratio of the vertical rise to the horizontal run. Referring to Figure 6–8, if the rise is 10'-0" and the run is 20'-0", the slope is 10 feet in 20 feet. But, since the slope is always expressed in terms of the rise of the roof per foot of run, and run is expressed as 12 inches,

Figure 6–7. Varying pitches.

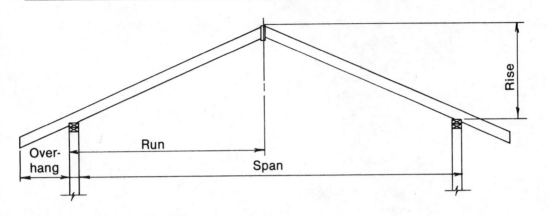

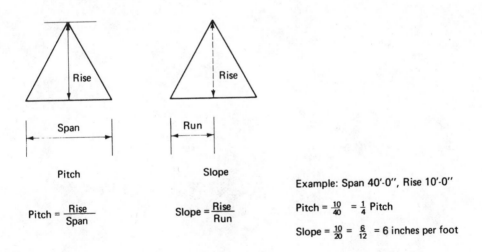

Example: Span 40'-0", Rise 10'-0"

Pitch = $\frac{10}{40}$ = $\frac{1}{4}$ Pitch

Slope = $\frac{10}{20}$ = $\frac{6}{12}$ = 6 inches per foot

Figure 6-8. Pitch and slope.

some conversion in numbers is required. A slope of 10 feet in 20 feet equals a slope of 120 inches in 240 inches, which is reduced to 6 in 12, the slope. The slope is noted on the drawings, usually elevations and sections, as a triangular symbol above the roof line.

Part II

Construction Assemblies and Materials

Chapter 7

Wood Frame Construction

This chapter covers the basics of wood frame construction (Figure 7-1), beginning with the foundation and progressing through the rough framing of a wood frame building.

7.1
FOUNDATIONS

The term *foundation* refers to the portion of the building which supports the main framing of the building. It transmits the loads of the building to the ground. Most commonly used foundations include:

1. Continuous exterior wall (Figure 7-2).
2. Slab on grade (Figure 7-3).
3. Piers (Figure 7-4).

Within each grouping of foundations there are several variations to also be considered.

7.2
CONTINUOUS FOUNDATIONS

Continuous exterior wall foundations utilize a concrete footing around the exterior of the building to transmit the load (weight) of the building down to the ground. This continuous footing may be all at one elevation (level) or may be stepped (Figure 7-2). Stepped footings are usually used on sloping sites. The continuous footing usually is sized as twice the width of the wall being placed on it and the same depth as the wall is wide (Figure 7-5). One of the basic rules for locating a footing of this type is that it must be placed *below* the frost line. Otherwise the freezing and thawing of the ground under the footing would cause the footing to shift and the building to move, causing cracks in the structure. The depth of the frost line varies considerably even within each state, and the building code and local building department should be checked.

Basements

In many northern climates the frost line is in excess of 3 feet deep, and often the builder will extend the excavation down and put a full or partial basement (Figure 7-6) under the building. Thus, for the cost of a few extra feet of excavation, the added materials for a higher wall, and a concrete floor, a sizable amount of square footage is added. Basements are also common on sloped sites where at least a portion of the space under the first floor can be excavated economically. Foundation walls commonly are concrete or block, with all-wood foundations popular in some areas.

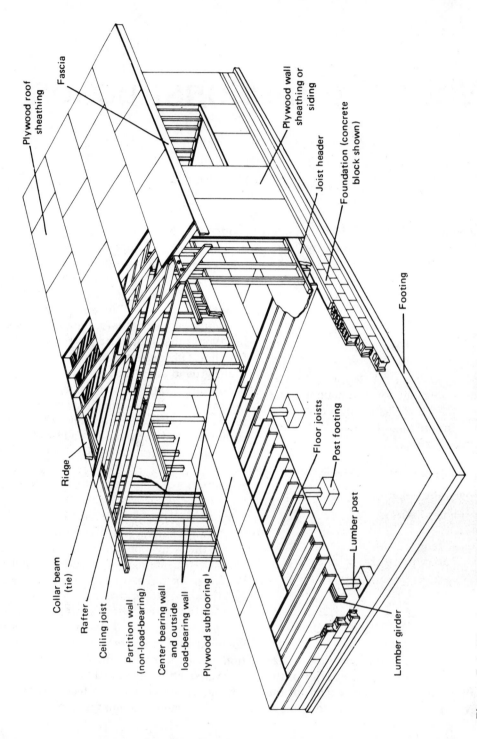

Plywood roof sheathing

Fascia

Plywood wall sheathing or siding

Joist header

Foundation (concrete block shown)

Footing

Ridge

Collar beam (tie)

Rafter

Ceiling joist

Partition wall (non-load-bearing)

Center bearing wall and outside load-bearing wall

Plywood subflooring

Floor joists

Post footing

Lumber post

Lumber girder

Figure 7-1. Wood construction framing. (Courtesy American Plywood Association)

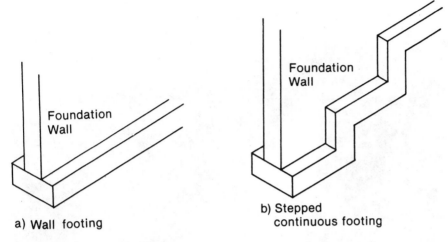

a) Wall footing

Foundation
Wall

b) Stepped
continuous footing

Foundation
Wall

Figure 7-2. Continuous Footing.

Figure 7-3. Slab on grade.

Figure 7-4. Pier foundation.

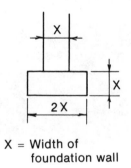

X = Width of
foundation wall

Figure 7-5. Typical footing size.

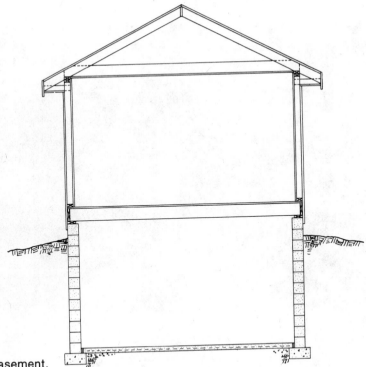

Figure 7-6. Basement.

Crawl Space

Where it is not economical to excavate for a basement the continuous exterior footing is used with a short foundation wall, usually of concrete or masonry. The bottom of the footing must still be below the frost line, but it is less costly than excavating for a basement. Crawl spaces are used extensively in mild climates but the high cost of construction has increased their usage even in cold climates. The clear space required in the crawl space varies, but as little as 18 inches is allowed in some locales.

There are several variables in the method of constructing the foundation wall. Two of the most common are the *continuous wall* (Figure 7-2) and the *pier* (Figures 7-4 and 7-7). The continuous wall may be of concrete or masonry, block and brick, or just brick. Pier construction is most commonly block, with a brick veneer placed around it later to cover the piers and hide the space under the house.

7.3
SLAB ON GRADE

Slab on grade construction is any method in which the floor of the residence is made of poured concrete. Slab on grade may be incorporated as part of the construction which uses a continuous wall footing (Figure 7-8) or the slab itself may turn and form its own footing (Figure 7-9). Reinforcing mesh is used to pro-

Figure 7-7. Piers.

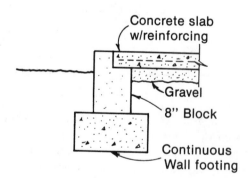

Figure 7-8. Slab on grade with continuous footing.

Concrete slab w/reinforcing

Gravel

8" Block

Continuous Wall footing

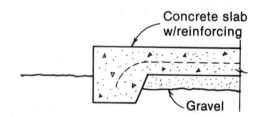

Figure 7-9. Slab on grade forming a foundation.

Concrete slab w/reinforcing

Gravel

Figure 7-10. Mechanical systems in slab.

vide strength to the slab as it expands and contracts due to temperature changes. Plumbing, heating, and electrical installations (Figure 7-10) must be carefully coordinated since they would be difficult to relocate if incorrectly installed.

7.4
WESTERN PLATFORM FRAMING

No matter what type of foundation is used, the western platform framing shown in Figure 7-1 is the most common form of wood frame construction used. (No floor joist is needed for slab on grade construction.)

Floor Framing

Typically, the floor framing for a wood frame residence consists of a sill, girder, floor joists, any required trimmers and headers, and the sheathing.

The *sill plate* (Figure 7-11) is placed on top of the foundation wall and serves as the nailing ledger for the floor joists which will be placed on it. The sill plate is usually a 2 x 6 or 2 x 8, and treated wood is recommended. A sill sealer (Figure 7-12) is usually installed on top of the foundation, and the sill plate is put on top to reduce the flow of air through the crack.

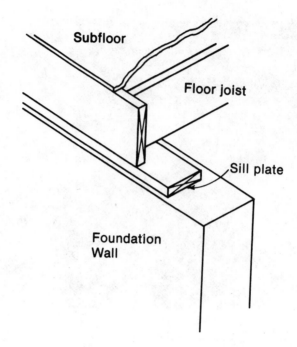

Figure 7-11. Sill plate.

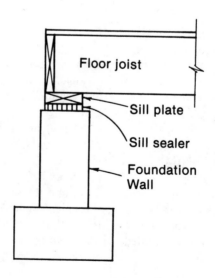

Figure 7-12. Sill sealer.

The sill plate is commonly anchored to the foundation to ensure the equal transfer of the building loads to the foundation and to offer extra strength against possible damage due to tornados or earthquakes. Anchorage usually consists of anchor bolts cast in the concrete foundation or set in the mortar placed in the cores of concrete block. Each length of sill plate should have at least two anchors, and the holes drilled in them should be at least ⅛ inch larger than the anchor to allow for shrinkage of the sill material.

Girder. When a building has a width greater than a floor joist span, a girder (beam) is used to reduce that span. The floor joist can then rest on, or tie into, the girder. The girder is often a built-up wood member (Figure 7–13). Typical sizes required are shown in Figure 7–14. Steel beams are also commonly used as girders, usually with a wood nailer along the top to which the floor joists are nailed (Figure 7–15). The girder should have a minimum of 4-inch bearing on the foundation wall. When untreated wood is used, allow a ½-inch air space at the end and sides of the girder (Figure 7–16) to allow air movement, keeping it drier.

Most codes restrict the clear span of girders supporting load-bearing walls to about 8 feet. This means the girder will have to be supported by a series of piers or posts. Round steel columns are usually used in wood frame construction with a basement. This steel column is cut to the required length and set on a footing (Figure 7–17). Adjustable steel columns are also available but, if used, are usually welded at the desired height so they cannot be tampered with. Piers of brick or block are commonly used in crawl spaces (Figure 7–18).

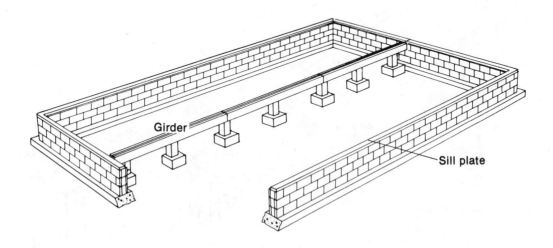

Figure 7-13. Girder.

Maximum Span for Wood Girders and Sills

Nominal Size in Inches	Maximum Clear Span	
	1 Story Dwelling	1½ to 2 Story Dwelling
4 × 6	5 ft.	4 ft.
4 × 8	6 ft. 4 in.	5 ft. 6 in.
6 × 8	8 ft.	7 ft.
4 × 10	8 ft.	7 ft.
6 × 10	9 ft.	8 ft.

Maximum Spans of Wood Girders For Non-Load Bearing Conditions:

Nominal Girder Size	Houses Up to 26 Ft. Wide	Houses 26-32 Ft. Wide
4 × 6	5 ft. 6 in.	
4 × 8	7 ft. 6 in.	7 ft. 0 in.
4 × 10	9 ft. 0 in.	8 ft. 6 in.
6 × 8	9 ft. 0 in.	8 ft. 6 in.
6 × 10	11 ft. 6 in.	10 ft. 6 in.
6 × 12	12 ft. 0 in.	11 ft. 6 in.

Figure 7-14. Girder—typical sizes.

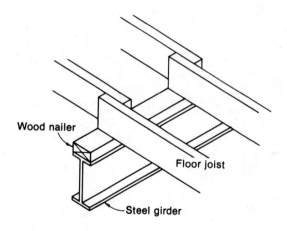

Figure 7-15. Steel girder.

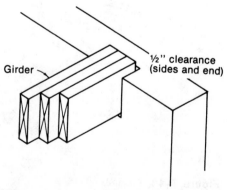

Figure 7-16. Girder air space.

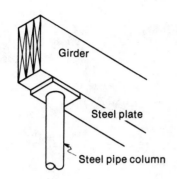

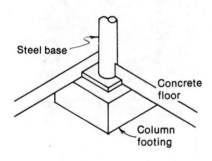

Figure 7-17. Steel column or pier.

NOTE: Occasionally wood trusses are used for floor joists and may span the entire width of the building. In such cases a girder, and its accompanying piers or posts and footings, are not required.

Floor Joists. The structural support for the floor are the floor joists (Figure 7-1). They are commonly a 2 x 8, 2 x 10, or 2 x 12, with sizing dependent on many factors (Section 10.9). Trusses utilizing lumber and plywood or lumber and lightweight metal tubing, are also used (Figure 7-19).

The floor joists bear directly on the sill and on, or into, the girder. A variety of girder-to-joist connections are used, and several common arrangements are shown in Figure 7-20).

To provide extra strength the floor joists are usually doubled when a wall runs parallel with them. Doubling of the floor joists is required under load-bearing walls. If the wall will be a "wet wall" (have plumbing pipes running in it), the joists are spaced apart with a wood block (Figure 7-21).

Openings in the floor joists (for stairs, fireplaces, etc.) are framed with trimmers running in the direction of the joist and headers, which support the cut-off "tail beams" of the joist. Typically, when the header is 4 feet or less single headers and trimmers may be used (Figure 7-22). For header lengths in excess of 4 feet most codes require that double headers and trimmers be used (Figure 7-23).

Bridging is often required by code between the floor joists (Figure 7-24). It is intended to provide the floor system with additional stability and strength by transferring the building loads to the foundation. While tests have shown that the bridging does not significantly assist in load transferring, many codes still require its use. Bridging may be solid or cross, and the cross may be wood or metal (Figure 7-24). Most codes do not require the use of bridging when plywood sheathing is glue-nailed to the floor joists. Sizing of floor joists is covered in Section 10.9.

Sheathing. The floor joists then have subfloor sheathing installed over them to form a working platform and as a base for the flooring (Figure 7-1). Plywood is commonly used and may be the two-layer (Figure 7-25) or single-layer flooring system (Figure 7-26). The two-layer system is most commonly used in residential construction and consists of a plywood subfloor and a separate underlayment to act as a finish floor base. This

Figure 7-18. Block or brick piers.

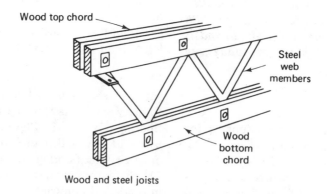

Figure 7-19. Floor joist trusses.

underlayment is usually a particle board or plywood. In the single-layer system the plywood serves as both the subfloor sheathing and the underlayment. The plywood panels should be installed with a ⅛-inch space at edge joints and a 1/16-inch space at end joints. (In humid and wet areas these spaces should be doubled).

Wall Framing

The wall framing consists of all the framing required to construct the exterior and interior walls of the building. Most walls are made of vertical studs, top and bottom plates, and headers. Wall framing is most commonly made with 2 x 4s, 16 inches on center, or 2 x 6s, 24 inches on center. To obtain the standard 8-foot ceiling height the assembly consists of a double top plate and a single bottom plate (Figure 7–1).

Studs. Wall studs are precut to a length of 93 inches to provide the 8-foot height. Several possible stud arrangements for corners are shown in Figure 7–27. Extra studs are required at these locations. The corners of the building are usually braced by the use of plywood sheathing (Figure 7–28), a 1 x 4 which is "let-in" (the studs are notched so that the outer edge of the 1 x 4 is flush with the outer edge of the stud), or metal straps.

Openings in the wall also require special framing. For openings 6 feet or less the header must extend over one stud on each end, and for openings over 6 feet the header must extend over two studs at each end (Figure 7–29). The horizontal piece below the opening is a sill and is the same size as the wall framing, while the short vertical pieces below the sill are referred to as cripples.

Headers. Headers are required to support the weight of the building over the openings. Headers for load-bearing walls (Figure 7–30) are often double 2 x 10 or 2 x 12 plywood box headers, or headers combining solid wood and cripples for high walls (Figure 7–31). For non-load-bearing walls a single 2 x 10 or 2 x 12, or cripples only, are placed above the opening (Figure 7–32).

Ceiling Assembly

The ceiling assembly typically consists of ceiling joists, headers, and trimmers, and is quite similar to the floor assembly (Figure 7–1). Ceiling joists are commonly a 2 x 6, 2 x 8, or 2 x 10 (size selection is covered in Section 10.9). Trusses are sometimes utilized as ceiling joists or as the entire ceiling/roof assembly.

The ceiling joists typically bear on the exterior wall and load-bearing interior wall. It is this load-bearing wall which should have a girder below it. Ceiling trusses may be designed to span from exterior wall to exterior wall, with no interior load-bearing walls. Trusses which are designed as the ceiling/roof assembly typically span from exterior wall to exterior wall.

Openings in the ceiling joists are handled with trimmers and headers in the same manner as floor joists (Figure 7–22). No bridging is required.

Roof Assembly

The roof assembly typically consists of rafters, collar ties, lookouts, plywood sheathing, and any special framing required for windows, such as dormers (Figure 7–1). The exact

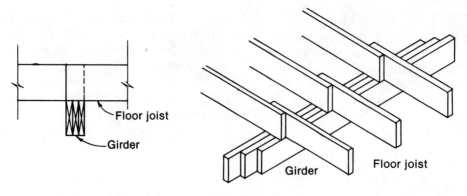

a) Floor joist on top of girder

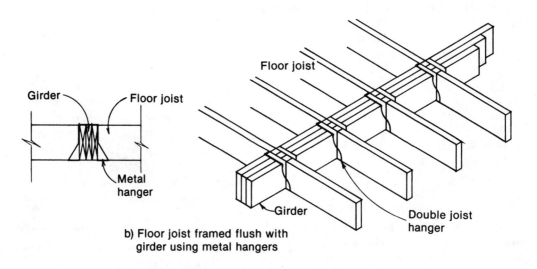

b) Floor joist framed flush with
girder using metal hangers

Figure 7-20. Girder to joist connections.

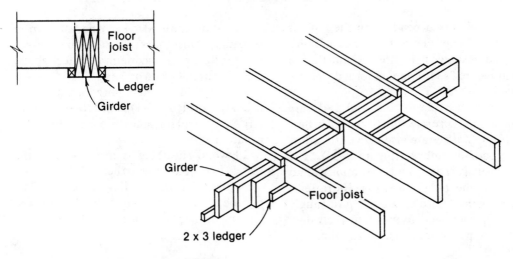

c) Floor joist framed with ledger

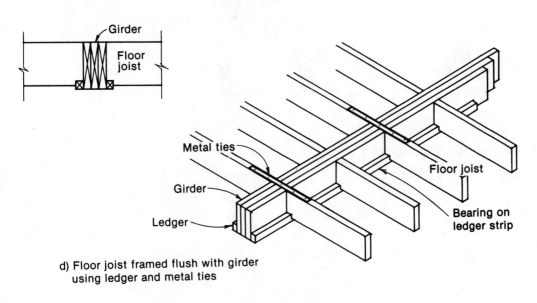

d) Floor joist framed flush with girder
using ledger and metal ties

Figure 7-20. (Continued)

framing will vary depending on the type of roof being built (Figure 6-1). The popular gable roof is shown in Figure 7-33 and the hip roof in Figure 7-34. Framing for a dormer is shown in Figure 7-35.

Rafter sizes of 2 x 6, 2 x 8, and 2 x 10 are most common (size selection is covered in Section 10.9). Trusses are also often utilized as ceiling/roof assemblies (Figure 7-36). The rafters usually span from the exterior wall to the ridge of the roof (Figure 7-1). The rafter is notched with a "bird's mouth" to fit snugly to the top plate and frames into the ridge, which is typically one size larger than the rafter.

Collar ties are used to keep the rafters from spreading. Most codes require collar ties every third rafter, or a maximum of 5 feet apart. They are usually 1 x 6 or 2 x 4.

Lookouts are often required when a soffit is boxed in (Figure 7-1). The lookout is usually a 2 x 4, and it frames into wood blocking or into the studs.

Plywood sheathing is installed over the rafters to tie it together structurally and to act as a base for the roofing materials. The plywood thickness is coordinated with the spacing of the rafters or trusses selected.

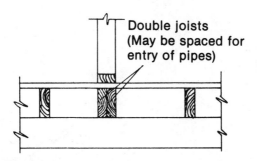

Figure 7-21. Doubled floor joist.

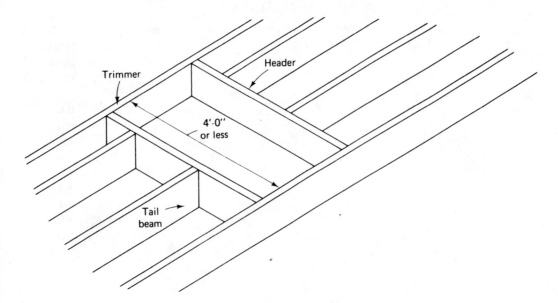

Figure 7-22. Headers, 4'-0" span or less.

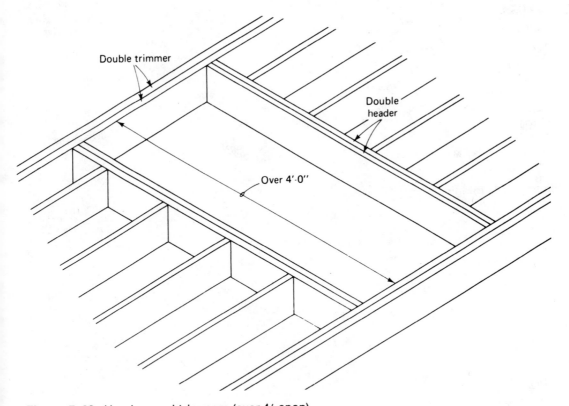

Figure 7-23. Headers and trimmers (over 4' span).

Bridging for floor and flat roof joists and beams

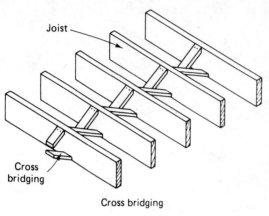

Cross bridging

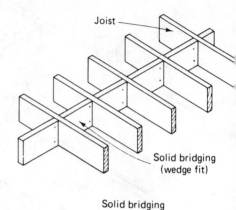

Solid bridging

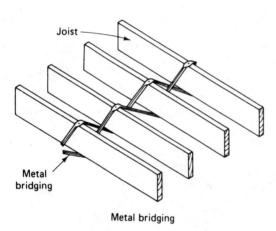

Metal bridging

Minimum sizes—wood cross bridging, 1 inch by 3 inches. Solid wood bridging, 2 inches thick, and same depth as members bridged. Metal bridging, minimum 18 gage.

Figure 7-24. Bridging.

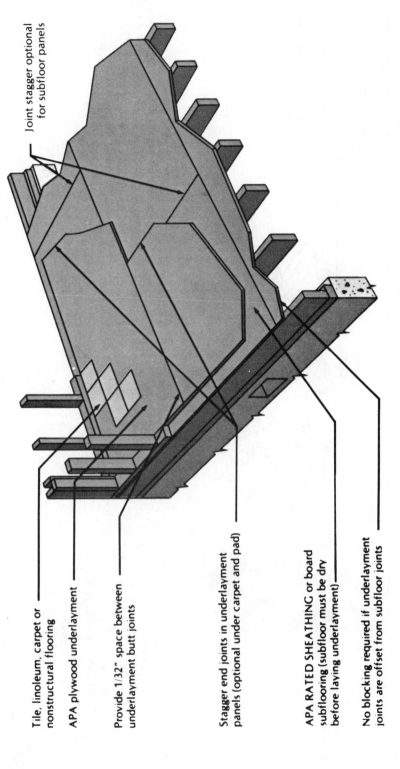

Joint stagger optional for subfloor panels

Tile, linoleum, carpet or nonstructural flooring

APA plywood underlayment

Provide 1/32" space between underlayment butt joints

Stagger end joints in underlayment panels (optional under carpet and pad)

APA RATED SHEATHING or board subflooring (subfloor must be dry before laying underlayment)

No blocking required if underlayment joints are offset from subfloor joints

Figure 7-25. Plywood sheathing (two-layer). (Courtesy American Plywood Association)

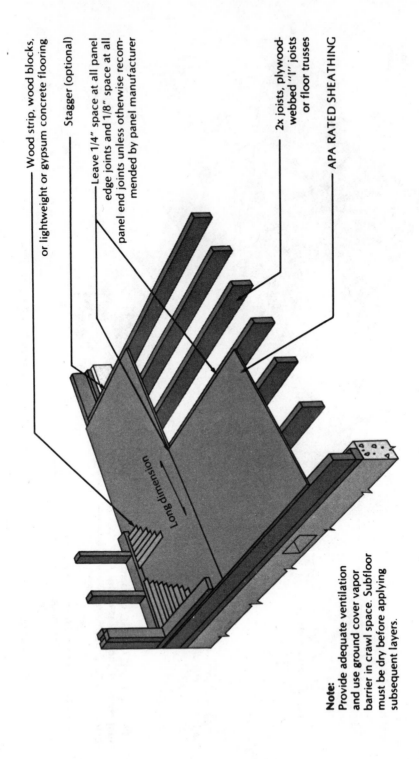

Wood strip, wood blocks, or lightweight or gypsum concrete flooring

Stagger (optional)

Leave 1/4" space at all panel edge joints and 1/8" space at all panel end joints unless otherwise recommended by panel manufacturer

2x joists, plywood-webbed "I" joists or floor trusses

APA RATED SHEATHING

Long dimension

Note:
Provide adequate ventilation and use ground cover vapor barrier in crawl space. Subfloor must be dry before applying subsequent layers.

Figure 7-26. Plywood sheathing (single-layer). (Courtesy American Plywood Association)

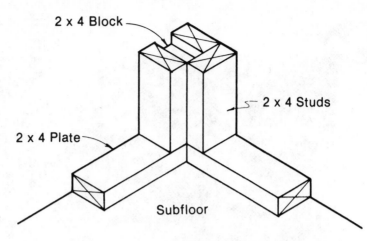

2 x 4 Block

2 x 4 Studs

2 x 4 Plate

Subfloor

a) Standard outside corner

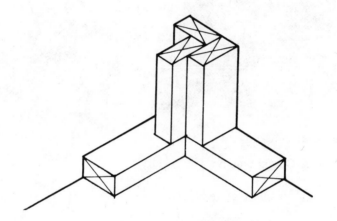

b) Alternate arrangement

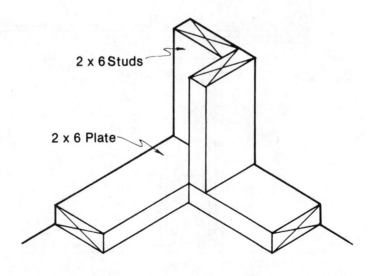

2 x 6 Studs

2 x 6 Plate

c) Typical corner using
2 x 6 studs

Figure 7-27. Corners.

Figure 7-28. Corner bracing.

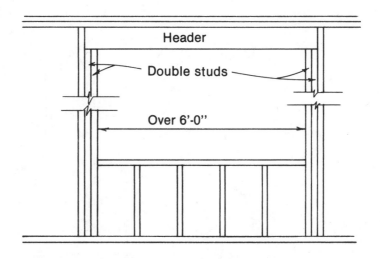

Header

Double studs

Over 6'-0''

Figure 7-29. Wall openings.

Figure 7–30. Headers.

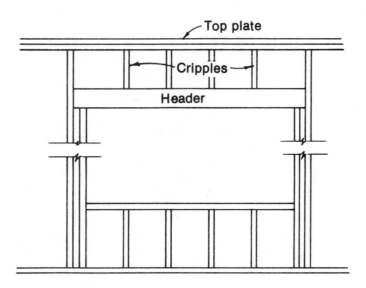

Figure 7–31. Headers.

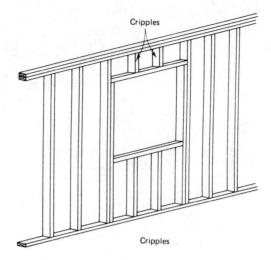

Figure 7-32. Headers, nonload-bearing.

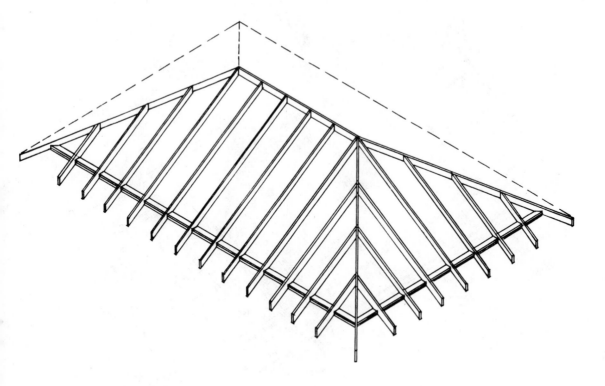

Figure 7-33. Gable roof framing.

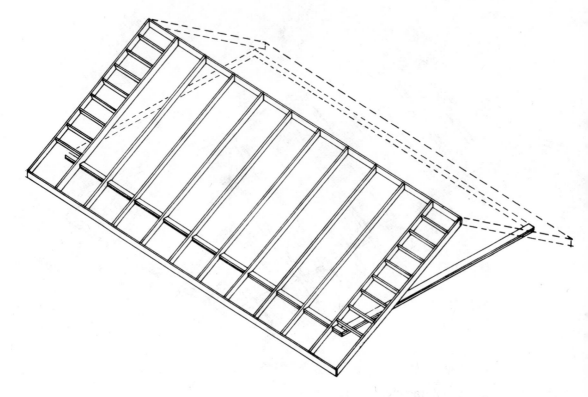

Figure 7-34. Hip roof framing.

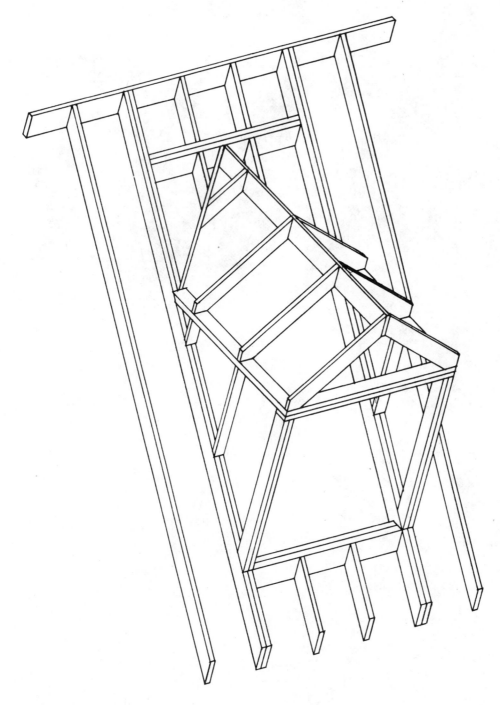

Figure 7-35. Dormer framing.

Figure 7-36. Truss.

Chapter 8

Concrete

Concrete is the only material brought onto the construction site for use in the building which can have its properties changed on the job. Other materials may be physically altered (cut, welded, burned, twisted, etc.) in a manner that would make them unsuitable, but only concrete offers the opportunity to change the properties of a material. This is why it is so important to learn all that is possible about the material. Workmen with 10 and 20 years' experience working with concrete do not necessarily understand what it is or how it should be handled. Misconceptions and misunderstandings are extensive, and the mistakes made become expensive. Organizations such as Portland Cement Association (PCA) and the American Concrete Institute (ACI) have many publications and technical bulletins which should be a part of your library.

The more that construction people know about concrete the better for the entire industry. This includes workers, contractors, laborers, architect-engineers, and inspectors. Too many times uninformed people from each category have been responsible for poor-quality concrete. The quality of the concrete may be affected at any time, from when the materials are being ordered, through mixing, placing, finishing, curing, and removal of forms.

The above are the reasons why the classrooms and laboratories of our technical schools and colleges are filled with people from the construction industry. Sponsors of such classes include the educational facilities themselves, building contractor associations, labor unions, construction industry advancement groups, and combinations of these.

8.1
CONCRETE

Concrete is made by mixing cement and water to form a cement paste binder that is mixed with materials such as sand, gravel, and crushed stone (referred to as aggregates).

The aggregates (sand, gravel, stone) are inert materials used as fillers. The basic relationship of cement to water will determine the strength of the concrete. Basically, the less water in the mix the stronger, more durable, and watertight will be the concrete, as long as the mixture is workable (usable, not so stiff it cannot be used). An excess of water will dilute the cement paste binder and the result will be a weaker and more porous concrete. The ratio of gallons of water to one bag of cement and corresponding strengths are shown in Figure 8-1. The proper water-cement ratio is the beginning of good concrete but it is not the

Total water (in the mix)	High-early strength		Regular	
Gal. per bag cement	Compressive strength*		Compressive strength	
	7 days	28 days	7 days	28 days
$4\frac{1}{2}$	6000	7000	4800	6200
5	5500	6400	4300	5600
$5\frac{1}{2}$	5000	5900	3800	5100
6	4500	5400	3300	4600
$6\frac{1}{2}$	4100	5000	2900	4100
7	3700	4500	2500	3700
$7\frac{1}{2}$	3300	4100	2200	3400
8	3000	3800	2000	3000
$8\frac{1}{2}$	2750	3500	1700	2700
9	2500	3250	1500	2400

*Compressive strength values are in lbs per sq in.

Approximate values, based on 6 x 12 inch cylinders, moist cured at 70° until tested.

Air-entrained concrete would have a compressive strength value about 80 percent that of the regular concrete.

Figure 8-1. Water-cement ratio.

only determinant. Other factors include aggregate type and proportions, proper curing of the concrete and adjustment of the water-cement ratio to damp or dry aggregates. A chemical reaction between the cement and water causes a hardening of the concrete referred to as hydration. This hydration of the concrete gives off heat, which is referred to as heat of hydration.

8.2
CEMENTS

There are several different types of cement available for use in the concrete mix; among the most common are those listed in Figure 8-2. As indicated, each has its own use, and several types are often required on any one

job. Cement is available in gray or white; however, the white may not be available in all of the types shown.

Air-Entraining Portland Cements

By adding small quantities of air-entraining materials, improved resistance to freeze-thaw action and to scaling caused by the chemicals which are applied to remove snow and ice is obtained. Minute air bubbles are well distributed throughout the concrete. Air-entrained concrete is also more workable, more sulfate resistant, abrasion resistant, and watertight. These advantages have led to widespread use of this type of cement. It is suggested for all exterior uses, especially where ice and snow are present, and is being increasingly used for interior concrete work.

All concrete has a certain amount of entrapped air, unavoidably present in any concrete regardless of how thoroughly it is compacted. Entrapped air is not intentionally retained in the concrete. By contrast entrained air is deliberately retained in the concrete. These bubbles are very small, with diameters of one to three thousandths of an inch. The bubbles do not connect with one another, each being individual, and are distributed evenly throughout the paste. The air voids

Type	Characteristics
I	All-purpose cement
II	Resists sulfates (moderate)
III	High-early strength
IV	Low heat of hydration
V	Resists heavy sulfate action

Air-entrained Portland cement is available in types IA, IIA, IIIA.

Figure 8-2. Types of portland cement.

caused by entrapped air amount to about one per cent of the volume, while air-entrained specifications most commonly require an air content of five to eight percent by volume.

Air-entraining materials may be added at the factory and are specified as Types IA, IIA, and IIIA, which correspond respectively with Types I, II, and III. White cement is also available with air entraining.

Miscellaneous types of cements which are available are plastic cements (used for making mortar, plaster, and stucco), waterproofed cements (add extra waterproofing qualities to the cements, available in gray or white), pozzolan cements (primarily for bridge piers and dams), as well as other special use varieties. It is important to determine what types of cement are readily available in your area, their cost, and how long it would take to obtain special types. Many times a minimum amount must be ordered to obtain the cement desired.

Shipment of Cement. The majority of portland cements used are shipped in bulk, loose, by railroad, truck, and barge. Cements are also shipped in paper bags; in the U.S. the standard bag weight is 94 pounds (1 cubic foot) with 376 pounds in a barrel, while the Canadian bag is 80 pounds. Bulk cement is often sold "by the barrel."

Storage. Portland cements will retain their qualities indefinitely provided they are kept dry. If the cement has been in contact with moisture, it will have less strength and set (harden) more slowly than dry cement. Bagged cement should be stored in a warehouse or shed, which should be as weathertight as possible. Bags should not be stacked on damp floors or against outside walls. Tarpaulins or plastic should be used to cover bags which will be stored for periods of time.

When no warehouse or shed is available, the bags should be stacked on raised wooden platforms and the bags of cement and platform covered with a waterproof covering.

Cement stored for long periods may harden. If the bags are "rolled" (turned back and forth), they will loosen up. If the cement is free of lumps and flows freely, it is still of good quality and was simply packed tight from long storage. If the lumps do not break up easily, or if there is any question relating to the quality of the cement, it should not be used, until it has been tested, especially where strength is important.

Bulk cements are usually stored in weathertight bins and may be stored for reasonable periods of time, the length of which depends on how weathertight the bin is. Bulk cement is usually used quickly, but if it becomes lumpy, or hard to move from the bin, it should be thoroughly checked. A word of caution on cement: different cements from different mills will vary in color from a light gray to a dark gray or "blue-gray" color. Use one type and make of cement, from one mill, on a project. If a change is necessary, slight color variations will not be noticeable if they are on different walls; for example, if the north wall is slightly darker than the east. It is noticeable when the color variation occurs on one wall.

8.3
AGGREGATES

The mineral aggregate comprises 65 to 75 percent of the total volume of the concrete. Basically, aggregate is broken down into two types, coarse and fine. The gradation of the particles (various sizes of the particles) is important in obtaining a dense, workable, and economical mix. The aggregates must be clean, strong, and well shaped.

Fine aggregates (such as sand) are those that are smaller than $3/16$ inch in diameter, while coarse aggregates (stone) are larger. Sieve analysis tests are made, in accordance with ASTM specifications, to determine the size groups of coarse material (¾, 1½, 2½ inches, etc.). A sieve analysis is also run on the fine aggregate to determine the gradation of the sample.

Normal weight concrete, 135 to 155 pounds per cubic foot, is obtained by the commonly used aggregates, such as sand, crushed stone, and gravel. Lightweight concretes, ranging from 80 to 115 pounds per cubic foot, use materials such as expanded shale, clay, and slag. The amount of moisture in the aggregates being mixed will affect the amount of water added to the mix. This cannot be determined ahead of time; it depends on whether the material being used is dry and will absorb water, or is wet and has excess "free" moisture, which requires that less water will be added.

The color of the concrete may be changed considerably by varying the cement and aggregates. Colors from pure white to buff to dark gray may be obtained in this manner.

8.4
ADMIXTURES

Any substance added to the portland cement, water, and aggregates (concrete) for the purpose of giving it some particular properties is called an admixture. Admixtures include those which are air-entraining, water-reducing, retarding, accelerating, pozzolans, workable and dampproofing. Many other types of admixtures are also available. Always find out as much as is possible about an admixture before you use it on a job. Read the manufacturer's technical bulletins and instructions before using the product. If possible, contact others who have used the product to determine what precautions, problems,

and results they had. Once you have decided to use a particular product, try it first in a sample batch with job materials and under the job conditions anticipated. As the use and control of admixtures can be tricky, questions which should be answered before using an admixture are:

1. What special mixing requirements are involved?
2. Are there corrosion hazards, especially when in contact with steel, aluminum, and galvanized steel?
3. Will it discolor the concrete (streaks)?
4. Will it change the color of the concrete, perhaps to lighter or darker shade?
5. What is its effect on strength of concrete?

Air-entraining may be added at the batcher (mixer) or at the factory. The admixture may be in powder or liquid form and is added either before or during the mixing. It is difficult to determine exactly what percent of the volume will be air-entrained and specifications allow some variation. The most common variation allowed is plus or minus one percent; in this case, if five percent air content is required, a range from four to six percent is acceptable. Closer tolerances in air content are not practical with admixtures and difficult even when the air-entraining material is added at the factory.

Retarding admixtures are used to retard (slow) the hardening of the concrete and increase the amount of time required for concrete to set. They may be used to counteract the effect hot weather has on the initial setting time of concrete and to delay initial set for any given reason, such as where continuous pours in foundations, piers, or abutments occur and there is the possibility of interruption. They are also used when the concrete is being pumped long distances.

Retarders are also used to obtain exposed aggregate finishes on sidewalks, panels, walls, and floors. The retarder is brushed, or sprayed, on either the form used for the finish face or the face of the freshly poured concrete (such as a sidewalk). Various types of retarders are available and will retard to different depths. Tests should be run to obtain desired results before they are used on the job and, if possible, enough retarder to complete the entire job should be purchased. The "pot life" (length of time it will stay good in the container) varies and will affect the amounts ordered. The retarder must be spread evenly over the surface since the thicker it is put on the deeper it retards.

Accelerating admixtures are used to speed the setting of the concrete and its development of strength. One of the most commonly used accelerators is calcium chloride. Many problems may occur with its use and it is suggested that other methods be used to achieve acceleration of the setting time and strength development. Among the alternate methods are the use of high-early-strength cements, increasing the temperature of curing and either reducing the water–cement ratio by decreasing the amount of water or increasing the amount of cement. It is the opinion of the author that calcium chloride should *not* be added to the concrete mix except when samples are made and tested. Great care should be taken in its use since excessive amounts will affect curing and strength.

Color is an admixture being used with increasing frequency as decorative concretes gain acceptance. When color pigments are required, samples should be made with job materials under anticipated job conditions and these samples should be submitted for the architect-engineer for approval. When possible, enough color for the entire project should be ordered at one time and be from one batch produced by the manufacturer. This batch should be checked to be certain that it will match the sample already made and approved. Any tendency for the color to wash out from rain and sun should also be checked.

The use of color can be tricky, and many firms have lost large sums of money due to unsatisfactory results. Care in the use of these admixtures must be extreme if uniform results throughout the project are required.

8.5
WATER

Water used in the concrete mix should be generally clean and free from any oil, acid, or alkali, with limited amounts of sulfates being acceptable. Water fit to drink, unless it has a high sulfate content, is acceptable for the concrete mix, but many sources of water not fit to drink are also acceptable. If any doubt exists, make samples using the water and have them tested, or take samples of the water and have it tested by an independent laboratory; quite often the cement company has a lab which will perform tests for you.

8.6
MIXING

The concrete required for the work may be obtained from job site facilities or from a central plant which supplies ready-mixed concrete in a given area. The proper proportion of materials must be accurately measured for each batch (quantity of materials to be used at one time) at these batching plants. The large majority of concrete used in building construction is purchased from the central plants, referred to as ready-mix companies.

With truck-mixing, the mixing water must be accurately controlled by the driver. Control of the water on a truck is difficult with the equipment in use and few drivers understand the importance of adding just the proper amount, not too little or too much. The consistency is checked by use of a standardized test, ASTM C172; this is referred to as the slump test. The amount of drop of the properly tamped fresh concrete, once the test cone is removed, is the slump.

Remixing of the concrete is not generally desirable and is not allowed by many project specifications. Concrete can be remixed provided that initial set has not occurred (initial set may occur up to two hours after placing), that as it is remixed it becomes plastic without any additional water, and that it may be compacted into the forms. The remixed concrete will harden rapidly and must be placed quickly.

8.7
HANDLING AND TRANSPORTING

There is a tendency for the concrete mix to have a segregation of materials, with the coarse aggregate settling to the bottom. This segregation may be minimized by careful handling, transporting, and placing of the mix. All concrete should be dropped vertically, straight down, since discharging at an angle tends to accumulate the mortar at the near side of the receiving container and the coarse aggregate at the far side. Once the concrete has segregated it can seldom be corrected in the handling and transporting process; instead it must be remixed.

Chutes

The chutes used to move concrete should be metal or metal-lined, with a round bottom. The chute should be large enough so that the concrete will not overflow. The slope of the chute should be set so that the concrete flows fast enough to keep the chute clean yet not so fast that there will be segregation of the materials. The slope of the chute for average mixes (about 3- to 5-inch slump) is usually from 1:3 to 1:2, but with stiffer, lower slump mixes a steeper slope may be used. The condition of the concrete mix as it leaves the chute is what should determine the slope; if the material has segregated it is too steep; if the mix will not flow easily it is too flat.

8.8
PLACING

After the concrete has been transported to its point of deposit it is ready to be placed. Before the concrete is placed the subgrade should be prepared and the forms and reinforcing set. Subgrades must be at the proper elevation with all loose soils removed or compacted, and in hot weather it may be necessary to moisten the area so that the mixing water needed for curing will not be absorbed into the soil. The forms should be tight, aligned, properly braced, and cleaned of sawdust, nails, and other debris. The inside surface of the forms may be oiled for easy form removal. Reinforcing should be free of loose rust, mill scale, hardened concrete, grease, and oil before the placing begins. The reinforcing should be securely tied in place so the placing of the concrete will not disturb it.

The actual placing of the concrete should be as close to its final position as possible. Large quantities of concrete should not be placed in one spot and worked, pulled, or allowed to run over long distances, since this tends to cause segregation of materials, often resulting in honeycombing. Honeycombing is the exposing of the coarse aggregate in the finished surface when there is insufficient paste (mortar) to cover this aggregate.

Concrete which drops freely for more than 4 feet also has a tendency to segregate. To guide the concrete into the form various types of chutes, made of metal or rubber, are used. The length of chute must vary as placing continues, and adjustable chutes or several lengths of chute must be used.

The concrete should be placed from the far end of the work, with each batch being placed directly against the concrete already in place. As the concrete is placed it is worked into place by vibrating. Vibration of the concrete consolidates each layer of concrete, embeds the reinforcing and any inserts or anchors while it is eliminating large air bubbles, and brings fine aggregates to the surface for finishing.

8.9
CURING

Proper curing is one of the principal factors in obtaining good concrete. The concrete is cured by the chemical combination between the cement and water, referred to as hydration. This hydration requires time and favorable temperatures and moisture; this time is referred to as the curing period.

The strength and watertightness of concrete will improve with proper curing, as will its resistance to weathering, freeze, and thaw. When placed on the job, in forms or as a slab, concrete has sufficient water for complete hydration, but it must be protected so that the moisture is not lost during the early stages of curing. Evaporation is one of the biggest factors in the loss of moisture, and on windy, dry days the problem is increased. Temperatures should be kept within a given range (50°–90° F) since as the temperature drops hydration slows, and high temperatures for curing are not acceptable unless there are precautions

taken to provide sufficient moisture for hydration.

Curing methods used to keep the concrete moist may be grouped into three types:

1. Those that prevent moisture loss from the concrete by sealing it in.
2. Those that supply additional moisture.
3. Those that supply heat and moisture to accelerate the strength gain of the concrete.

Watertight covers such as waterproof paper and polyethylene films are often used to seal in moisture on floors and other horizontal areas. These covers should not be applied until after the concrete has sufficiently hardened and surface damage may be prevented. Covers applied too soon will damage the surface finish which was applied to the concrete.

The covers used should be as wide as possible, overlapped several inches, and sealed tightly with pressure sensitive tape, sand, wood planks, or glue. Wood planks are commonly used since they are inexpensive, can be reused, are easy to apply, and easy to remove. Sand is inexpensive, but if the sand gets on the concrete it must be removed, a time consuming job. Removing the sand from the cover may also be time consuming and awkward. Pressure sensitive tape and glue are time consuming to apply and once applied are not easily removed. However, the paper can be easily cut for reuse. There is less possibility of the moisture escaping with this type of seal if it is applied correctly. The sides and ends of all covers should be sealed so that moisture will not escape at these points.

Covers must not stain the concrete, should be inspected for rips and tears, and the upper surface should be white for use during hot weather. Commonly used materials include plastic sheets (polyethylene) which are available in thicknesses of 2, 4, 6, 8, and 10 mils, lengths of 50, 100, 200, and 300 feet (depending on thickness and width, the 100-foot length is most common), and widths from 3 to 40 feet (not all sizes in all thicknesses). The sheets are available in black or clear; white sheets are available in 4 mil, 100-foot lengths, with 12- and 24-foot widths. Always check with manufacturers and suppliers for exactly what is available and what may be ordered. Tape is available in 1- and 2-inch thicknesses, in rolls 100 feet long.

Curing compounds provide a seal against the loss of water used in the concrete for a period of 7 days or more. The compounds are usually applied right after the concrete surface has been finished and is still moist. The compounds may be applied by use of power-driven or hand-operated spray equipment. On large paving projects power-driven spray equipment, with spray nozzles, is used. There are both one- and two-coat applications in clear (or translucent), white, light grey, and black. The white is used when it is necessary to reflect the heat of the sun, thus preventing a temperature rise which might cause cracking and spalling of the concrete. The colored compounds also provide for visual inspection by the applicator, and inspector, to insure uniform coverage of the compound.

Care in selecting the type of compound used is important since some compounds prevent the bonding of fresh and hardened concrete, and some affect the adhesion of paint and adhesives (such as required for resilient floors) to the concrete. The manufacturer of the curing compound should be consulted to determine what problems may occur. The compounds may be removed by grease removers, power wire brush buffers, or sandblasting, but these remedies are expensive and it is far less expensive to select the proper curing compound at the outset. Some curing compounds are formulated so that they will

become brittle and disintegrate with age to facilitate easy removal.

Forms left in place will provide protection against the loss of moisture as long as certain precautions are taken. The top, exposed surface of the concrete must also be kept moist. Either watertight covers, compounds, wet coverings, or sprinkling may be used to keep this exposed surface moist. During hot, dry weather the wood forms must be kept moist (sprinking is the most common method). The forms should be kept in place as long as is practical unless they are not kept moist. In this case, the forms should be immediately removed and some other form of curing used.

Sprinkling the concrete with a continuous spray of water is one of the most effective methods of curing. The water should be applied continuously through a system of nozzles, and care must be taken that the finish of the concrete is not damaged by too strong a spray too early in the curing process. If the sprinkling is to be performed at intervals, take care that the concrete does not dry out between applications, since this tends to increase the possibility of crazing, the fine cracks which may occur in the surface of new concrete as it hardens. This method is more expensive than the other methods and requires careful supervision and an adequate supply of water.

Wet covering by moisture-retaining fabrics, such as burlap, hay, or straw, is sometimes used for curing. The covers must be kept constantly moist and care must be taken to cover the entire surface, including any exposed sides and edges. The covers should be placed as soon as possible but not until the concrete has sufficiently hardened, so that no surface damage will occur. There is a possibility of discoloration with burlap, hay, and straw and this may be a major disadvantage. However, nonstaining burlap is available as is white burlap-polyethylene sheeting. Most new burlap must have any soluable substances in it removed by rinsing in caustic soda; this also makes the burlap more absorbent.

Ponding is a curing method which may be used on flat surfaces such as sidewalks, pavements, and floors. Dikes of earth or sand are built up around the perimeter of the surface and the enclosed area retains a pond of water. This method is effective at preventing the loss of moisture and also is effective in controlling the temperature of the concrete in hot weather. However, care must be taken in regard to the following items:

1. If the area is flooded too soon the concrete may be weakened.
2. The surface must remain ponded.
3. Be sure that there is no possibility of freezing weather.
4. If any soil gets on the concrete surface it will be difficult to remove.

Considerable labor and supervision are involved, thus raising the cost. Ponding is most commonly used for small areas.

8.10
COLD WEATHER CONCRETING

Successful winter concreting is possible provided that precautions are taken when the temperature, during placing and the early curing period, is 40°F or lower. During these periods the fresh concrete must be protected against freezing. Air-entrained concrete is commonly used in winter construction since it reduces the possibility of damage from any freeze-thaw which may occur.

To obtain concrete which will attain its required properties care must be taken to allow the fresh concrete to cure properly. Commonly used methods involve a combination of heating the materials used, providing a cover or enclosure and maintaining the proper temperature by using some type of heater. At the time of placing, the temperature of the cement should be between 50 and 70°F (temperatures above 70°F result in lower strength). The application of dry heat has a tendency to draw moisture out of the concrete and it may be necessary to add moisture (sprinkling) to reduce the possibility of cracking and crazing.

The use of low water-cement ratio, air-entrained, high-early-strength concrete will reduce curing time and the tendency of the concrete to scale from freezing and thawing. This allows quicker reuse of forms, quicker removal of forms, savings in heat costs, and early completion of concrete work.

Job batching presents a few added problems to cold weather concreting. In cold weather the aggregates, which generally contain moisture, become frozen and ice is often present when there are freezing temperatures. These frozen aggregates must be heated so the ice will melt, since if it thaws after the concrete is placed a void will be left where the ice was, and if it melts during mixture it will add excessive water and alter the desired water-cement ratio.

The aggregates may be heated by stockpiling them over pipes through which steam is circulated. Live steam may be used; however, high free moisture may occur which must be considered in the mix design. Aggregates, for small jobs, may be heated by stockpiling them over large metal pipes and then building fires in the pipes. This may result in a low free moisture content for the aggregates closest to the fire. All aggregate stockpiles which are being heated should have tarpaulin (canvas) over them. This serves to distribute the heat, as well as retain it, and also acts to prevent any icing.

Heating the water used in the mix is the easiest and most practical approach to heating the mix. This approach may be used provided all other ingredients are above 32°F. Water stores about five times as much heat, per pound, as the other materials, can be easily heated, easily handled, and the cost is low. As a precaution, do not heat the water above 180°F since it might cause quick setting of the concrete. There are times when both the aggregates and the water may be heated.

Before concreting, certain preparations must take place:

1. If the subgrade is frozen, it must be thawed before placing the concrete. If it is not thawed, when it does thaw uneven settlement may occur which may (and probably will) cause the concrete to crack.

2. The placing of warm concrete on a frozen subgrade will draw heat from the concrete, which will slow the hydration process and possibly freeze the lower part of the concrete member.

At the time concrete is placed, the insides of forms, reinforcing steel, and accessories must be free of ice and snow. If not, one of two things will occur:

1. The ice and snow will melt as the concrete is being placed and this increased water will weaken the concrete in that area.

2. The ice and snow will not melt during placing of the concrete, thus taking up space which should have been filled with concrete, preventing bonding of concrete to reinforcing bars and accessories and leaving voids when it does melt.

Frozen subgrade may be thawed by burning straw, steaming, or covering with an enclosure and applying heat.

Temperatures must also be maintained during the curing process. Heat for curing can use either live steam or unit heating devices in an enclosure which may be made from tarpaulins, wood, building boards, plastic sheets, or waterproof paper. The enclosure should be flameproof and the unit heaters vented. Heaters may be powered by kerosene, gas, oil, or coal; all produce a dry heat and care must be taken to prevent drying of the concrete, in particular near the heating elements. Do not place heaters directly on a concrete floor slab; instead they should be elevated and damp sand placed on the concrete near them.

Heat may be more effectively retained in the enclosure by use of insulation placed on the enclosure. For recommendations and amounts of insulation required for various temperatures refer to ACI's "Recommended Practice for Cold Weather Concreting" (ACI 306). Factors in determining the amount of insulation required include type and design of forms, temperature, wind, and amount of construction enclosed. Leave the forms in place as long as possible since even within the heated enclosure they help prevent drying, distribute the heat more evenly, and help prevent localized overheating.

8.11
HOT WEATHER CONCRETING

While it is generally easier to get good quality concrete during hot weather than during cold weather, several precautions must be taken for successful concreting. The keys to success in hot weather concreting are keeping the temperature of the fresh concrete down and preventing rapid evaporation of water or moisture from the concrete.

Fresh concrete temperatures are best obtained by controlling the temperatures of the materials to be used. The water and aggregates should be kept as cool as possible. Water is the easiest material to cool, and cold water may be obtained by refrigeration or adding ice. Crushed or flaked ice is sometimes used as part of the mixing water since it is more effective than water in reducing the temperature (at the moment that the ice melts it absorbs extra heat). The ice must be completely melted by the time the mixing is complete. Since the aggregates represent 60 to 80 percent of the total weight of the concrete mix, their temperature will also have an effect on the temperature of the fresh concrete. Simple methods which will keep the aggregates cool include shading them from the sun and spraying them with water, which cools them as the sprayed water is evaporated.

Rapid evaporation of the water or moisture may be controlled by keeping the subgrade damp so that it won't absorb water from the concrete, wetting the wood forms, discharging the concrete immediately after mixing, protecting the concrete with temporary covers (such as polyethylene or wet burlap); immediately after striking off the concrete, finish only small areas at a time and then protect them with a covering immediately after finishing. In extreme cases, windbreaks may be required, sun shades erected, and fog nozzles spraying water used. The concrete should be cured as soon as possible, but you must be certain that the curing method will allow sufficient moisture to be retained in the concrete for hydration.

High temperatures accelerate the hydration of the concrete and extra water is required to maintain a desired consistency. If water is added to maintain the consistency it is important that cement also be added so that the water-cement ratio will not be changed. A higher water content may cause cracks to develop in hardened concrete due to shrinkage.

Cracks may also occur due to the rapid evaporation of water from the hot concrete and from volume changes as the concrete cools down from its initial high temperatures.

Higher temperatures also require an increase in the amount of air-entraining to produce a given air content and accelerate the setting time, thus leaving less time for the handling and finishing of the concrete. The concrete should be kept plastic for a sufficient time to allow each layer being placed to be continuous with no cold joints. A cold joint occurs when fresh concrete is placed against hardened concrete, since the joint is not continuous.

An upper temperature limit for the placing of fresh concrete is about 90°F. Satisfactory concrete can be placed at high temperatures but it requires careful adjustment of the mixture and extra care in placing and curing of the concrete.

8.12
REINFORCING

Concrete has significant compressive strength but very little tensile strength. Steel is placed in the concrete to provide:

1. Tensile strength.

2. Increased compressive strength.

3. Resistance to the stresses from temperature changes which cause expansion and contraction and from shrinkage during the curing period.

4. Distribution of the loads through the concrete and over the soil bearing area.

The location of the steel in the concrete will vary depending on what the steel is required to do.

The types of reinforcing discussed in this chapter include reinforcing bars and welded wire mesh.

Reinforcing Bars

These bars are available in #2 through 11, 14 and 18 (Figure 8–3); the most common length is 20 feet. All reinforcing bars, except the ¼-inch, are deformed bars; that is, they have a raised pattern. Deformed bars provide a greater surface area for bonding to the concrete than plain bars. The bars are deformed in accordance with ASTM A305, which specifies the size, position, and spacing of the projections. The bars are numbered in eighths, from #3 through #8; for example, a #7 bar designates a deformed bar averaging ⅞

Bar no.	Nominal size — inches		Area sq. in.	Weight lbs/ft.
2	$\frac{1}{4}''$	ϕ	0.05	0.167
3	$\frac{3}{8}''$	ϕ	0.11	0.376
4	$\frac{1}{2}''$	ϕ	0.20	0.668
5	$\frac{5}{8}''$	ϕ	0.31	1.043
6	$\frac{3}{4}''$	ϕ	0.44	1.502
7	$\frac{7}{8}''$	ϕ	0.60	2.044
8	1.0''	ϕ	0.79	2.670
9	1.13''	ϕ	1.00	3.400
10	1.27''	ϕ	1.27	4.303
11	1.41''	ϕ	1.56	5.313
14	1.693''	ϕ	2.41	8.18
18	2.257''	ϕ	3.98	13.52

No. 2 bar is a plain round bar. All other bars are deformed.

Figure 8-3. Reinforcing bars.

inch in diameter. Bars #9, #10, and #11 replace bars, 1, 1¼, and 1⅛ inch, respectively, and have the same area and weight as the bars replaced. Bars #14 and #18 are used only in the heaviest construction and usually must be specially ordered.

The primary use of reinforcing bars in footings is to help distribute the loads over the area of soil required. To accomplish this, several smaller bars often work better than a smaller number of larger bars. Four #4 bars have about the same area as two #6 bars but the four #4 bars will do a better job of distributing the load. The placement steel ties the reinforcing bars together, making them easier to place in the form and allowing better control of spacing (distance between bars) and cover; #3 bars 2 feet on center are usually adequate as placement steel.

Welded Wire Mesh

This material is made of steel wires at right angles to each other and secured at the intersections. An economical reinforcing for floors and driveways, it is commonly used as temperature reinforcing as well as reinforcing for beam and column fireproofing. The wires which run lengthwise are referred to as longitudinal, while the transverse wires are those which run in the short dimension. The longitudinal wire is generally heavier than the transverse wire when the mesh is being used for other than temperature reinforcing. Longitudinal spacing varies from 2 to 16 inches in ½-inch increments while transverse wires vary from 1 to 18 inches in 1-inch intervals. Almost any spacing is available but it may have to be specially ordered. Wire mesh is designated by a combination of wire size and spacing. Mesh designated as 4 x 8 6/12 means that the longitudinal wires are spaced 4 inches on center while the transverse wires are spaced 8 inches on center; the longitudinal wire is 6 gauge while the transverse is 12 gauge. Mesh is available in rolls 5 feet wide and 150 feet long or in flat sheets (of a variety of sizes, usually specially ordered). The finish may be either plain or galvanized (often specially ordered). The mesh is usually lapped one square and tied with wire.

Once the concrete hardens, any stresses placed on the concrete are transmitted to the reinforcing by the bond between the concrete and the reinforcing. The reinforcing used must be free from oil, paint, or loose, flaking rust so the concrete and reinforcing will bond. Excessive rusting may actually reduce the cross-sectional area of the bar but will not otherwise affect the use of the bar as long as it is removed. This scaly rust is usually removed with a wire brush. Another method of removal is with a solution of hydrochloric acid and water (1:3 ratio), which should be thoroughly washed off with running water.

Masonry

Masonry may be concrete, clay, or stone. Concrete masonry includes hollow and solid block, concrete brick, and a variety of specialty units. Clay masonry includes brick and hollow tile, while stone masonry may use any of the natural stones such as granite and marble, which are used to construct walls.

The various types of masonry walls used are referred to as single wythe, cavity, bonded, and composite (faced) walls (Figure 9–1). A single wythe wall is a wall with only one thickness of material. A cavity wall is composed of an inner and outer wythe with an air space (with or without insulation) between them. Composite walls have two wythes of different materials that are bonded together by masonry units.

_____ 9.1 _____
CONCRETE MASONRY

Concrete masonry includes all of the molded concrete units used in the construction of a building such as concrete brick and hollow and solid concrete, as well as specialty units such as split-face, slump, fluted, and screen block. Historically most concrete masonry units are manufactured on the local level, and industry standards are not always adhered to.

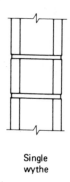
Single
wythe

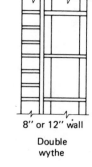
8″ or 12″ wall
Double
wythe

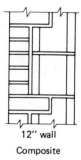
12″ wall
Composite

Figure 9–1. Typical walls.

Variations in shape, size, and surface texture are typical.

There is no complete standardization of sizes for concrete block, and sizes must be checked in each locality. The most common block size available has an actual face dimension of 7 5/8 inches high by 15 5/8 inches long. A 3/8-inch mortar joint on each side provides a nominal face dimension of 8 inches by 16 inches. Thickness available includes 3, 4, 6, 8, 10 and 12 inches nominal thickness. (The actual dimension is 3/8 inch less). Most commonly available sizes and shapes are shown in Figure 9-2. In the metric system the basic module of 100 mm is used, and with a standard block length of 390 mm and a 10 mm mortar joint, the nominal length is 400 mm (Figure 9-3).

Concrete block units are usually specified by the labels *hollow load-bearing, solid load-bearing,* and *hollow nonload-bearing.* These standards set specifications for moisture content, absorption, thicknesses of face-shell and web, and the allowable variations in dimensions.

Units with a core area (holes) that amounts to more than 25 percent of the gross cross-sectional area are referred to as hollow, while solid units have less than a 25 percent core area. Many concrete block units have from 40 to 50 percent core area; such information is available from the manufacturer. Concrete block may be available with three cores or two cores, depending on the manufacturer. The two-core units weigh about 10 percent less and provide more space for the vertical placement of electrical conduit, other utilities, and reinforcing bars.

Face shells and webs increase in thickness from the bottom to the top of the unit. The thickness referred to is the narrowest point, not an average top and bottom. This taper of the core allows for the easy removal of the form during manufacture and provides a larger bedding area for the mortar. Many manufacturers provide mortar grooves on the end flanges of block units, which provides a mortar key to make a more watertight joint.

Concrete blocks are available in normal weight or lightweight units. Both types use portland cement, but normal weight units use aggregates such as sand, gravel, and crushed stone, while the lightweight units incorporate lightweight aggregates such as expanded shale, slag, clay, and pumice. Since aggregates amount to about 90 percent of the block by weight, the choice of aggregates has a significant effect, and lightweight units may be as much as 30 percent lighter than normal weight units.

The *compressive strength* often referred to for concrete masonry is based on the gross area of the unit, that is, the total area including any core spaces. The primary factors influencing the compressive strength of a unit are the type and amount of cement, type and gradation of aggregate, degree of compaction, amount of water used, and moisture content and temperature of the units when tested. Lightweight units generally have lower compressive strengths than the normal weight units. Unit strengths may range from 300 pounds per square inch for nonload-bearing blocks to over 1,000 pounds per square inch for load-bearing units.

9.2
SURFACE FINISH

The texture of the surface varies from fine to coarse depending on the gradation of aggregates used by the manufacturer. The surface may be smooth or raised or have other special designs (Figure 9-4). The plain (smooth) faced units are the most widely used where economy is most important, where the

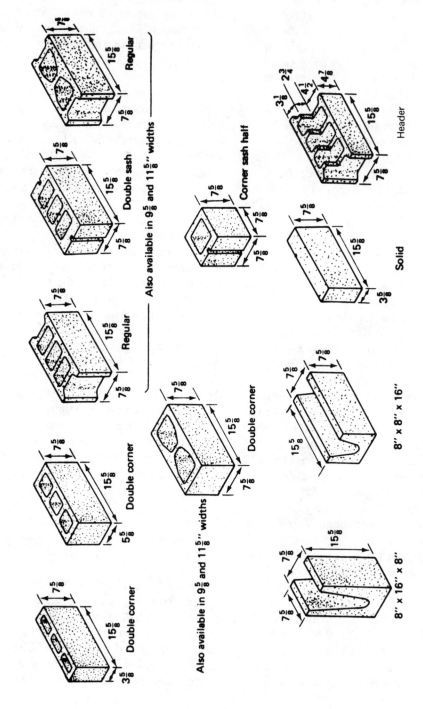

Figure 9-2. Typical block sizes (all sizes are actual).

110

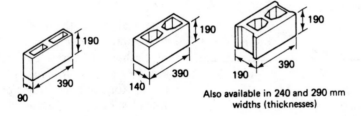

Figure 9-3. Metric block sizes.

Figure 9-4. Surface finishes of block.

111

unit will not be exposed to view, or where the block will back up other materials.

9.3
COLOR

A range of colors from off-white to gun-metal gray and from brownish shades to light tan are all available with the commonly used aggregate and portland cements. The standard color available in any given area will depend on the aggregates available locally. Manufacturers may specially blend natural aggregates to obtain colors within the ranges above, and white units may be made by using white portland cement and white fine and coarse aggregates. Variations in the color of the aggregates as they are quarried may cause some color variation in the units, and care must be taken to obtain uniformly colored structures.

A wide color range can also be obtained by using pure mineral oxide pigments in the mix. Colors such as browns, reds, and grays are easily obtained and give the best results. Blue and green are also available, but these colors have a tendency to fade. Color and special mixes cost extra, but the savings on painting and future maintenance often outweight the additional initial cost.

9.4
BLOCK USES

The most commonly available sizes and shapes of concrete block are shown in Figure 9–2. However, no concrete block manufacturer is likely to have all sizes and shapes at all times. Many manufacturers have only a limited number of the various sizes and shapes. For this reason it is important to know what units are available in a project's locale. This can be accomplished easily by visiting local manufacturers.

Wall and Partition Blocks

The straight lengths of exterior walls and interior walls are commonly made of regular block shapes (Figure 9–2), often referred to as *regular block* or *stretchers*. The corners are usually formed out of single-corner block and L-corner block. Bullnose corners are sometimes used, although they are not as readily available. The 4-inch and 6-inch units are available only as double corners, and these are considered regular block for those widths.

Bond-Beam Blocks

The block walls of a building may be strengthened by using bond-beam blocks, reinforcing bars, and concrete to form a continuous beam around the outside walls of the building. Bond beams are often used in warehouses and manufacturing facilities where there is a possibility of a vehicle or piece of equipment hitting the walls, in areas where soil conditions are unstable, and in earthquake zones. In each case the bond beam will provide added strength to prevent an entire wall or building from collapsing. The bond-beam block may also be used to form beams and lintels to span openings.

There are two types of bond-beam blocks in general use (Figure 9–5), the channel and low-web. The channel block is most commonly used. This U-shaped block is available

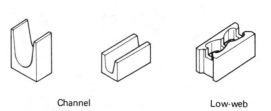

Channel   Low-web

Figure 9-5. Bond-beam blocks.

is two basic heights (8-inch and 16-inch) and a variety of widths.

Lintel Blocks

While bond-beam block may be used to span openings, there are specially shaped blocks available for window lintels. There are two common lintel block shapes available. The lintel block shape in Figure 9–6(a) is designed for use with either wood or metal sash (window frames). The wood window frames fit against the L-shaped portion of the block. There is also a lintel block available for use with certain types of commonly used metal sash. This block has a groove to receive the top edge of the window frame.

The lintel block must be supported (usually by wood shoring) as it is laid. Once it is in place reinforcing bars are placed and concrete is poured into the block. Once the concrete begins to harden (3 to 7 days) it will be strong enough to remove the forms.

Jamb Blocks

The sides of any opening in a block wall are referred to as the jamb. When windows will be installed in the opening, quite often special block, referred to as *jamb block,* is used on the sides of the opening to provide support for the window frame. There are two basic jamb block shapes, one to complement each of the lintel shapes. The jamb block shown in Figure 9–6(b) is designed for use with wood and metal sash. This shape allows the windows to be installed after the block wall is built, and the window frame fits against the L-shaped portion of the block.

The jamb block shown in Figure 9–6(c) is designed for use with certain types of commonly used metal sash. This block is similar to a corner unit, but it has a vertical groove in it that will receive the metal frame. When this style jamb block is used the window frame must be installed as the wall is built. There are several advantages to being able to install the window later in the construction process, after the walls are built and probably after the roof is on and the interior walls are built. Among these advantages are:

1. There is less chance of damage to the window frame during the construction process.

2. Windows may be delivered to the job at a later date. (Having to install the window frames as the wall is built may result in construction delays if the windows are not delivered when needed.)

3. Damaged windows can be removed and replaced more easily whether they are damaged during or after the construction process.

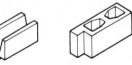

(a) Lintel blocks   (b) Jamb block   (c) Jamb block
                               (wood windows)   (steel windows)

Figure 9-6. Specialty block.

Decorative Blocks

There are a wide variety of blocks that may be decorative as well as structural or that may be only decorative. Many of these decorative units are available only on special order, and they may be specially ordered in a wide range of colors and textures.

Structural/Decorative Blocks.
Structural/decorative blocks are usually available in two basic groups, those with shapes cast on the face of the unit (Figure 9–7) such as fluted, convex (pyramid), and scored, and those that have been split to expose the aggregate (Figure 9–8) such as split and split-fluted blocks.

Pierced Blocks.
Pierced blocks (Figure 9–9) are used as decoration or as sunscreens to provide partial privacy. When they are used as walls in front of windows or at an angle to windows they act as a sunscreen to keep at least some of the sun's rays out of a building and provide partial privacy. They may also become part of the security system, which uses them to keep out possible intruders and to reduce the possibility of damage from rocks or bottles that might be thrown at the glass. Pierced blocks may be used as a decorative facing (cover) over other building materials such as concrete block. They are also commonly used as decorative garden walls and to provide partial privacy when used in fences.

Specialty Blocks

There are a wide variety of concrete masonry units available for special uses. Local manufacturers should be checked for availability of these products.

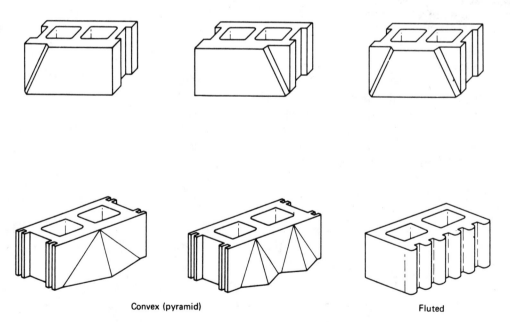

Convex (pyramid)   Fluted

Figure 9-7. Decorative block.

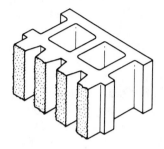

Figure 9-8. Split-fluted block.

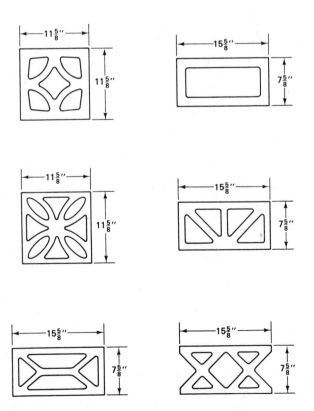

Figure 9-9. Decorative pierced sunscreen block.

Chimney Blocks. All fireplaces and furnaces that use gas, coal, oil, or wood as their fuel must have the smoke and fumes vented to the exterior of the building. Quite often they are vented vertically through terra cotta flues (Figure 9–10). These flues are not structural; they must be surrounded by a structural material—usually brick concrete block or chimney block—for support (Figure 9–11). Chimney blocks are available in a variety of sizes, as solid block or with small vertical cores to reduce weight. These blocks are laid up several courses (usually 3 courses), and then the flue is inserted in them. Specially shaped units to allow for the installation of the clean-out door and the furnace exhaust vent are also available.

Header Blocks. Header blocks are L-shaped units (Figure 9–2), which are used in composite walls of concrete block and brick as shown in Figure 9–1. As the blocks are laid every other course is header block; this allows the bricks to be laid lengthwise into the block. Then, as other block and brick courses are added above, the interlocking effect creates a single structural wall of block and brick. This also results in a decorative brick bond pattern (Figure 9–12).

Half-Height Blocks. Half-height blocks are block units that are half the height of the regular units and may also be available in colors. Half-height blocks must usually be specially ordered and are primarily used for the decorative aspect of a unit 3 ⅝ inches high and 15 ⅝ inches long. They are structural and can be used for load bearing walls; however, their cost per 100 square feet of wall area is much higher than the cost for regular sized units. Half-height blocks are sometimes used where the coursing of the block is based on a 4-inch module. For example, a concrete block wall 8 feet-4 inches high would be laid using twelve courses of the regular 8 inch high block, which would result in a wall 8 feet high. Then a half-height unit would be used to raise the wall to 8 feet-4 inches.

Cap Blocks. Cap blocks are solid units, usually 3 ⅝ inches thick and 15 ⅝ inches long. They are also used to add a half course to the top of a wall. They are often used on the top of block piers since they will more evenly distribute their building load to the pier.

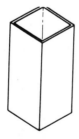

Figure 9-10. Flue.

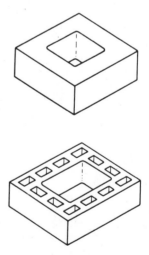

Figure 9-11. Chimney block.

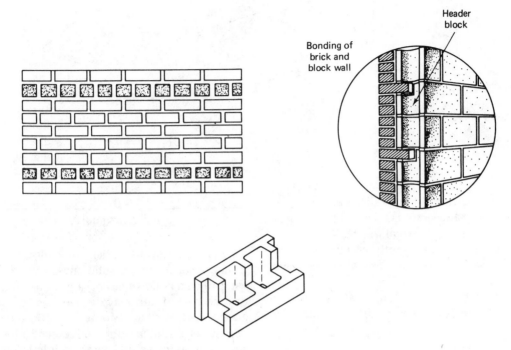

Figure 9-12. Header block.

Miscellaneous block shapes (Figure 9–13) include curved shapes for manholes, septic tanks, sidewalk pavers, and catchbasins. There are many other shapes available for special uses.

9.5
ADHESIVE MORTAR

The conventional mortar used to lay masonry units is discussed later in this chapter. In concrete masonry construction an adhesive mortar is occasionally used instead of the conventional mortar. This very strong adhesive mortar is usually made of epoxy binders, which form an organic mortar that is very strong. It is applied to the masonry unit with a caulking gun in a thin bead about 1/16 inch

thick. When this type of mortar is used the block used must be square and true for best results. In addition, this thin mortar joint requires slightly larger block units if the basic 4-inch module will be followed. (Remember that the standard unit is 7 5/8 inches high, leaving 3/8 inch for mortar joint and a nominal size of 8 inches.)

9.6
SURFACE BONDING

Another development has been a special coating which can be applied in a thin coat over a block wall that has been laid up without mortar. This special cement coating is reinforced with glass fibers and applied directly to each side of the block wall in thicknesses of 1/16

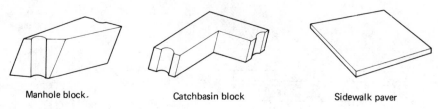

Manhole block.   Catchbasin block   Sidewalk paver

Figure 9-13. Specialty shapes.

to ⅛ inch. The resulting wall has an attractive stucco-like appearance and is just as strong in flexure as a conventional wall built with mortar, but it does have slightly less (about 10 percent less) resistance to vertical compressive loads.

The special coatings may be troweled or sprayed on, and color pigments may be added.

9.7
BLOCK BONDS (PATTERNS)

The pattern (design) of the masonry units in the wall is referred to as the *bond*. Typical bonds used for block are shown in Figure 9-14. Variations on possible bonds are limited only by the designer's imagination. Many designs of bonds have been modified by the projection or recession of some units or the omission of units to provide pierced walls.

The most common bond used for concrete block is the running bond. The running bond is also the most economical in terms of labor costs. Walls of stack bond and basket weave bond are not as strong as walls of running bond. This is because loads are basically transmitted straight down while the overlapping of the units in the running bond provides for the spreading out of any imposed loads.

9.8
CLAY MASONRY

Clay masonry includes brick and hollow tile. Brick is considered a solid masonry unit, while tile is a hollow masonry unit. The faces of each type of unit may be given a ceramic glazed finish or a variety of textured finishes.

Brick is a solid masonry unit even if it has cores in it, as long as the core area is less than 25 percent of the cross-sectional area of the unit. Bricks are available in a large variety of sizes (Figure 9-15). They are usually classified by material, manufacture, kind (common or face), size, and texture or finish on the face. Special giant bricks with a face dimension of 4 x 16 inches (nominal) and thicknesses of 4, 8, and 12 inches are also available. Typical bonds are shown in Figure 9-16.

9.9
COLOR

The color of the brick can be varied by the clay used. The most common color is red, but brown, cream, gray, and white are also available. Fire markings are introduced on the brick during burning.

All brick needed for an entire job should be ordered at one time, for it is very difficult to

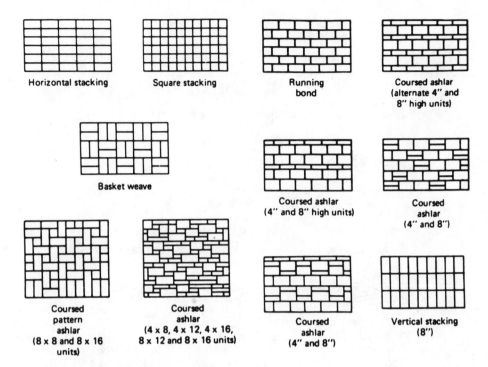

Figure 9-14. Typical concrete block patterns.

get an exact match of colors from batch to batch in the plant. As the clay varies so will the color of the brick.

water-struck, sand-struck, and stippled. A smooth, slick finish is obtained from the glazed finishes, which are available in a wide range of colors.

9.10
TEXTURES

Textures are produced on brick by the dies or molds used in forming. Smooth textures result from the steel die. Attachments may be applied to cut, roll, scratch, brush, or otherwise roughen the surface as the clay is extruded from the die. Textures range from fine through coarse, the principal textures being smooth, matt, rug, bark, sand-molded,

9.11
STONE MASONRY

Stone masonry is used primarily as a veneer (attached to a backing material) for interior and exterior walls. It is also used for walkways and for such trim on buildings as sills, copings, stools, and steps.

The types of stone most commonly used are granite, marble, limestone, sandstone,

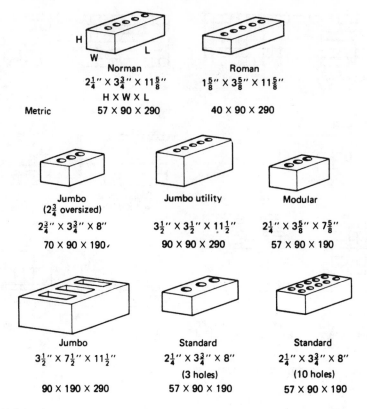

Figure 9-15. Brick sizes.

and slate. Other types of stone such as rock with fossils in it and rocks of sea shells are sometimes used. Stone is generally available in random irregular sizes for rubble masonry and other rustic work; sawed-bed stone is used primarily for veneers on the interior and exterior of buildings; and cut stone (smooth cut pieces) is used for veneer.

Granite is the hardest and most difficult to work (shape) of all stones. The stone takes a high polish, and its color range includes white, green, pink, gray, and red. It is used primarily as a thin veneer over other construction and for steps, trim, and platforms.

Marble itself is white, but other substances in the stone give it the color patterns and variations that make it so popular as a decorative feature. Widely used as a veneer for the most monumental buildings, it may be used on both interior and exterior. Decorative uses include panels, mantels, and hearths. Floor tiles and steps of marble should not be used where heavy foot traffic will occur unless special marble with high wear-resistance is used. A high polish finish is available in marble.

Limestone is used extensively as cut stone and sawed-bed stone for veneer, trim, cop-

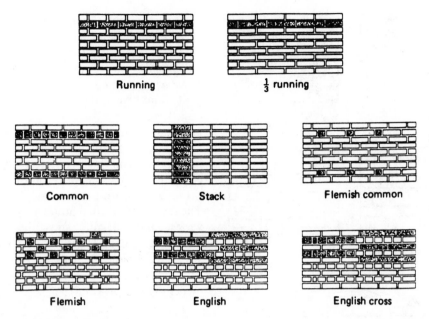

Figure 9-16. Brick patterns.

ings, sills, stools, and walkways. Colors available range from buff to gray, and the pieces are graded according to the defects on the surface. This type of stone is easily carved if decorative effects are desired. Floors of limestone may be used in light traffic areas. Limestone that has a texture of small irregular cavities and a mixture of color tones is called *travertine.* It is used primarily for decoration, but may even be used for floors in areas that receive very light traffic.

Sandstone is used primarily as sawed-bed stone veneer on buildings and cut stone for sills, copings, wall caps, mantels, and hearths. Colors range from light gray and buff, pink, and light tan, to brown and dark variegated colors.

Slate is used as cut stone veneer on buildings, sills, copings, wall caps, mantels, flooring, and even for roofing. The color most commonly used is blue-gray, but red and green slate is also available.

9.12
STONE BONDS (PATTERNS)

The pattern (design) of the stone in the wall is referred to as the bond. Typical bonds used for stone are shown in Figure 9-17. Many times these designs are modified by the projection or recession of some of the individual pieces of stone.

9.13
MORTAR

The mortar used to lay up the masonry units is mixed on the job site using mortar cement, sand, and water.

Figure 9–17. Stone patterns.

Type of mortar	Average compressive strength 2-inch cubes at 28 days (psi)
M	2500
S	1800
N	750
O	350

Figure 9–18. Mortar types.

Mortar strength requirements are shown in Figure 9–18. The mortar may be made with portland cement, lime, fine aggregates (sand), and water, or a specially formulated masonry cement with fine aggregates and water. Another possibility is a combination of portland and masonry cements. Proportions of the mix, by volume, are shown in Figure 9–19. The types of mortar suggested for use in given situations are shown in Figure 9–20.

Mortar joint thicknesses range from ¼ inch up. A ⅜-inch joint is generally used with concrete masonry and clay brick, while hollow clay tile commonly has ½-inch joints.

Masonry cements are available in gray, white, and black. Natural colors may be obtained by the use of natural materials, aggregates, and the masonry cements to obtain a range of colors from white to gray, buff, and black. Once a mix has been selected for a project the same materials must be used throughout to obtain a continuity of color. The color of each manufacturer's cement varies; some are lighter and some are darker.

Colored mortar is used extensively in construction. Colors such as red, brown, and green may be obtained by adding color pigments to the mortar mix.

_____ 9.14 _____

MORTAR JOINT SHAPES

The most common shapes of mortar joints are illustrated in Figure 9–21. The joints are first struck off flush with the edge of the trowel, and as soon as the mortar has partially set the joints are formed. Tooled joints are molded and compressed with a rounded or V-shaped joint tool. Raked joints are formed by raking or scratching out the mortar to a given depth (usually about ⅜ inch). The edge of the trowel is used to form weathered and struck joints.

Proper construction of the mortar joint is important to keep water from penetrating the wall. Only certain joints should be used on the exterior of a building. Watertight joints require proper shaping and good workmanship; the mortar must be made smooth and dense by the application of considerable pressure on the tool used to form them. The mortar used for the joints in exterior walls should completely fill the joint space to reduce moisture penetration. The weathered joint is the

MORTAR PROPORTIONS BY VOLUME*

Mortar type	Portland cement cu. ft.	Masonry cement cu. ft.	Hydrated lime or lime putty cu. ft.	Aggregate measured in damp loose condition cu. ft.
M	1	None	$\frac{1}{4}$	
	1	1- type 2	None	
S	1	None	Over $\frac{1}{4}$ to $\frac{1}{2}$	Not less than $2\frac{1}{4}$ and not more than 3 times the sum of the volumes of cement and lime used.
	$\frac{1}{2}$	1- type 2	None	
N	1	None	Over $\frac{1}{2}$ to 1	
	None	1- type 2	None	
O	1	None	Over 1 to 2	
	None	1-type 1 or type 2	None	

*The weight of one cubic foot of the respective materials used shall be considered to be as follows:

Portland cement . . . . 94 pounds
Masonry cement · · · · · · · · · · · · · · · · · · weight printed on bag
Hydrated lime . . . . 40 pounds
Lime putty (quicklime) . . . . 80 pounds
Sand, damp and loose . . . . .80 pounds of dry sand

Figure 9-19. Mortar proportions. (Permission to reprint granted by the American Society for Testing and Materials, copyright ASTM)

TYPES OF MORTAR REQUIRED

Type of Masonry	Types of Mortar Permitted
Foundations: (below grade masonry)	
Footings	M or S
Walls of Solid Units	M, S or N
Walls of Hollow Units	M or S
Hollow Walls	M or S
Masonry Other Than Foundation Masonry	
Piers of Solid Masonry	M, S or N
Piers of Hollow Units	M or S
Walls of Solid Masonry	M, S, N or O
Walls of Hollow Masonry	M, S or N
Hollow Walls and Cavity Walls	
(a) Design Wind Pressure Exceeds 20 psf.	M or S
(b) Design Wind Pressure 20 psf or less.	M, S or N
Glass Block Masonry	M, S or N
Non-Bearing Partition and Fireproofing	M, S, N, O or Gypsum
Gypsum Partition Tile or Block	Gypsum
Fire Brick	Refractory Air Setting Mortar
Masonry Other Than Above	M, S or N

Figure 9-20. Mortar uses. (Permission to reprint granted by the American Society for Testing and Materials, Copyright ASTM)

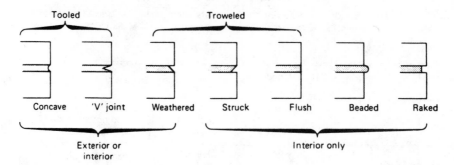

Figure 9-21. Mortar joint shapes.

troweled joint most acceptable for exterior use, but often not enough mortar is forced against the brick above the joint and moisture is allowed to seep in.

While some of the other joint shapes may seem more attractive or desirable, it is important to remember that water will enter the wall unless the appropriate joint shape is selected and the workmanship is good. The struck and raked joints provide a shelf on which water can collect and work its way into the wall. A flush joint is formed by cutting the mortar off with the trowel; this action pulls the mortar away from the brick, leaving a hairline crack through which water can enter. Furthermore, with this type of joint the mortar is simply cut off and is not compressed into the joint; this leaves a porous mortar that will readily absorb and transmit water through the wall.

Rain penetration should be handled from two aspects: (1) to keep moisture out, and (2) to get out any moisture that does get into the masonry. Moisture can best be kept out by use of flashing, caulking, tight mortar joints, parging (covering with mortar), dampproof coatings, and providing adequate slope overhangs and drip edges on all sills and copings.

Moisture accumulation by capillary action from contact with the ground is prevented by dampproofing and waterproofing. Moisture accumulation by condensation of water vapor often occurs where the walls are uninsulated. The application of a water-emulsion asphalt paint to the wall surface will act as a vapor barrier. Insulated masonry often does not require a vapor barrier since there is much less possibility of water vapor condensation.

9.15
MOISTURE CONTROL

Rain is the primary source of moisture in the masonry. Moisture may also accumulate through capillary action from contact with the ground and condensation of vapor in the masonry.

9.16
MASONRY WALL REINFORCING

Reinforcing bars are often used in masonry walls to create multistory bearing wall construction and in conjunction with bond-beam blocks. The reinforcing bars are placed in the cores and surrounded by concrete or mortar.

Sometimes steel reinforcing is placed in

continuous strips in the mortar joints. It is used primarily to minimize shrinkage and temperature and settlement cracks, and also to provide a shear force transfer to the reinforcing steel. This type of reinforcing is also used to tie the inner and outer wythes (thicknesses) of wall together in cavity wall construction (Figure 9–22). The most common spacing of reinforcing is 16 inches on center. Suggested reinforcing layouts for various conditions are shown in Figure 9–23. The reinforcing rods may be deformed, plain, or a combination of the two. Finishes may be plain, galvanized, or stainless steel. Cross rods, which serve as metal ties in cavity or multi-wythe walls, should be galvanized. At splices and corners the side rods should be lapped at least 6 inches. Special corner and tee sections are available from some manufacturers. The reinforcing used is 2 inches less than the nominal thickness of wall (or wythe) and should be placed to insure a minimum of $5/8$-inch mortar cover on the exterior face and $1/2$-inch mortar cover on the interior face. Often wall reinforcings are used in conjunction with wall ties.

9.17
WALL TIES

These are also used to tie the inner and outer wythes of wall together and also to tie single wythe walls to back-up materials such as concrete, wood, and steel. Typical shapes of ties are shown in Figure 9–24, and typical installations are shown in Figure 9–25. Wall ties permit the mason to construct one wythe of the wall to any given height without working on the other wythe, resulting in increased productivity. Adjustable wall ties may be used where the coursing of the inner and outer wythes does not line up. Most common spacing of the rectangular type of adjustable wall

ties should be one for every 2.66 square feet of wall area (about 16 inches vertically and 24 inches horizontally). Z-type adjustable wall ties should be spaced to provide one tie per 1.33 square feet of wall area. Ties should also be located within 8 inches of each side of vertical supports and control joints. The codes governing the project should also be checked. Many of the ties have moisture drips, which reduce any possibility of moisture traveling along the tie from the outer wythe to the inner wythe. Ties may be available in uncoated steel, zinc coated steel, hot dipped galvanized steel, copper coated steel, zinc steel, 25 percent copper steel, and stainless steel, depending on the manufacturer. Uncoated steel, zinc coated, and hot dipped galvanized are the most commonly used.

9.18
CONTROL JOINTS

In order to minimize shrinkage and cracks and to control cracks at locations where the concentration of stresses or points of weakness are expected, continuous vertical joints are built into the masonry wall. They are referred to as *control joints* (Figure 9–26). When properly constructed and placed, they permit slight movements in the wall to occur without cracking the masonry units, and they also seal the joints against weather.

Control joints should be located at all intersections of nonload-bearing and load-bearing walls, where changes in wall heights occur at the intersection of walls and piers, at pilasters and columns attached to or built into walls, and on walls weakened by openings. Suggested locations and spacing of control joints are shown in Figure 9–27.

Masonry below grade that must be watertight should not have control joints. For such cases a bond beam (Figure 9–28) should be

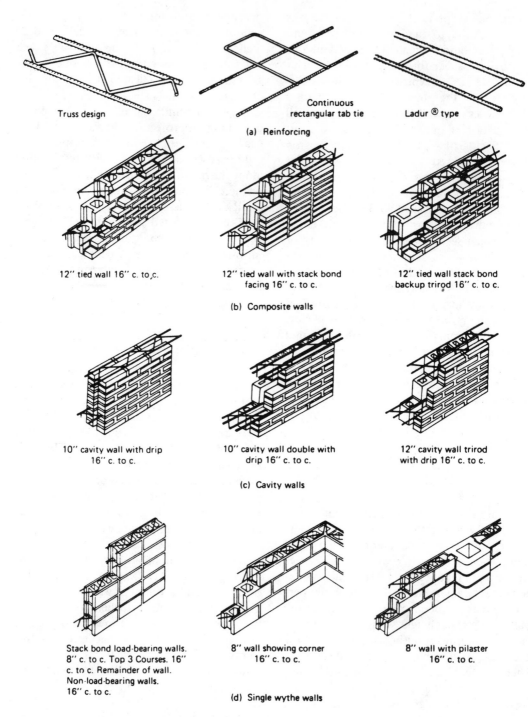

Truss design

Continuous rectangular tab tie

Ladur ® type

(a) Reinforcing

12" tied wall 16" c. to c.

12" tied wall with stack bond facing 16" c. to c.

12" tied wall stack bond backup trirod 16" c. to c.

(b) Composite walls

10" cavity wall with drip 16" c. to c.

10" cavity wall double with drip 16" c. to c.

12" cavity wall trirod with drip 16" c. to c.

(c) Cavity walls

Stack bond load-bearing walls. 8" c. to c. Top 3 Courses. 16" c. to c. Remainder of wall. Non-load-bearing walls. 16" c. to c.

8" wall showing corner 16" c. to c.

8" wall with pilaster 16" c. to c.

(d) Single wythe walls

Figure 9-22. Horizontal masonry reinforcement.

Wall openings. Reinforcing should be installed in the first and second bed joints, 8 inches apart immediately above lintels and below sills at openings and in bed joints at 16-inch vertical intervals elsewhere. Reinforcement in the second bed joint above or below openings shall extend two feet beyond the jambs. All other reinforcement shall be continuous except it shall not pass through vertical masonry control joints.

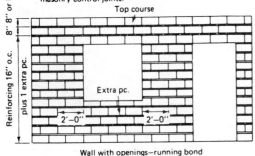

Wall with openings—running bond

Single wythe walls. Exterior and interior. Place reinforcing 16″ o.c. and in bed joint of the top course.

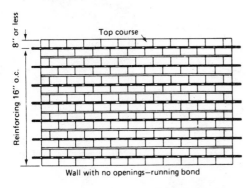

Wall with no openings—running bond

Foundation walls. Place reinforcing 8″ o.c. in upper half to two-thirds of wall.

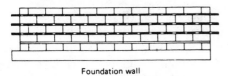

Foundation wall

Figure 9-23. Reinforcement layout.

Basement walls. Place reinforcing in first joint below top of wall and 8″ o.c. in the top 5 bed joints below openings.

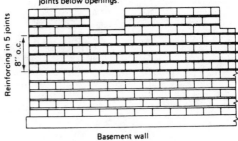

Basement wall

Stack bond. Reinforcing should be placed 16″ o.c. vertically in walls laid in stack bond except it shall be placed 8″ o.c. for the top 3 courses in load bearing walls

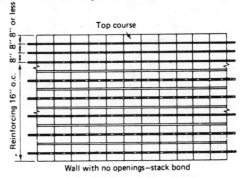

Wall with no openings—stack bond

Control joints. All reinforcement should be continuous except it shall not pass through vertical masonry control joints.

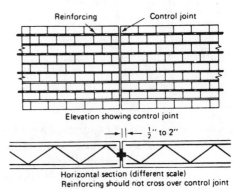

Elevation showing control joint

Horizontal section (different scale)
Reinforcing should not cross over control joint

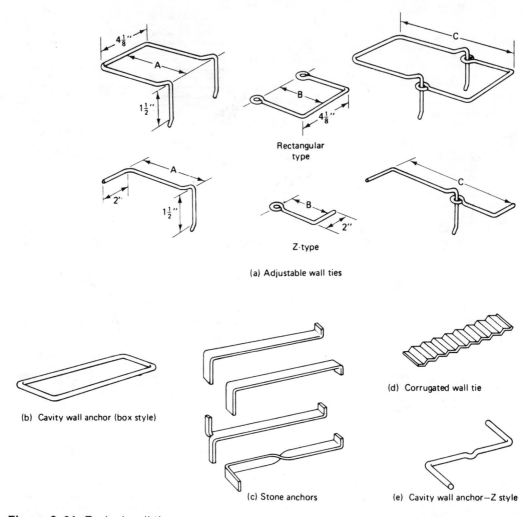

Rectangular
type

Z-type

(a) Adjustable wall ties

(b) Cavity wall anchor (box style)

(c) Stone anchors

(d) Corrugated wall tie

(e) Cavity wall anchor—Z style

Figure 9-24. Typical wall ties.

formed at the top of the foundation wall, and the control joint should then be placed in the masonry above grade, extending from the top of the bond beam to the top of the masonry wall.

The most satisfactory results are achieved when preformed control joints of neoprene rubber are used (Figure 9-29). Caulking is required with all types of control joints. Caulking requires maintenance, so proper selection of caulking is important; the control joint manufacturer's recommendation should be followed.

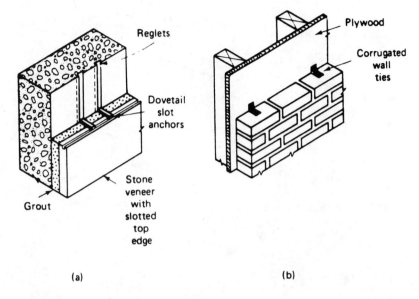

(a) (b)

Figure 9-25. Wall-tie installation.

Figure 9-26. Control joints.

Recommended control joint spacing (ft)			
Wall height (ft)	No joint reinforcing	Joint reinforcing 16″ O.C. vertically	Joint reinforcing 8″ O.C. vertically
Up to 8	20′	25′	30′
8 to 12	25′	30′	35′
Over 12	30′	35′	40′

Control joints also recommended at

(a) Changes in wall height
(b) Building corners

Figure 9-27. Control joint location.

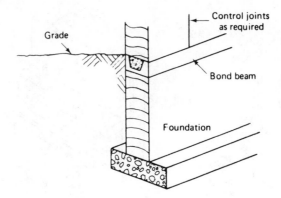

Figure 9-28. Control joint above bond beam.

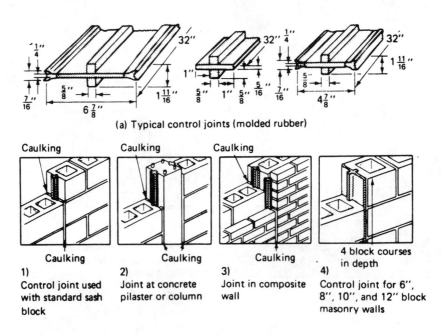

(a) Typical control joints (molded rubber)

Caulking Caulking Caulking

Caulking Caulking Caulking

1)
Control joint used
with standard sash
block

2)
Joint at concrete
pilaster or column

3)
Joint in composite
wall

4)
4 block courses
in depth

Control joint for 6″,
8″, 10″, and 12″ block
masonry walls

Note: Distance from sash groove to outside face of wall is same for 8″, 10″ and 12″ masonry units.

(b) Applications

Figure 9-29. Control joints. (Courtesy Dur-o-waL)

9.19
LINTELS

A lintel is the horizontal member that supports the masonry above a wall opening. It spans the entire opening from one side to the other. Materials used as lintels include steel angle irons, composite steel sections, lintel and bond-beam blocks with reinforcing bars and filled with concrete, and precast concrete lintels. Some of the various types of lintels and their installation details are shown in Figures 9–30 and 9–31.

The ends of all lintels must have adequate bearing on the walls so that the loads imposed on the lintels are transmitted to the masonry walls. For steel lintels a minimum of 4 inches of bearing is required, while a minimum of 6 inches of bearing is required of lintel block and precast lintels. Bearing information is usually contained in the specifications or on the lintel schedule.

Lintel sizes should be determined on the basis of structural design for the loads involved. At their best the rule-of-thumb methods that generally result in over-sizing of the lintel are uneconomical. At their worst such methods may result in actual failure of the lintel. Inadequate sizing usually results in excessive deflection of the lintel and structural cracking. When steel angles are used as lintels the blocks surrounding the steel must be cut, and all steel lintels require painting.

9.20
SILLS AND STOOLS

The exterior members at the bottom of masonry openings are called *sills;* the interior members at the bottom of windows are called *stools.* Materials used include brick, stone, tile, precast concrete, aluminum, bronze, and stainless steel. Sills and stools may be classified as *slip* or *lug* units. Slip sills and stools are those that are slightly smaller than the width of the opening, and they may be installed after the masonry work is completed.

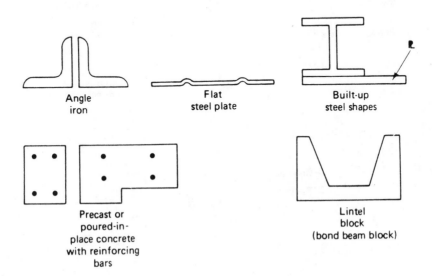

Angle iron

Flat steel plate

Built-up steel shapes

Precast or poured-in-place concrete with reinforcing bars

Lintel block (bond beam block)

Figure 9–30. Lintels.

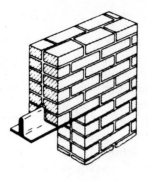

(a) Angle iron lintel

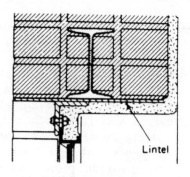

Lintel

(b) Built-up steel lintel

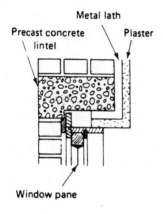

Metal lath

Precast concrete lintel

Plaster

Window pane

(c) Precast concrete lintel

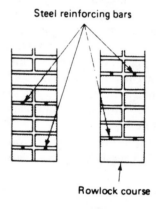

Steel reinforcing bars

Rowlock course

(d) Reinforced brick masonry lintels

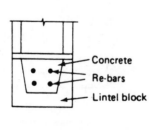

Concrete

Re-bars

Lintel block

(e) Lintel block

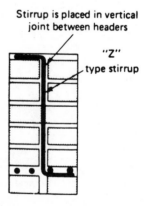

Stirrup is placed in vertical joint between headers

"Z" type stirrup

(f) Reinforced brick masonry beam

Figure 9-31. Lintel details.

132

Lug sills and stools extend into the masonry at each end and must be built into the masonry as the job progresses. Typical installations are shown in Figure 9–32.

The exposed materials, both inside and outside the building, will probably require cleaning while the concealed masonry, such as block used as a back-up, generally recieves no cleaning.

Clay Masonry

For brick work, no attempt to clean should be made until a minimum of 48 hours after completion of the wall. After the minimum time, soap powder (or other mild solutions) with water and stiff brushes may be tried. When cleaning unglazed brick and tile, the first at-

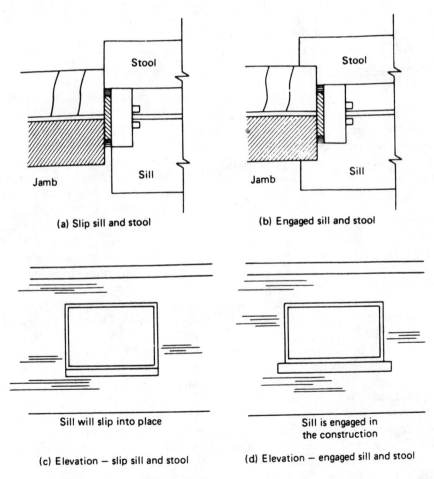

(a) Slip sill and stool

(b) Engaged sill and stool

(c) Elevation — slip sill and stool

Sill will slip into place

(d) Elevation — engaged sill and stool

Sill is engaged in the construction

Figure 9-32. Sill and stool installation.

tempt is made with plain water and stiff brushes. If these solutions do not work, the surface should be thoroughly wet down with clear water, scrubbed with a solution of acid and water, and thoroughly rinsed down. The acid solution should always be tried on an inconspicuous area prior to using it on the entire wall. Acids should not be used on glazed facing tile.

Concrete Masonry

Acid is not used on concrete masonry. If mortar droppings fall on the units they should be allowed to dry before removal to avoid smearing the face of the unit. Once the droppings are dry they can be removed with a trowel, and a final brushing will remove most of the mortar.

The better the workmanship on the job the less cleaning is required. When the mortar is a different color from the masonry unit the droppings must be cleaned off to get an unblemished facing.

Stone Masonry

Stone masonry should be cleaned with a stiff fiber brush and clear water—although soapy water may be used if necessary—then rinsed with clean water. This will remove construction and mortar stains. Machine cleaning processes should be approved by the stone supplier before they are used. Wire brushes, acids, and sandblasting are not permitted for cleaning stonework.

Masonry should not be laid if the temperature is 40°F and falling or less than 32°F and rising unless adequate precautions against freezing or thawing are taken. The masonry is required to be protected from freezing for at least 48 hours after being laid. Any ice on the masonry materials must be thawed off before they are used.

Mortar also has special requirements: its temperature should be between 70° and 120°F. During cold weather construction it is common practice to heat the water used in order to raise the temperature of the mortar. Any moisture present in the sand also will freeze unless it is heated, and it must be thawed before it can be used.

Cold weather construction is more expensive than warm weather construction. The increased costs include the cost of temporary enclosures so the masons can work, higher frequency of equipment repair, thawing materials, and expenses for temporary heat.

_____ 9.22 _____
STRENGTH

Masonry units are used together as an "assembly" of materials. Wall assemblies must be considered in terms of appearance, strength, durability, watertightness, fire resistance, sound resistance, and resistance to heat transfer and radiation transfer. Any number of these considerations may determine the details of the assembly used in a given situation. The masonry units, workmanship, and mortar must all be of a good quality for best results, since an assembly is no better than its weakest point. All of the elements discussed in this chapter must be considered for good construction. Details and sections showing what is required should be carefully drawn by the draftsman and checked by the contractor.

All masonry walls, load-bearing and nonload-bearing, require support or bracing of some type. This required bracing is referred to as _lateral support,_ and it may be vertical or horizontal, such as a cross wall (vertical) or a roof (horizontal), but does not have to be both. The limits of unsupported masonry lengths and heights allowed are shown in Figure 9–33. Lateral support may be obtained by use of cross walls, piers (pilasters), or columns when the limiting distance is measured horizontally, and by roofs and floors when the limiting distance is measured vertically. The masonry walls must be sufficiently bonded or anchored to the lateral support so they can resist all applied forces.

The working stresses allowed for unreinforced masonry construction are governed by the material and type of mortar used. The building codes give the type of mortar to be used and allowable compressive strengths for the construction (Figure 9-34).

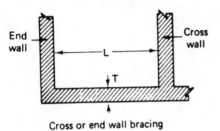

Cross or end wall bracing

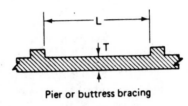

Pier or buttress bracing

(a) Vertical bracing

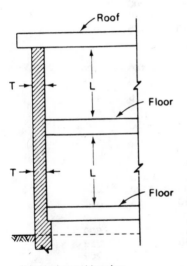

(b) Horizontal bracing
(where floors and
roofs serve as bracing)

Maximum ratio of unbraced height or length (L) to thickness (T)	
Type of masonry	Ratio
Solid masonry (except rubble stone) type M, N, S, O mortar	20
Grouted solid masonry (except rubble stone) type M, N, S mortar and grout	22
Hollow walls and walls of hollow masonry units type M, N, S mortar	18
Cavity walls, type M, N, S mortar	14 *
Plain concrete 2000 psi	22
Rubble stone: coursed and bonded	18
Rubble stone: random	14

*Based on sum of inner and outer parts of wall.

Example: Assuming a wall (T) of 8″ hollow masonry units with type N mortar, what maximum spacing of vertical or horizontal bracing will be required? The spacing is determined by multiplying the wall thickness times the applicable ratio.

8″ x 18 = 144″ or 12′–0″ maximum spacing

Figure 9-33. Lateral support.

ALLOWABLE COMPRESSIVE STRESSES IN UNIT MASONRY

Construction: grade of unit	Allowable compressive stresses gross cross-sectional area (except as noted)			
	Type M Mortar	Type S Mortar	Type N Mortar	Type O Mortar
Solid masonry of brick and other solid units of clay or shale; sand-lime or concrete brick:	psi	psi	psi	psi
8,000 plus, psi................ 400	350	300	200	
4,500 to 8,000, psi 250	225	200	150	
2,500 to 4,500, psi 175	160	140	110	
1,500 to 2,500, psi 125	115	100	75	
Grouted solid masonry of brick and other solid units of clay or shale; sand-lime or concrete brick:				
4,500 plus, psi................. 350	275	200	—	
2,500 to 4,500, psi 275	215	155	—	
1,500 to 2,500, psi 225	175	125	—	
Solid masonry of solid concrete masonry units:				
Grade A 175	160	140	100	
Grade B 125	115	100	75	
Masonry of hollow units 85	75	70	—	
Piers of hollow units, cellular spaces filled, as in Section 1405.6 105	95	90	—	
Hollow walls (cavity or masonry bonded)[2] Solid units:				
Grade A or 2,500, psi plus 140	130	110	—	
Grade B or 1,500 to 2,500, psi .. 100	90	80	—	
Hollow units 70	60	55	—	
Stone ashlar masonry:				
Granite 800	720	640	500	
Limestone or Marble 500	450	400	325	
Sandstone or cast stone 400	360	320	250	
Rubble stone, coursed, rough or random 140	120	100	80	

Figure 9-34. Compressive strengths of masonry. (Permission to reprint granted by the American Society for Testing and Materials, copyright ASTM)

Chapter 10

Wood

Nine out of ten homes built in America are wood. While research has brought about the substitution of other products for wood for many uses, this same surge of exploration and technology has brought about increased and more economical use of wood, as well as increased knowledge of what species (types) of wood are best suited for particular uses.

The long popularity of wood may be traced to several different factors. The easy accessibility and workability tends to keep it economical in comparison to other materials. Its easy workability eliminates the need for some of the expensive manufacturing and refining equipment required for materials such as steel, and reduces the amount of equipment required to install it. Wood is easily cut, sawed, or planed to a desired shape, unlike most materials. The characteristics that keep it free from corrosion and rust, comparatively light weight, and more shock resistant than steel are other reasons that the use of wood continues to increase.

Perhaps the most important quality that wood has is its beauty. The warm feeling of wood around us is, in some way, comforting. Perhaps it is the feeling that nature is with us, or that no two trees are alike. No matter the reason, it is simply a warm, friendly material. This durable, strong, easy to maintain, and good insulating material will be in use for a long time.

10.1
DEFINITIONS

While the terms *wood, lumber* and *timber* are often used synonymously, they have distinct meanings in the building industry and are not interchangeable. *Wood* is the hard, fibrous material that makes up the tree under the bark. *Lumber* is wood that has been sawed or planed, but not manufactured further than sawing, resawing, passing lengthwise through a standard planing machine, crosscutting to length, and working. *Timber* is lumber that has as its smallest dimension at least 5 inches (125mm). *Millwork* is the term used for wood building materials that are manufactured in millwork plants and planing mills. Such products include window frames, shutters, doors, interior trim, mantels, stairways, and moldings; but items such as flooring, siding, and ceiling materials are not millwork.

10.2
TREE CLASSIFICATION

The two classes of trees used for construction purposes are the *needle-leaved conifers* such as pine, spruce, fir, and hemlock and the *broad-leaved* trees such as maple, oak, poplar, and walnut (Figure 10–1).

HARDWOODS	SOFTWOODS
Ash	Cedar
Basswood	Cyprus
Beech	Fir
Birch	Hemlock
Cherry	Pine
Elm	Redwood
Hickory	Spruce
Maple	
Oak	
Poplar	
Walnut	

Figure 10-1. Typical tree classifications.

10.3
MOISTURE CONTENT

Moisture content (m.c.) of wood is a measure of the amount of moisture contained in the wood. The moisture content is the weight of the water the wood contains expressed as a percentage of the weight of the wood when it is oven-dry. For example, if a sample of wet wood weighs 3.3 pounds (1.5 kg) and is then oven-dried and weighs 2.2 pounds (1 kg), the water amounted to 1 pound. Then the moisture content would equal:

$$\frac{(\text{original weight} - \text{oven-dry weight}) \times 100}{\text{oven-dry weight}}$$

$$= \frac{(3.3 - 2.2) \times 100}{2.2} = 50\%$$

$$\text{Metric:} \frac{(1.5 - 1) \times 100}{1} = 50\%$$

Determining the moisture content by oven-drying the wood sample may require several days. Instantaneous moisture content readings may be obtained by using a moisture meter. It is often used in the field, where quick

analysis is required and oven-drying may not be practical.

Moisture present in wood may be *free* or *hygroscopic*. Free moisture is found in the cavities (spaces) between the fibers. Hygroscopic moisture is the water that has been absorbed by the wood fibers, and this moisture is present in the wood throughout its life as a tree. When the tree is cut, it has a high percentage of moisture in it—from 35 percent to 200 percent depending on the type of tree (wood species)—and it is considered to be in a *green condition*.

Once the tree is harvested (cut) it begins to dry, and the free moisture begins to evaporate. The free moisture will continue to evaporate until it is all gone. It is only after all the free moisture has evaporated that any of the absorbed moisture in the wood fibers will evaporate. The point at which all the free moisture in the cells has evaporated but the wood fibers are still fully saturated is referred to as the *fiber saturation point*. In most wood species, the fiber saturation point occurs at about 30 percent m.c. (moisture content). During the evaporation of the free moisture the wood remains at its original size. No shrinkage of the wood occurs until the cell cavities have lost their free moisture. Once the fiber saturation point is reached further drying causes the absorbed water in the cell fibers to evaporate and the fibers begin to constrict, causing the wood to shrink.

The total amount of shrinkage that can occur takes place between its fiber saturation point, about 30 percent m.c., and 0 percent, the theoretical point at which all moisture is dried out of the wood. The rate of shrinkage is uniform during this drying out process. This means if 30 percent m.c. was the fiber saturation point, at 15 percent m.c. one half of the total possible shrinkage has occurred. In actual use, wood rarely reaches a moisture content of 0 percent, since it absorbs moisture from the surrounding air.

Thus, once a tree is harvested its moisture content begins to drop, and after it is cut into lumber this drying out continues. The lumber must be dried out further before it can be used in construction. The lumber may be *air-dried*, a process in which it is stacked in perpendicular rows (rows at 90 degree angles to each other), with spaces left between the pieces so that air will circulate through the stack to dry it out. This lumber should not be used until it has a moisture content of 19 percent or less. Above 19 percent m.c. it is classified as *green,* or *green lumber,* and lumber with a 19 percent m.c. and below is classified as *S-dry,* or *dry lumber.* The process of air-drying the lumber may take several months.

To speed up the drying process the lumber may be put in a kiln, which uses heat to evaporate the moisture. The lumber should be dried to 15 percent m.c. and may be designated *KD* or *kiln-dried.*

Once placed in service the wood will take on or give off moisture to the surrounding air until there is an equilibrium (balance) between the surrounding air and the moisture in the wood. This balance is the *equilibrium moisture content.* Since the temperature and humidity of the surrounding air will vary, the moisture content of the wood will also vary as it seeks equilibrium with the surrounding air. Using the chart in Figure 10–2 it is possible to estimate the approximate equilibrium moisture content the wood will have once put into service. For example, if the surrounding air has a temperature of 70°F and 50 percent relative humidity, the equilibrium moisture content will be about 9 percent. For the least shrinkage it is desirable that when the wood is fabricated and installed it have a moisture content as close to the equilibrium moisture content as possible. Some recommended moisture contents for various lumber uses are shown in Figure 10–3. Unless specially ordered, the lumber used for the structural framing on a project has a higher moisture

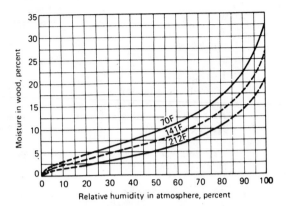

Figure 10–2. Relation of the equilibrium moisture content of the wood to the relative humidity of the surrounding atmosphere at three temperatures. (Courtesy U.S. Forest Products Laboratory)

content than the recommendeded moisture content, and the higher cost of specially ordered lumber usually makes it impractical.

The effects of shrinkage on a piece of lumber are shown in Figure 10–4. The amount of shrinkage and distortion on the lumber is affected by the direction of the annual ring, with the tangential shrinkage (parallel to the growth rings) about twice as much as the radial (across the growth rings). The approximate shrinkage for a piece of 2 x 10 (38 x 235mm) lumber as it seasons from green condition to the theoretical 0 percent moisture content is shown in Figure 10–5.

―――――――――― 10.4 ――――――――――
WOOD GRAIN

The process of sawing lumber from logs results in a visible pattern showing the annual rings on the end of the lumber or the pattern of longitudinal fibers on the surface. Theoret-

| USE OF LUMBER | MOISTURE CONTENT (%) | | | | | |
| | Dry Western Areas | | Damp & Coastal Southern Areas | | Remainder of United States | |
	Average	Ind. Pieces	Average	Ind. Pieces	Average	Ind. Pieces
Exterior trim, siding, sheathing and framing[1]	8–10	7–12	11–13	9–14	11–13	9–14
Interior trim, woodwork and softwood flooring	5–7	4–9	10–12	8–13	7–9	5–10
Hardwood flooring	6–7	5–8	10–11	9–12	7–8	6–9

[1]Framing lumber of higher moisture content is commonly used in construction because lumber with the moisture content recommended may not be available except on special order.

Figure 10-3. Recommended approximate moisture contents.

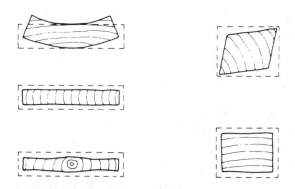

Figure 10-4. Shrinkage as affected by direction of annual rings.

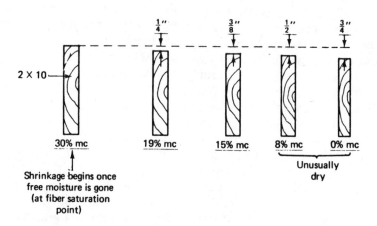

Figure 10-5. Approximate lumber shrinkage.

140

ically, all lumber is cut from a log either tangentially to the annual growth rings or radially to the rings (which is parallel to the rays) (Figure 10–6). However, not all of the pieces of lumber cut from a log will fit exactly into these two definitions.

In an attempt to classify the wood graining directions possible, several typical wood grain situations have been identified. There are no exact definitions or parameters for these groupings, but the most commonly used are included here.

Edge-Grain

Lumber cut from logs radially to the growth rings will have the annual rings run at about a right angle (90°) to the face (Figure 10–7). This is referred to as *edge-grain* if it is a softwood, and *quarter-sawed* if it is a hardwood. Generally, any piece with the grain at more than a 45 degree angle is referred to as edge-grain or quarter-sawed.

Angle-Grain

In the actual cutting of the lumber radially to the growth rings, some will be at less than a 45 degree angle from the face. This is referred to as *angle-grain* in softwoods and *plain-sawed* in hardwoods.

Flat-Grain

When the lumber is cut tangentially to the growth rings from the logs, the growth rings run about parallel to the surface. This is referred to as *flat-grain*.

Close-Grain

In some woods such as poplar and most softwoods, the wood fiber bundles are packed closely together and are referred to as *close-grain* woods. In such woods no open spaces or pores are visible along the surface.

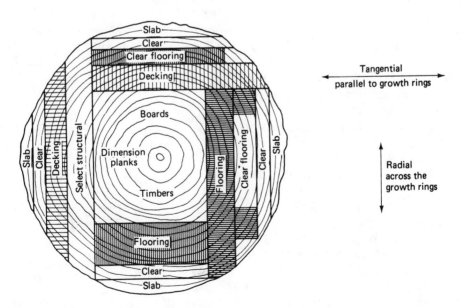

Figure 10-6. Lumber cut from logs.

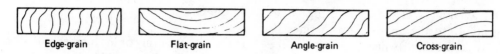

Figure 10-7. Grain patterns.

Open-Grain

Other woods, such as oak, do not have their wood fiber bundles as closely packed together, and open spaces or pores may be visible along the surface. This is referred to as *open-grain* wood.

Cross-Grain

When the wood fibers do not run parallel to the length of the board, it is referred to as *cross-grain*. This means that somewhere along the surface the ends of the fibers have been cut off, leaving a rough area that will be difficult to finish neatly.

10.5
STRUCTURAL PROPERTIES

The fibers in the wood give the wood its strength. The amount of strength depends on the thickness of the walls of the wood fiber and the direction of the fibers in relation to the loads applied to it. These fibers are parallel with the vertical axis of the tree, which is also parallel to the wood grain (Figure 10-8). The strength of the wood is much higher when tested parallel to the grain than when tested perpendicular to the grain.

The basic structural properties of wood for bending, tension, shear, and compression must be known, since whenever a piece of wood is used in the structural framing of a building it will be affected by these stresses.

Bending

When a load is applied to a beam, as in Figure 10-9, the beam begins to deflect (bend), which creates bending stresses in the top fibers of the beam. The beam can support the load as long as the load is not so great that it exceeds the fiber-bending strength of the wood. If this occurs the beam will fail (break) from the bending stresses. The further a fiber is from the neutral axis (center) of the beam, the greater the bending stress that is produced in the fiber. For example, a fiber 2 inches (50mm) from the central axis has a bending stress twice that of a fiber 1 inch (25 mm) from the central axis. In the beam (Figure 10-9), the maximum bending stresses occur in the fibers furthest from the central axis, which are at the top and bottom of the beam. These outer fibers are referred to as the *extreme* fiber. The allowable extreme fiber bending stress for several species and grades of wood is shown in

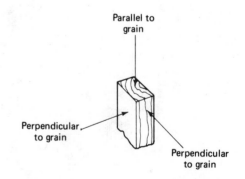

Figure 10-8. Compression strengths.

Figure 10–10 and is expressed in pounds per square inch or psi (MPa). This is referred to as the "f" value.

Tension

When a load is applied to a beam, as in Figure 10–9, the beam tends to deflect, which creates a stress in which the fibers below the neutral axis are pulled in opposite directions, tending to pull them apart. This stress is called *tension,* and it is usually parallel to the grain of the wood. Wood has great tensile strength, but it is difficult to measure since no testing device is available that can develop tension across the entire cross section to be tested. So, the allowable extreme fiber bending stress (f) is used for the tensile strength value, parallel to the grain. Wood members are rarely used in such a manner that tension perpendicular to the grain must be considered.

Horizontal Shear

As a load is applied to a beam the wood fibers above the neutral axis tend to move horizontally with respect to the fibers below the neutral axis (Figure 10–9). This movement results in stresses in the fibers referred to as *horizontal shear.* The maximum horizontal shear stresses occur at the end of the beam along the neutral axis. The allowable horizontal shear for several species and grades of wood is shown in Figure 10–10. This is expressed in pounds per square inch or psi (MPa) and is referred to as the "H" value.

Vertical Shear

As the load is applied to the beam the wood fibers at the point where the beam and support meet (Figure 10–9) have a stress put on

them that tends to cause them to separate and move vertically with respect to the fibers that rest on the support. This movement results in stresses in the fibers referred to as *vertical shear.* The wood's allowable horizontal shear is much lower than its vertical shear, and wood beams would be more likely to fail due to horizontal stresses than to vertical. Because of this, vertical shear values are rarely given in typical wood strength lists.

Compression

When a wood member is used as a column (Figure 10–8) the load applied presses the wood fibers together, creating compression stresses in the wood fibers. The load may be applied in any one of these directions:

1. Parallel to the grain.
2. Perpendicular to the grain, radially.
3. Perpendicular to the grain, tangentially.

There is little difference in the strength of the wood for loads perpendicular to the grain, either radial or tangential. However, the

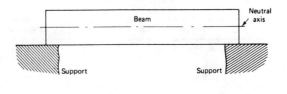

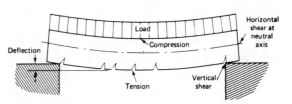

Figure 10–9. Forces in a beam.

		Allowable Unit Stresses in psi				Millions psi
SPECIES Grades	Grading Authority 2 Size Classification	**Extreme fiber in bending "f" and tension parallel to grain "t"**	**Horizontal shear "H"**	**Compression perpendicular to grain "c⊥"**	**Compression parallel to grain "c"**	**Modulus of elasticity "E"**
PINE, SOUTHERN	SPIB					
Dense Structural 86KD	2" thick only	3,000	165	455	2,250	1.76
Dense Structural 72KD	2" thick only	2,500	150	455	1,950	1.76
Dense Structural 65KD	2" thick only	2,250	135	455	1,800	1.76
Dense Structural 58KD	2" thick only	2,050	120	455	1,650	1.76
No. 1 Dense KD	2" thick only	2,050	135	455	1,750	1.76
No. 1 KD	2" thick only	1,750	135	390	1,500	1.76
No. 2 Dense KD	2" thick only	1,750	120	455	1,300	1.76
No. 2 KD	2" thick only	1,500	120	390	1,100	1.76
Dense Structural 86	2", 3", 4" thick	2,900	150	455	2,200	1.76
Dense Structural 72	2", 3", 4" thick	2,350	135	455	1,800	1.76
Dense Structural 65	2", 3", 4" thick	2,050	120	455	1,600	1.76
Dense Structural 58	2", 3", 4" thick	1,750	105	455	1,450	1.76
No. 1 Dense	2", 3", 4" thick	1,750	120	455	1,550	1.76
No. 1	2", 3", 4" thick	1,500	120	390	1,350	1.76
No. 2 Dense	2", 3", 4" thick	1,400	105	455	1,050	1.76
No. 2	2", 3", 4" thick	1,200	105	390	900	1.76
No. 1 Dense SR	3" & 4" thick	1,750	120	455	1,750	1.76
No. 1 SR	3" & 4" thick	1,500	120	390	1,500	1.76
No. 2 Dense SR	3" & 4" thick	1,400	105	455	1,050	1.76
No. 2 SR	3" & 4" thick	1,200	105	390	900	1.76
Select	Decking	1,500	120	390	1,350	1.76
Select No. 1	Decking	1,200	105	390	900	1.76
No. 2	Decking	1,200	105	390	900	1.76
PINE, IDAHO WHITE, LODGEPOLE PONDEROSA & SUGAR	WWP					
Selected Decking	Decking	900	—	350	—	1.10
Commercial Decking	Decking	700	—	350	—	1.10
REDWOOD	RIS					
Dense Structural	J. & P., B. & S.	1,700	110	320	1,450	1.32
Heart Structural	J. & P., B. & S.	1,300	95	320	1,100	1.32
SPRUCE, ENGELMANN	WWP					
Selected Decking	Decking	750	—	215	—	1.10
Commercial Decking	Decking	600	—	215	—	1.10
SPRUCE, SITKA	WCLB WWP					
Select Dex	Decking	1,100	—	305	—	1.32
Commercial Dex	Decking	850	—	305	—	1.32

1. The allowable unit stresses are for normal loading conditions. See provision of the National Design Specification of the National Forest Products Association for adjustments of the tabulated allowable unit stresses.
2. Abbreviations:
MC 15—Lumber of 2" thickness surfaced to size at ·15% moisture content.
RIS—Redwood Inspection Service.

SPIB—Southern Pine Inspection Bureau.
WCLB—West Coast Lumber Inspection Bureau.
WWP—Western Wood Products Association.
f—Allowable unit stress—extreme fiber in bending.
L.F.—Light Framing.
J. & P.—Joists and Planks.
KD—Kiln Dried.
B. & S.—Beams and Stringers.

Figure 10-10. Allowable wood stresses. (Courtesy Southern Forest Products Association)

strength of the wood is five to ten times greater in compression parallel to the grain than in compression perpendicular to the grain. The allowable compression values for parallel to the grain (C) and perpendicular to the grain (C⊥) for several species and grades of wood are shown in Figure 10–10 and are expressed in pounds per square inch (MPa).

Modulus of Elasticity

When a load is applied to a beam, as in Figure 10–9, the beam begins to deflect. The wood's ability to resist this deflection indicates the wood's stiffness and is measured by its modulus of elasticity. The higher a material's modulus of elasticity, the greater its stiffness and the less deflection it will undergo under any given loading situation. The modulus of elasticity for several species and grades of wood is shown in Figure 10–10 and is expressed in pounds per square inch (MPa) and referred to as the "E" value.

When a load is applied to a column there is a tendency for the column to buckle outward. The ability of the wood column to resist buckling is proportional to the modulus of elasticity of the wood used and the width and depth dimensions of the column.

To obtain the maximum unit stresses and an accurate modulus of elasticity it is necessary to test small, perfect samples of the species of wood. All testing is done in accordance with the procedures outlined in ASTM D2555, *Methods for Establishing Clear Wood Strength Values*. To determine the maximum unit stresses it is necessary to test the samples until they fail. The maximum values are determined by those stresses at which the samples failed, since they can withstand none higher.

The actual load on the member must be lower than the maximum, to be certain it will not fail. For this reason a factor of safety must be used. The amount of the safety factor depends on the material being tested and its various defects. For wood the allowable unit stresses are determined using ASTM D245, *Methods for Establishing Structural Grades for Visually Graded Lumber, as Basic Stresses for Clear Lumber Under Long-Time Service at Full Design Load.*

Using ASTM D245 it is necessary to determine the location, magnitude, and condition of the defects that would reduce the strength and stiffness of the wood member. This is used to determine the allowable unit stresses for a given piece of wood. In this manner the stress grades of lumber are determined by a visual inspection by an experienced lumber grader.

Lumber 2 inches (38mm) or less in thickness can now be tested by an electromechanical stress grader at the sawmill. The grader tests the lumber for strength and automatically marks it. This has allowed the manufacturers to standardize the grading process and guarantee the strength of each piece of lumber. Under the visual grading process the grader had a tendency to assign lower grading values to the wood, just to be on the safe side. Tests have shown that at least half of the visually graded lumber should have been graded at twice the grading given.

As the piece of lumber is fed through the machine it is subjected to bending in two directions, and the modulus of elasticity (E) is measured at 6-inch intervals. Then the machine takes into account the growth rate, density, moisture content, slope of grain, and knots to electronically compute the stress grade. The allowable fiber stress in bending (f) and modulus of elasticity (E) is then stamped on the piece of lumber as it leaves the machine. The lumber may be graded into any one of nine stress grade categories that have been established (Figure 10–11). A typical

EXTREME FIBER STRESS IN BENDING (PSI)	MODULES OF ELASTICITY (MILLION PSI)
900	1.0
1200	1.2
1500	1.4
1800	1.6
2100	1.8
2400	2.0
2700	2.2
3000	2.4
3300	2.6

Figure 10-11. Stress grades. Lumber is mechanically graded into these nine categories.

grademark that would be stamped on a piece of lumber is shown in Figure 10-12.

The allowable stress for a piece of lumber establishes its maximum span when used as a structural wood framing such as floor, ceiling and wood joists, and roof rafters.

When the lumber will be exposed to view, such as in exposed beams and columns, it will also be visually graded for appearance.

_____ 10.6 _____
LUMBER QUALITY AND DEFECTS

The quality of the lumber cut from the logs varies considerably due to the natural growth characteristics of the tree and manufacturing process. The lumber is divided into various lumber grades. Any defects may affect the quality of the lumber in terms of its strength or its appearance. When appearance is important even relatively small defects may lower its grade considerably. The grading associations have rules limiting the various defects for each wood species, and they are available from the associations.

Imperfections Due to Natural Characteristics

The most common natural characteristics that may affect the quality of the lumber are the following:

1. *Knots,* which occur when the lumber is cut at cross or longitudinal sections to a branch. The knots interrupt the basic grain direction of the wood, resulting in a reduction of its strength, so the grading rules restrict knot size and location. In addition, the knots may affect the appearance of the wood, especially in the *select* grades. A knot that is free from decay is called "sound"; if the knot has decayed, it is called "unsound." When the knot is firmly in place, it is "tight," but if it is loose or may not stay in place, it is

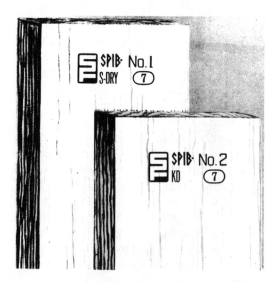

Figure 10-12. Typical Southern Pine grademark. (Courtesy Southern Forest Products Association)

"loose." Types and sizes are shown in Figure 10–13.

2. *Shakes* are separations in the wood that occur between the annual rings. These lengthwise separations reduce the allowable shear stress of the wood, but affect only slightly its tensile and compressive strengths. The separations make the wood undesirable for use when appearance is important.

3. *Pitch pockets* are the accumulations of tree sap that may occur between the annual rings. The number, size, and location of the pitch pockets will determine whether they have any effect on strength.

Decay of the wood affects both strength and appearance of the wood, and it is carefully controlled in the grading rules.

Imperfections Due to Manufacturing Process

Lumber quality may also be affected by the manufacturing process. Following are some typical imperfections that occur.

1. *Wanes* occur when the bark or other soft material is left on the edge of the piece or when there is no wood on the edge. This has no effect on the strength, but does reduce the amount of nailing surface and affects the appearance of the wood.

2. *Torn grain* occurs when the machine tears out small bits or portions of the wood during the dressing operation. This does not reduce its strength, but does affect its appearance.

3. A *skip* is any area that is not smoothed as the piece passes through the planer. While

Spike knot
(lumber sawn
parallel to the
branch)

Oval knot
(lumber sawn
diagonally to
the branch)

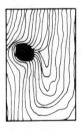

Round knot
(lumber sawn
perpendicular
to the branch)

Types of knots

Pin knot, less than $\frac{1}{2}$ in.

Small knot, less than $\frac{3}{4}$ in.

Medium knot, less than $1\frac{1}{2}$ in.

Large knot, more than $1\frac{1}{2}$ in.

Knot

Knots are measured by averaging the maximum and minimum diameters.

Figure 10-13. Knots.

a skip does not affect the wood's strength, it does affect its appearance.

4. *Warp* is any distortion or deviation that occurs in the flat, plane surface of the piece. The various types of warping are shown in Figure 10–14. A "cup" is any deviation across the width of the piece; a "bow" is any deviation along the flat of the piece; a "crook" is any edgewise deviation; and a "twist" is any deviation or twist in one of the corners. Warp is always measured at its point of greatest deviation from a straight line.

5. A *check* is a separation that occurs lengthwise in the wood across the annual ring. When present on only one surface, it is called a *surface check;* if it extends through the piece, it is a *through check*. It is measured by depth and length, and primarily affects appearance, unless it is large.

6. *Machine burn* occurs when the machine rollers cause an area to darken due to overheating. This has no effect on strength, but does affect the appearance of the wood.

7. *Seasoning* of the lumber may also result in lowering lumber quality (de-grading it).

This is usually a result of unequal shrinkage and can usually be avoided with proper care during seasoning, in the lumber yard and on the construction site. Generally, lumber that dries too quickly has a tendency to split and check, while lumber dried slowly at unfavorable temperatures may decay or stain.

10.7
HARDWOOD LUMBER

While hardwood accounts for about 25 percent of lumber production, it is seldom used for the wood framing members of residential and small commercial buildings. Instead it is used primarily for wood floors (oak and maple), wood trim (beech, birch, oak, cherry, mahogany, maple, and walnut), and various types of plywood veneers, doors, paneling, and furniture.

Size and grading rules of hardwood sawed lumber are published by the National Hardwood Lumber Association (NHLA). Since hardwood is primarily used to be reshaped or remanufactured into other shapes, the size and grade rules are of primary interest to the

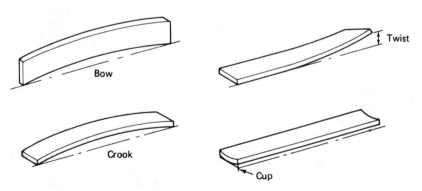

Figure 10–14. Types of warps.

lumber industry and not to the construction industry. Hardwood used for construction is usually specially ordered, with the size, stress, and appearance requirements spelled out and agreed to by the buyer and seller.

10.8

SOFTWOOD LUMBER CLASSIFICATIONS

Softwood lumber is classified according to how it will be used, the size of the lumber, and the amount of manufacturing, and is graded according to appearance, stress grading, and moisture content.

Use

The three broad categories of softwood lumber use are yard, structural, and factory and shop lumber. *Yard lumber* is commonly used for general building construction uses. These include the stress grades, sizes, and appearance grades shown in Figure 10–15.

Structural lumber is commonly used in construction where strength is of primary importance and the visual appearance of the wood is less important, such as for wood trusses. This type of lumber has all four longitudinal surfaces checked for strength before a stress grade is assigned to the member. Structural lumber is often referred to as "stress-graded" lumber and must be 2 inches (38mm) or more in thickness and width.

Factory and shop lumber is commonly used for windows, doors, frames, trim pieces, and other millwork items. It is graded by the amount of clear wood that can be cut from a given piece of lumber, recut or reshaped, and used in the factory or shop.

Size

Softwood lumber is classified by size into boards, dimension, and timbers. *Boards,* Figure 10–15, are any pieces that are less than 2 inches (38mm) thick and more than 1 inch (19mm) wide. Boards are typically used as trim, siding, and panelling, and occasionally are used as wall and roof sheathing and subflooring. Boards that are less than 6 inches (140mm) in width are often classified as *strips.*

Dimension lumber, Figure 10–15, is any piece with a thickness of 2 inches (38mm) or over, but less than 5 inches (113mm), and a width of 2 inches (38mm) or more. Dimension lumber is commonly used for the structural framing in wood frame structures such as residences and may be used as floor and ceiling joists, roof rafters, and wall studs. With this type of usage the strength is the most important consideration and appearance is secondary.

Timbers, Figure 10–15, are any pieces with a minimum dimension of 5 inches (112.5mm) in both width or thickness. Timbers are used for heavy framing such as columns, beams, and girders. Since it is difficult to get a low moisture content in such thick cross sections, timbers tend to have higher amounts of shrinkage than thinner members. Many times laminated timbers are used when seasoned timbers are required. The smaller pieces of wood, which are glued together to make laminated timbers, can be easily kiln-dried so they will have less shrinkage than the timber sections.

Manufacture

Softwood lumber is also classified by the amount of processing it undergoes during the sawing and milling process. The classifica-

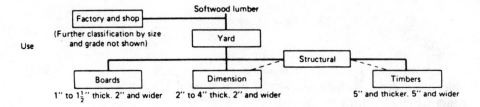

PRODUCT	GRADE	CHARACTER OF GRADE AND TYPICAL USES
DIMENSION **Structural Light Framing** 2″ to 4″ thick 2″ to 4″ wide	*Select Structural *Dense Select Structural	High quality, relatively free of characteristics which impair strength or stiffness. Recommended for uses where high strength, stiffness and good appearance are required.
	No. 1 No. 1 Dense	Provide high strength, recommended for general utility and construction purposes. Good appearance, especially suitable where exposed because of the knot limitations.
	No. 2 No. 2 Dense	Although less restricted than No. 1, suitable for all types of construction. Tight knots.
	No. 3 No. 3 Dense	Assigned design values meet wide range of design requirements. Recommended for general construction purposes where appearance is not a controlling factor. Many pieces included in this grade would qualify as No. 2 except for single limiting characteristic. Provides high quality and low cost construction.
STUDS 2″ to 4″ thick 2″ to 6″ wide 10′ and Shorter	Stud	Stringent requirements as to straightness, strength and stiffness adapt this grade to all stud uses, including load-bearing walls. Crook restricted in 2″ x 4″ — 8′ to ¼″, with wane restricted to 1/3 of thickness.
Structural Joists & Planks 2″ to 4″ thick 5″ and wider	*Select Structural Dense Select *Structural	High quality, relatively free of characteristics which impair strength or stiffness. Recommended for uses where high strength, stiffness and good appearance are required.
	No. 1 No. 1 Dense	Provide high strength, recommended for general utility and construction purposes. Good appearance, especially suitable where exposed because of the knot limitations.
	No. 2 No. 2 Dense	Although less restricted than No. 1, suitable for all types of construction. Tight knots.
	No. 3 No. 3 Dense	Assigned stress values meet wide range of design requirements. Recommended for general construction purposes where appearance is not a controlling factor. Many pieces included in this grade would qualify as No. 2 except for single limiting characteristic. Provides high quality and low cost construction.
**Light Framing 2″ to 4″ thick 2″ to 4″ wide	*Construction	Recommended for general framing purposes. Good appearance, strong and serviceable.
	*Standard	Recommended for same uses as Construction grade, but allows larger defects.
	*Utility	Recommended where combination of strength and economy is desired. Excellent for blocking, plates and bracing.
	*Economy	Usable lengths suitable for bracing, blocking, bulkheading and other utility purposes where strength and appearance not controlling factors.

Figure 10-15. Softwood lumber classification. (Product and grade information courtesy Southern Forest Products Association)

150

tions are rough, dressed, and worked lumber.

Rough lumber has been sawed, edged, and trimmed, and the rough marks from the sawing are visible on the surface of the lumber.

Dressed (or *surfaced*) lumber has been put through a planing machine, which gives it a smooth surface and uniform size. It is further designated by the amount of surfacing it receives. For example, if it is surfaced on one side it is designated S1S (*surfaced one side*), and when it is surfaced on two sides it is S2S. When one edge has been surfaced it is S1E (*surfaced one edge*); when two edges have been surfaced it is S2E. There may even be a combination of sides and edges surfaced such as two sides and one edge (S2S1E) or one side and one edge (S1S1E). When both sides and both edges are surfaced it is designated S4S (*surfaced four sides*); this is the type of lumber commonly used for wood studs, rafters, and joists in wood frame construction.

Worked lumber, Figure 10–16, has been dressed and then matched, shiplapped, or patterned. *Matched* lumber has a tongue on one edge and a groove on the opposite edge. The use of tongue and grooved pieces together provides tight-fitting, interlocking construction joints. The ends may also be tongue and groove to provide ''end-matching'' (tongue and groove joint at the ends). *Shiplapped* lumber has been rabbeted (grooved) on one or both edges. When two pieces are fitted together they give a close-fitting, lapped joint. *Patterned* lumber has been dressed and either matched or shiplapped, and is then shaped to a determined pattern. Typically, this includes wood siding and millwork, such as the various moldings used.

Grades

Softwood lumber is classified into select (appearance or finish) grades, common grades, and stress grades.

Select grades are based on the appearance and finish of the piece and are used where the wood will be left natural, stained, or painted with a high quality paint. The select grades are A, B, C, and D, with A the highest grade representing the best appearance. Many times grades A and B are combined and sold under the grading designation ''B and Better'' (B & Btr.). These grades are commonly used where the appearance of the wood is most important—probably in locations where it will be left natural or where the finish selected (e.g., stain, varnish, or sealer) will be used to high-

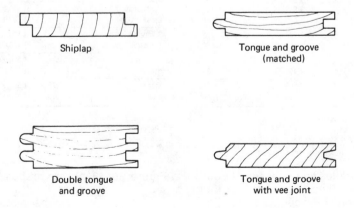

Shiplap

Tongue and groove
(matched)

Double tongue
and groove

Tongue and groove
with vee joint

Figure 10–16. Worked lumber.

light the natural beauty of the wood. This high-quality material is often used for interior and exterior trim, flooring, paneling, ceilings, and siding. Grades C and D have excellent finishing surfaces and provide an excellent base for high quality painting at much less cost than the higher grades. They are generally considered undesirable for use where a natural wood finish is preferred.

Common grades of lumber are used for general construction and utility purposes where moderate stresses are involved. This lumber is graded by numbers (No. 1, No. 2, No. 3, No. 4) or by names (stud utility, standard, construction). The type of grading identification used varies with the grading association. The grading for southern pine and its allowable stresses have been determined by the Southern Forest Products Association

and are illustrated in Figure 10-17. The stud and construction grades, which roughly correspond with Number 3 grade, are commonly used in wood frame construction.

Stress grades of lumber have been visually inspected or machine tested for their structural properties. Stress grades include medium and dense grain, structural, and select structural. These and others are listed in Figure 10-18. They are commonly used in wood frame construction when high stress values are required.

| GRADE | 2-4" THICK, 5" & WIDER | | | 2-4" THICK, 2-4" WIDE | |
	Extreme Fiber In Bending* "Fb" psi	"1.15Fb" psi	Modulus of Elasticity "E" psi	Extreme Fiber In Bending* "Fb" psi	"1.15Fb" psi
No. 1 KD	1600	1850	1,800,000	1850	2150
No. 1	1450	1650	1,700,000	1700	1950
No. 2 KD	1300	1500	1,600,000	1550	1800
No. 2	1200	1400	1,600,000	1400	1600
No. 3 KD	750	850	1,500,000	850	975
No. 3	700	800	1,400,000	775	900
Construction KD			1,500,000	1100	1150
Construction			1,400,000	1000	1050
Standard KD			1,500,000	625	725
Standard			1,400,000	575	650
Utility KD			1,500,000	275	325
Utility			1,400,000	275	325
Stud KD	800**	925	1,500,000	850	975
Stud	725**	900	1,400,000	775	900

*See section entitled " Design Values"
**Applies to 5" and 6" widths only.

Terms and abbreviations: Sel. Str. means select structural; KD means kiln dried to a moisture content of 15% or less; and where KD is not shown the material is dried to a moisture content of 19% or less.

Figure 10-17. Grades and stresses of Southern Pine. (Courtesy Southern Forest Products Association)

| GRADE | 2-4" THICK, 5" & WIDER | | | 2-4" THICK, 2-4" WIDE | |
	Extreme Fiber In Bending* "Fb" psi	"1.15Fb" psi	Modulus of Elasticity "E" psi	Extreme Fiber In Bending* "Fb" psi	"1.15Fb" psi
Dense Sel Str KD	2200	2550	1,900,000	2500	2900
Dense Sel Str	2050	2350	1,800,000	2350	2700
Sel Str KD	1850	2150	1,800,000	2150	2450
Sel Str	1750	2000	1,700,000	2000	2300
No. 1 Dense KD	1850	2150	1,900,000	2150	2450
No. 1 Dense	1700	1950	1,800,000	2000	2300
No. 1 KD	1600	1850	1,800,000	1850	2150
No. 1	1450	1650	1,700,000	1700	1950
No. 2 Dense KD	1550	1800	1,700,000	1800	2050
No. 2 Dense	1400	1600	1,600,000	1650	1900
No. 2 KD	1300	1500	1,600,000	1550	1800
No. 2	1200	1400	1,600,000	1400	1600
No. 3 Dense KD	875	1000	1,500,000	1000	115u
No. 3 Dense	825	950	1,500,000	925	1075
No. 3 KD	750	850	1,500,000	850	975
No. 3	700	800	1,400,000	775	900
Construction KD			1,500,000	1100	1150
Construction			1,400,000	1000	1050
Standard KD			1,500,000	625	725
Standard			1,400,000	575	650
Utility KD			1,500,000	275	325
Utility			1,400,000	275	325
Stud KD	800**	925	1,500,000	850	975
Stud	725**	900	1,400,000	775	900

*See section entitled " Design Values"
**Applies to 5" and 6" widths only.

Terms and abbreviations: Sel. Str. means select structural; KD means kiln dried to a moisture content of 15% or less; and where KD is not shown the material is dried to a moisture content of 19% or less.

Figure 10-18. Stress grades of wood. (Courtesy Southern Forest Products Association)

10.9
LUMBER SIZES

The thickness and width of lumber used as framing material is referred to as its *nominal size*. This nominal size represents the approximate end dimensions of the piece of lumber *before* it is surfaced. Most framing lumber is surfaced on four sides (S4S) and dried in some manner. The *actual size* is the dimension of the lumber *after* it is dressed (surfaced) and dried. The nominal and actual sizes for various framing member sizes are shown in Figure 10–19 and are based on the American Softwood Lumber Standard (PS 20–70). So, while a 2 × 8 framing member is actually only 1½ by 7¼ inches, it is still called a 2 × 8. The corresponding metric measurement for the 2 × 8 is 38 × 138 millimeters, its actual size.

Lumber Size Selection

The grading of the lumber is important in the proper selection of lumber sizes for construction purposes. For light wood frame construction the sizes of the floor joists, ceiling joists, and rafters must be determined. The sizes required will be determined based on:

1. Span (distance between supports).
2. Loading requirements.
3. Grade of lumber used.

The span is usually determined by reviewing the layout of the project. For example, for the building in Appendix B the following information is accumulated:

1. Overall dimensions, 24'-0" (7.3m) (width) × 50'-0" (15.2m) (length).

Metric Equivalents

Nominal Size (Inches)	Lumber Metric (MM) Nomenclature (Actual Size)	Actual Size (Inches)
2 × 2	38 × 38	1½ × 1½
2 × 4	38 × 89	1½ × 3½
2 × 6	38 × 140	1½ × 5¼
2 × 8	38 × 184	1½ × 7¼
2 × 10	38 × 235	1½ × 9¼
2 × 12	38 × 286	1½ × 11¼

STUD SPACING

12"	300mm
16"	400mm
24"	600mm

A 92¼", precut stud will become 2305 mm

Figure 10–19. Metric equivalents.

2. Girder (3'-2" × 12') (38 × 286mm) and a row of piers down the center of the building.

With this information a sketch through the building, along with some dimension calculations, will determine the actual span required for the floor joists. In this layout the actual span is one-half the building width minus the width of the sill (5½ inches actual) plus one-half the actual size of the built-up girder (Figure 10–20). So the actual span of the floor joists is 11'-4½" (3,468mm).

NOTE: The actual *span* of the floor joists is 11'-4½" (3,468mm), but the floor joist used will be longer than the span calculated, since it rests on top of the sill and on top of the girder.

Loading is reviewed next. Many building codes have two basic loading conditions for wood frame residences: (a) 30 pounds per square foot (207 kPa) for all sleeping rooms and attic floors; and (b) 40 pounds per square foot (276 kPa) for all rooms except sleeping rooms and attic floors. It is most important to determine what the building code requirements are in the locale in which the building will be constructed. Some codes require 40 pounds per square foot (276 kPa) for all residential floor joists, with no reduction for sleeping rooms. When the building use (type of occupancy) will be other than residential, for example a bank, offices, or store, it will be necessary to determine the live load requirements per square foot of floor area. Most building codes set live loads for various types of occupancies (Figure 10–21), and these limits are used for the design of the floor joists.

For the 24' × 50' (7.3 × 15.2m) building, the following information has now been accumulated:

1. Span = 11'-4½" (3,468mm)
2. Live load = 40 pounds per square foot (276 kPa) (residence).

A typical floor joist selection table is shown in Figure 10–22 for southern pine. The floor joist selection table shows the following variables involved in the actual decision on what size to select:

1. Grade of lumber. The various grades are grouped along the top of the table.
2. Various sizes of floor joists typically used. These are listed along the left edge of the table.

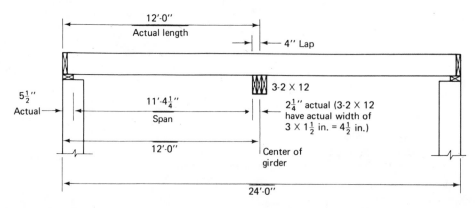

Figure 10-20. Floor joist.

3. Spacing. The distances between the floor joists most commonly used are shown just to the right of the sizes.

The actual selection must take all of these things into consideration. The first step is to determine what grade of lumber is "stock," or readily available at the local lumber companies. Stock materials are usually less costly than ordering specially graded lumber for one small building. If the local lumber supplier has southern pine, No. 2 KD, the selection of any other grade will usually result in higher costs.

Assuming No. 2 KD is available, the design selection would proceed by using the table. This is done by first finding the grade along the top and then moving down vertically to a figure that is the same as or higher than the ac-

tual span, in this design 11′-4¼″. As selected from Figure 10–22, a 2 × 8 (38 × 184mm), 19.2 inches (480mm) on center is the selection. However, 19.2 inch (480mm) spacing is not common in residential construction, and a 16 inch (400mm) on center spacing is chosen since it will be within the allowable design limits.

A review of the table shows that using higher grades of lumber would not reduce the size of floor joist required and that grades as low as No. 3 Dense KD may be used 16 inches (400mm) on center. However, if No. 3 KD were to be used, the floor joists would have to be 2 × 8 (38 × 184mm), 12 inches (300mm) on center or increase a size to 2 × 10 (38 × 235mm), 16 inches (400mm) on center.

Ceiling Joists

Ceiling joist sizes are selected in a similar manner, but different loading requirements come into effect. In this building, the ceiling joists run in the same direction as the floor joists. They bear on an exterior wall and an interior bearing wall, which runs down the center of the building. This results in an actual span of 11′-7¾″ (3,550mm), as taken from the sketches shown in Figure 10–23.

Choosing the appropriate ceiling joist table involves a consideration of (a) the type ceiling to be used (plaster or drywall) and (b) the live load requirements. It is assumed that there will not be any sleeping spaces above. If there is attic storage, then a 20 pounds per square foot (138 kPa) live load is used; when there is no attic storage, a 10 pounds per square foot (69 kPa) live load is used. When there will be sleeping spaces above the ceiling joists, they should be sized using the appropriate floor joist table [using 30 psf (207 kPa)]. In this building, assume there is attic storage and the live load would be 20 pounds per square foot (138 kPa).

Occupancy	Live load (psf)
Educational	
Classrooms	40
Corridors	100
Businesses	
Offices	80
File rooms	125
Mercantile	
Wholesale	125
First floor, rooms	100
Upper floor	75
Residental	
Dwellings:	
First floor	40
Upper floors	30
Attics (uninhabitable)	20
Multifamily:	
Private apartments	40
Corridors	60
Public rooms	100

Figure 10–21. Typical floor live loads.

TABLE NO. 2. FLOOR JOISTS—40 psf live load. All rooms except sleeping rooms and attic floors. (Spans shown in light face type are based on a deflection limitation of $l/360$. Spans shown in color, bold face type are limited by the recommended extreme fiber stress in bending value of the grade and includes a 10 psf dead load.)

Size and Spacing in	Grade in. o.c.	Dense Sel Str KD and No. 1 Dense KD	Dense Sel Str, Sel Str KD, No. 1 and No. 1 Dense and No. 1 KD	Sel Str, No. 1 and No. 2 KD	No. 2 Dense, No. 2 KD and No. 2	No. 3 Dense KD	No. 3 Dense	No. 3 KD	No. 3
2 x 5	12.0	9-3	9-1	8-11	8-9	8-3	8-0	7-8	7-4
	13.7	8-11	8-9	8-7	8-5	7-9	7-6	7-2	6-11
	16.0	8-5	8-3	8-2	8-0	7-2	6-11	6-7	6-5
	19.2	7-11	7-10	7-8	7-6	6-6	6-4	6-0	5-10
	24.0	7-4	7-3	7-1	7-0[1]	5-10	5-8	5-5	5-3
2 x 6	12.0	11-4	11-2	10-11	10-9	10-1	9-9	9-4	9-0
	13.7	10-10	10-8	10-6	10-3	9-5	9-2	8-9	8-5
	16.0	10-4	10-2	9-11	9-9	8-9	8-6	8-1	7-10
	19.2	9-8	9-6	9-4	9-2	8-0	7-9	7-4	7-1
	24.0	9-0	8-10	8-8	8-6[1]	7-1	6-11	6-7	6-4
2 x 8	12.0	15-0	14-8	14-5	14-2	13-3	12-11	12-4	11-11
	13.7	14-4	14-1	13-10	13-6	12-5	12-1	11-6	11-1
	16.0	13-7	13-4	13-1	12-10	11-6	11-2	10-8	10-3
	19.2	12-10	12-7	12-4	12-1	10-6	10-2	9-9	9-5
	24.0	11-11	11-8	11-5	11-3[1]	9-5	9-1	8-8	8-5
2 x 10	12.0	19-1	18-9	18-5	18-0	16-11	16-5	15-8	15-2
	13.7	18-3	17-11	17-7	17-3	15-10	15-5	14-8	14-2
	16.0	17-4	17-0	16-9	16-5	14-8	14-3	13-7	13-1
	19.2	16-4	16-0	15-9	15-5	13-5	13-0	12-5	12-0
	24.0	15-2	14-11	14-7	14-4[1]	12-0	11-8	11-1	10-9
2 x 12	12.0	23-3	22-10	22-5	21-11	20-7	20-0	19-1	18-5
	13.7	22-3	21-10	21-5	21-0	19-3	18-9	17-10	17-3
	16.0	21-1	20-9	20-4	19-11	17-10	17-4	16-6	16-0
	19.2	19-10	19-6	19-2	18-9	16-3	15-10	15-1	14-7
	24.0	18-5	18-1	17-9	17-5[1]	14-7	14-2	13-6	13-0

1. The span for No. 2 grade, 24 inches o.c. spacing is: 2x5, 6-10; 2x6, 8-4; 2x8, 11-0; 2x10, 14-0; 2x12, 17-1.

Figure 10–22. Floor joist spans. (Courtesy Southern Forest Products Association)

Using the design information accumulated—a span of 11′-7¾″ (3,550mm); a 20 pounds per square foot live load; and a lumber grade of Number 2, kiln-dried (No. 2, KD)—the ceiling joist selection would be 2 × 6 (38 × 140mm), 19.2 inches (480mm) on center, with a final selection of 2 × 6 (38 × 140mm), 16 inches (400mm) on center.

Rafters

Sizing the rafters requires a little extra care in determining the actual span. The rafter span can be determined by first finding the horizontal distance from the ridge to the inside of the exterior wall, as shown in Figure 10–24. The actual span distance is the sloping

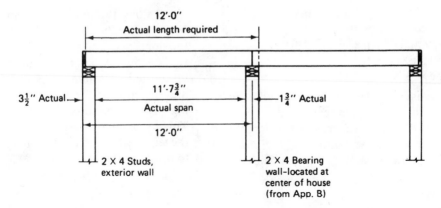

Figure 10-23. Ceiling span.

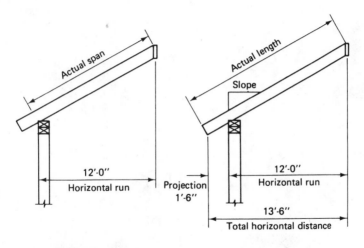

1. Determine horizontal run
2. Determine roof slope
3. Determine actual length using graph in Fig. 1.58 and illustrated in Fig. 1.33.

(a) Determining actual span

1. Determine total horizontal run
2. Determine roof slope
3. Determine actual length using graph Fig. 1.58.

(b) Determining actual length

Figure 10-24. Rafter span.

distance from the ridge to the inside of the exterior wall, and this sloping distance is found by:

1. Determining the horizontal distance from the ridge to the inside of the exterior wall.

2. Determining the pitch or slope of roof. Typically the pitch or slope is shown on either the elevations or the wall sections. Referring to Appendix B, the slope is shown as 3 in 12. This means that the roof slopes up at a rate of 3 inches vertically for

every 12 inches of horizontal distance.
Accumulated information:
Horizontal distance = 11′-7¾″
(3550mm).
Slope = 3 in 12.

3. The actual span can now be determined by using the conversion chart for rafters (Figure 10-25). First, locate the horizontal distance along the bottom of the chart (11′-7¾″) (3550mm), and then find the

slope along the upper right curved line. Move vertically up until it intersects with the slope, which is brought down from the right. From this point (1), follow the curve up to the left, where the sloping distance may be estimated. In this design, the sloping distance is about 12′-0″ (3658mm). This is the actual span of the rafter, but not the length of the rafter, which would have to consider any overhangs.

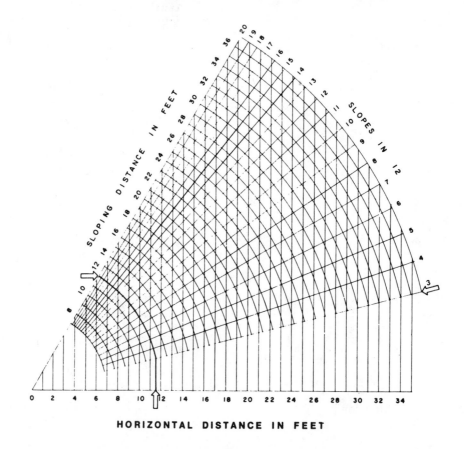

HORIZONTAL DISTANCE IN FEET

To use the diagram select the known horizontal distance and follow the vertical line to its intersection with the radial line of the specified slope, then proceed along the arc to read the sloping distance. In some cases it may be desirable to interpolate between the one foot separations. The diagram also may be used to find the horizontal distance corresponding to a given sloping distance or to find the slope when the horizontal and sloping distances are known.

Figure 10-25. Rafter conversion diagram. (Courtesy Southern Forest Products Association)

The rafter span may also be determined by a multiplying factor that represents the particular slope or pitch. These factors can be found in Figure 10–26. For a 3 in 12 slope, the factor used is 1.03, so the horizontal distance of 11'-7¾" (3550mm) is multiplied for a span of just about 12'-0" (3657mm).

Selecting the rafter chart to use takes some careful reading and consideration, since there are several variables including:

1. High slope (over 3 in 12).
2. Low slope (under 3 in 12).
3. Plaster ceiling.
4. Drywall ceiling.

Pitch of roof	Slope of roof	For length of rafter multiply length of run by
$\frac{1}{12}$	2 in 12	1.015
$\frac{1}{8}$	3 in 12	1.03
$\frac{1}{6}$	4 in 12	1.055
$\frac{5}{24}$	5 in 12	1.083
$\frac{1}{4}$	6 in 12	1.12

Note: Run measurement should include any required overhang.

Example:
For a 24'-0" span with a 2'-0" overhang on each side, $\frac{1}{8}$ pitch, what length rafter is required?

$$\frac{\text{Span \& Overhang}}{2} = \frac{24 + 4}{2} = \text{Run} = 14'\text{-}0''$$

Rafter length = 14'-0" × 1.03 = 14.42' = 14'- $5\frac{1}{32}$'

Figure 10–26. Rafter length.

5. No finished ceiling.
6. Amount of live load (20, 30, or 40 psf).
7. Amount of dead load (for snow, wind and roofing).

The drawings and basic design will answer several of these questions, and the building code should provide the rest of the required information. For this project there is a 3 in 12 slope with no finished ceiling. Since this project is to be built in South Carolina it will require only a 20 pounds per square foot live load, and it has a light roof (standard asphalt shingles), so only 7 pounds per square foot dead load need be allowed. (By contrast, the same project built in New York State would require a live load of 40 psf.) By using the appropriate table a rafter selection of 2 × 6 (38 × 140mm), 19.2 inches (480mm) on center is made, and then a final selection of 2 × 6 (38 × 140mm), 16 inches (400mm) on center is made, to conform with standard wood framing practices.

NOTE: In many areas wood framing is installed using 24-inch (600mm) centers. Of course this often will require larger structural members. For example, for this project the framing would be:

10.10
GRADING AUTHORITY

There are three basic lumber producing regions in the United States. The species of wood harvested in each region, the lumber association, and the grading authorities are listed in Figure 10–27. Grading authorities set the minimum standards for the various grades of lumber produced, as well as for the test procedures to be used to verify the grading. Lumber that is graded under their guidelines

	16" (400 MM) O.C.	24"(600MM) O.C.
Floor joists	2 × 8 (38 × 184 mm)	2 × 10 (38 × 235 mm)
Ceiling joists	2 × 6 (38 × 140 mm)	2 × 8 (38 × 184 mm)
Rafters	2 × 6 (38 × 140 mm)	2 × 8 (38 × 184 mm)

is stamped with the appropriate identification markings. A knowledge of the various grades and where they are best used is important in selecting and using the most economical grade for each proposed use.

10.11
PLYWOOD

Plywood is an engineered wood panel that is made by bonding (gluing) together thin wood sheets. An odd number of sheets are usually used, and each sheet is placed so that its grain is at right angles to the one next to it (Figure 10-28). Reversing the grain in this manner

REGION	SPECIES	LUMBER ASSOCIATION	GRADING AUTHORITY
Western Wood Region	Douglas fir Ponderosa pine Western red cedar Western hemlock White Fir Spruce Pine	Western Wood Products Association	Western Wood Products Association West Coast Inspection Bureau
Redwood	Redwood Douglas fir	California Redwood Association	Redwood Inspection Bureau
Southern Pine Region	Longleaf pine Shortleaf pine	Southern Forest Products Association	Southern Forest Products Inspection Bureau

Figure 10-27. Lumber regions and associations.

and using an odd number of sheets (or plies) of wood gives a stronger, more dimensionally stable product that is less susceptible to warping or cupping. Each layer or sheet of wood used in the plywood sheet is called a *veneer.* The surface veneer of the plywood, which will be exposed, is called the *face,* while the hidden side of the plywood sheet is the *back* or *back face* veneer. The center sheet is the *core,* and where more than three plies are used, the plies between the core and face veneers are the *crossbands.*

Plywood may be manufactured from softwoods or hardwoods. The softwood plywoods are used primarily where strength is important. They are commonly used as construction material, but they are sometimes used for decorative paneling. Hardwood plywoods are used primarily for decorative purposes, such as wall paneling, furniture, and cabinet work.

Softwood Plywood

The American Plywood Association (APA) is the primary trade association for softwood plywoods. This trade organization provides quality testing and inspection of association member firms. It also conducts research and provides promotional information for the association. All approved plywoods from association members have grade-trademarks, which provide a variety of information to the plywood user. Softwood plywood must meet the requirements of the U.S. Products Standard, which contains performance specifications for softwood plywood and a simplified, uniform basis for its use.

Softwood plywood is manufactured using wood veneers from softwood trees. This plywood is manufactured in two basic types, *exterior* and *interior*. Exterior plywood is bonded with a glue that is waterproof. Only exterior plywood should be used in any installation where it will be permanently exposed to the weather or moisture, such as siding on a building. Exterior plywood has higher grade veneers used for its backs and inner plies than those used in interior plywood. The interior plywood is bonded with a glue that is highly water resistant, but not waterproof. It may be installed in any location where it will not be permanently exposed to the weather or moisture, and is commonly used for floor and ceiling sheathing. Both ex-

terior and interior plywood are available in a variety of appearance and structural grades. Interior type plywoods, with their usual backs and inner plies, are also manufactured with exterior glue as the bonding agent. This is used where the plywood will be exposed to the weather during long construction delays. This interior plywood is not for use in any conditions where it would be permanently exposed to the weather or moisture. In such cases, exterior plywood must be used. However, if the plywood subfloor must be installed and it will be exposed to the weather for a while before the roof is on, or if it will be a long time before the house is completely dried in (windows and doors), the use of interior plywood with exterior glue may be appropriate. The APA grade-trademarks (Figure 10–29) clearly indicate the type of panel being used.

Veneer Grades

A variety of grades of plywood are available in both exterior and interior types. These grades are usually based on the quality and appearance of the veneer used for the face and back of the panel. The veneer grades range from N, the highest, through A, B, C, C-plugged, and D, which is the lowest. The grade selected is based on the use of the plywood sheet. The veneer may have knotholes, pitch pockets, and splits repaired with patches or oval-shaped plugs to raise its quality. Allowable defects are listed in Figure 10–30.

N Grade.   A select veneer, free of any open defects, with only limited, well-matched repairs allowed. It must be specially ordered and is seldom used. The veneer is either all heartwood or sapwood and is either left natural or finished to highlight its natural beauty.

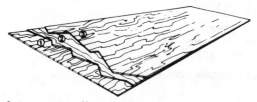

3 ply construction (3 layers of 1 ply each)

5 ply construction (5 layers of 1 ply each)

Figure 10–28. Plywood panel construction. (Courtesy Southern Forest Products Association)

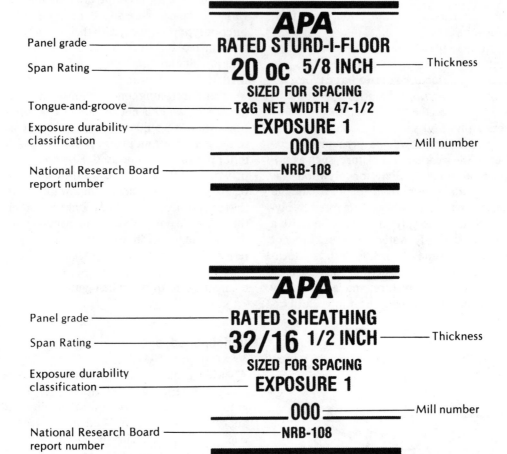

Panel grade

Span Rating

Thickness

Tongue-and-groove

Exposure durability
classification

Mill number

National Research Board
report number

Panel grade

Span Rating

Thickness

Exposure durability
classification

Mill number

National Research Board
report number

Figure 10-29. APA grade-trademarks explained. (Courtesy Southern Forest Products Association)

A Grade.   Primarily used where the plywood will be painted. A restricted number of patches, plugs, and repairs are allowed, but no splits or knotholes are allowed.

B Grade.   Used where smooth-surfaced plywood is required, such as concrete formboards. This grade has a larger number of neatly made patches, plugs, and repair work.

Veneer Grades

N	Smooth surface "natural finish" veneer. Select, all heartwood or all sapwood. Free of open defects. Allows not more than 6 repairs, wood only, per 4 x 8 panel, made parallel to grain and well matched for grain and color.
A	Smooth, paintable. Not more than 18 neatly made repairs, boat, sled, or router type, and parallel to grain, permitted. May be used for natural finish in less demanding applications.
B	Solid surface. Shims, circular repair plugs and tight knots to 1 inch across grain permitted. Some minor splits permitted.
C Plugged	Improved C veneer with splits limited to 1/8-inch width and knotholes and borer holes limited to 1/4 x 1/2 inch. Admits some broken grain. Synthetic repairs permitted.
C	Tight knots to 1-1/2 inch. Knotholes to 1 inch across grain and some to 1-1/2 inch if total width of knots and knotholes is within specified limits. Synthetic or wood repairs. Discoloration and sanding defects that do not impair strength permitted. Limited splits allowed. Stitching permitted.
D	Knots and knotholes to 2-1/2 inch width across grain and 1/2 inch larger within specified limits. Limited splits are permitted. Stitching permitted. Limited to Interior (Exposure 1 or 2) panels.

Figure 10-30. Veneer grade summary. (Courtesy American Plywood Association)

C Grade. Often used as the face on floor, ceiling, and sheathing sheets and as the core and back face of higher grade sheets. No veneer lower in grade than C is allowed on *exterior* plywood. Allowable defects include tight knots to 1 inch, knotholes to 1 inch (1½ inches under certain conditions), and splits, which may be up to ½ inch and taper to a point. A variation on the C grade is the *C-plugged,* on which the defects have been repaired, patched, or plugged.

D Grade. Used as the core and back face of higher grade sheets. It is the lowest grade and may only be used on *interior* plywood. Allowable defects include knotholes to 2½ inches (3 inches under certain conditions), pitch pockets to 2½ inches wide, and splits to 1 inch wide and tapered to a point.

Grading and Finishes

A plywood panel is graded by the two letters that indicate the grade of veneer used for the face and back face of the panel. For example, plywood graded A-C has an A face and a C back. Typical grade-trademarks are shown in Figure 10-29. For most construction framing purposes where the panel will not be exposed to view or will be painted, it is simply sanded smooth by running the finished panel through large drum sanders.

Decorative finishes on the face are often desired when the plywood panel will be used as an exterior or interior finish material. Finishes available range from grooves in various widths to a large variety of surface textures such as rough-sawn and brushed. When the plywood will be painted, it is available with a special resin-treated surface that is hot-bonded to the panel (APA calls it MDO, for Medium Density Overlaid Plywood).

Plywood may also be used as a base for other materials such as plastics, metals, and natural aggregates. The illustrations show a variety of installations of APA graded ply-

wood. In Figure 10–31, the exterior plywood is Texture 1–11 (T1–11). In Figure 10–32, an MDO plywood is used for lap siding. Figure 10–33 shows an example of APA 303 siding with a reverse board and batten finish.

Figure 10-31. Exterior plywood—Texture 1–11.

Figure 10-33. APA 303 siding. (Courtesy American Plywood Association)

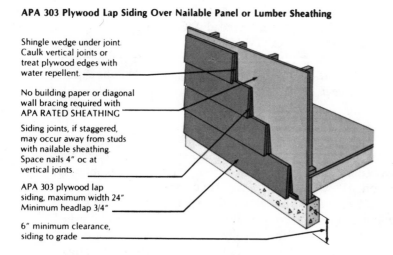

APA 303 Plywood Lap Siding Over Nailable Panel or Lumber Sheathing

Shingle wedge under joint. Caulk vertical joints or treat plywood edges with water repellent.

No building paper or diagonal wall bracing required with APA RATED SHEATHING

Siding joints, if staggered, may occur away from studs with nailable sheathing. Space nails 4" oc at vertical joints.

APA 303 plywood lap siding, maximum width 24" Minimum headlap 3/4"

6" minimum clearance, siding to grade

Figure 10-32. MDO plywood.

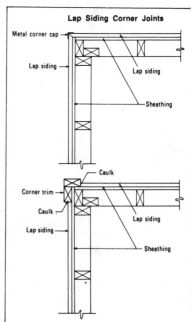

Lap Siding Corner Joints

Metal corner cap
Lap siding
Lap siding
Sheathing
Caulk
Corner trim
Caulk
Lap siding
Lap siding
Sheathing

Species

The various species of softwoods available for use in softwood plywood have different strength and stiffness characteristics. They are broken into five groups (Figure 10–34), based on stiffness, in accordance with Product Standard PS-1-66. Group 1 species has the greatest stiffness; Group 5 the least. The group species that is used is what determines the ultimate strength of the plywood panel. For example, ½ inch (13mm) thick plywood used as roof sheathing may have its supports (usually roof rafters) 32 inches (800mm) on center if the panel face and back are made of Group 1 species, but only 24 inches (600mm) on center for panels of Group 2 or 3 species. The species group number is often a part of the grade-trademark placed on the plywood panel (Figure 10–35). If it is not included it will have the suggested maximum support spacing (referred to as its identification index) such as the grade-trademark illustrated in Figure 10–35b. The relationship between plywood panel thickness, group species, and identification index is shown in Figure 10–36.

Classification of Species

Group 1	Group 2	Group 3	Group 4	Group 5
Apitong	Cedar, Port	Alder, Red	Aspen	Basswood
Beech,	Orford	Birch, Paper	Bigtooth	Poplar,
American	Cypress	Cedar, Alaska	Quaking	Balsam
Birch	Douglas	Fir,	Cativo	
Sweet	Fir 2(a)	Subalpine	Cedar	
Yellow	Fir	Hemlock,	Incense	
Douglas	Balsam	Eastern	Western	
Fir 1(a)	California	Maple,	Red	
Kapur	Red	Bigleaf	Cottonwood	
Keruing	Grand	Pine	Eastern	
Larch,	Noble	Jack	Black	
Western	Pacific	Lodgepole	(Western	
Maple, Sugar	Silver	Ponderosa	Poplar)	
Pine	White	Spruce	Pine	
Caribbean	Hemlock,	Redwood	Eastern	
Ocote	Western	Spruce	White	
Pine, South.	Lauan	Engelmann	Sugar	
Loblolly	Almon	White		
Longleaf	Bagtikan			
Shortleaf	Mayapis			
Slash	Red			
Tanoak	Tangile			
	White			
	Maple, Black			
	Mengkulang			
	Meranti,			
	Red(b)			
	Mersawa			
	Pine			
	Pond			
	Red			
	Virginia			
	Western			
	White			
	Spruce			
	Black			
	Red			
	Sitka			
	Sweetgum			
	Tamarack			
	Yellow-			
	Poplar			

(a) Douglas Fir from trees grown in the states of Washington, Oregon, California, Idaho, Montana, Wyoming, and the Canadian Provinces of Alberta and British Columbia shall be classed as Douglas Fir No. 1 Douglas Fir from trees grown in the states of Nevada, Utah, Colorado, Arizona and New Mexico shall be classed as Douglas Fir No. 2

(b) Red Meranti shall be limited to species having a specific gravity of 0.41 or more based on green volume and oven dry weight

Figure 10–34. Species group classification. (Courtesy American Plywood Association)

APA

UNDERLAYMENT

GROUP 1

INTERIOR

—————— 000 ——————

PS 1-74 EXTERIOR GLUE

(a)

APA

RATED SHEATHING
48/24 3/4 INCH

SIZED FOR SPACING

EXPOSURE 1

—————— 000 ——————

NRB-108

(b)

Figure 10–35. Grade-trademarks: (a) group species; (b) identification index. (Courtesy American Plywood Association)

APA RATED STURD-I-FLOOR[a]

Span Rating (Maximum Joist Spacing) (in.)	Panel Thickness[b] (in.)	Fastening: Glue-Nailed[c]			Fastening: Nailed-Only		
		Nail Size and Type	Spacing (in.)		Nail Size and Type	Spacing (in.)	
			Panel Edges	Intermediate		Panel Edges	Intermediate
16	19/32, 5/8, 21/32	6d deformed-shank[d]	12	12	6d deformed-shank	6	10
20	19/32, 5/8, 23/32, 3/4	6d deformed-shank[d]	12	12	6d deformed-shank	6	10
24	23/32, 3/4	6d deformed-shank[d]	12	12	6d deformed-shank	6	10
	7/8, 1	8d deformed-shank[d]	12	12	8d deformed-shank	6	10
48 (2-4-1)	1-1/8	8d deformed-shank[e]	6	(f)	8d deformed-shank[e]	6	(f)

(a) Special conditions may impose heavy traffic and concentrated loads that require construction in excess of the minimums shown. See page 22 for heavy duty floor recommendations.

(b) As indicated above, panels in a given thickness may be manufactured in more than one Span Rating. Panels with a Span Rating greater than the actual joist spacing may be substituted for panels of the same thickness with a Span Rating matching the actual joist spacing. For example, 19/32-inch-thick Sturd-I-Floor 20 oc may be substituted for 19/32-inch-thick Sturd-I-Floor 16 oc over joists 16 inches on center.

(c) Use only adhesives conforming to APA Specification AFG-01, applied in accordance with the manufacturer's recommendations. If non-veneered panels with sealed surfaces and edges are to be used, use only solvent-based glues; check with panel manufacturer.

(d) 8d common nails may be substituted if deformed-shank nails are not available.

(e) 10d common nails may be substituted with 1-1/8-inch panels if supports are well seasoned.

(f) Space nails 6 inches for 48-inch spans and 10 inches for 32-inch spans.

Figure 10–36. APA STURD-I-FLOOR. (Courtesy American Plywood Association)

Grades

Plywood is also available in several specialized grades, which are often used for wood frame construction such as roof sheathing, subflooring, and wall sheathing (Figure 10–37).

Standard Grade, Interior. This panel is Standard C-D with interior glue and should be used where it will not be exposed to high moisture for any length of time. It is also available with intermediate glue and with fully waterproofed glue. When fully waterproofed glue is used on this interior panel, the word *Exterior* is put below the grade-trademark (Figure 10–38). An interior panel with exterior glue is often used where the panel will be exposed to the weather for a period of time before it is covered with other materials or protected by the roof, or where long construction delays are expected. Woods from species groups, 1, 2, 3, or 4 may be used in the manufacture of this panel.

C-C Exterior Grade. This panel has only C or better veneers and a fully waterproof glue. It is used where the panel will be permanently exposed to moisture or to the weather, such as exterior siding, exposed overhangs, and boxed-in soffits (Figure 10–39). Wood from species groups 1, 2, 3, or 4 may be used in the manufacture of this grade panel. When an open soffit is used exterior grade plywood should be used as roof sheathing whenever it will be exposed to the outside weather. The typical plywood panel layout for a building is shown in Figure 10–40. For open-soffit construction those panels shaded in should be exterior grade.

Structural I and II Grades. These unsanded panels are designed to provide maximum strength and are primarily used as a part of a structural diaphragm (Figure 10–41) used for roof construction, box beams, gusset plates, and stressed skin panels. They are available both as interior with exterior glue and exterior panels. Structural I (Figure 10–42) is stronger than Structural II since it is manufactured from the stronger species groups.

Underlayment Grade. Applied over the structural subfloor (Figure 10–43), these panels provide a smooth surface for the application of resilient floor coverings. Underlayment panels have a C-plugged face veneer and are available as interior, interior with exterior glue, and exterior panels.

Subfloor/Underlayment Combination Panel. The APA refers to this as STURD-I-FLOOR, and this one thicker plywood panel (Figure 10–44) may be used in place of the two layers usually used in subfloor/ underlayment construction. These combination panels have a C-plugged face veneer and are available as interior with exterior glue and exterior panels. These combination panels are also available in a 1⅛ inch (28mm) thickness, which can be used on 32 and 48 inch (800 and 1200mm) spans. These combination panels are available in tongue and groove or square edges. When the combination panel with square edge is used, most codes require that all edges of the panel be supported. This means it will be necessary to support the edges between joists with 2 × 4 (38 × 89mm) blocking. This blocking is not required with the square edge combination panel if it is used under structural finish flooring, such as wood strip flooring, or if a separate underlayment panel or sheet is to be applied over it. The tongue and groove panels do not require any blocking.

Fire-retardant treated plywood is often used as a roof deck in wood frame construction. The fire-retardant treated plywood must be a minimum of ¾ inch (19mm) thick and is manufactured with exterior glue. In addition, each panel treated is identified by an Underwriters Laboratories® label.

Guide to APA Performance-Rated Panels[1][2]

APA RATED SHEATHING

TYPICAL
TRADEMARK

APA
RATED SHEATHING
32/16 1/2 INCH
SIZED FOR SPACING
EXPOSURE 1
000
NRB-108

Specially designed for subflooring and wall and roof sheathing. Also good for broad range of other construction and industrial applications. Can be manufactured as conventional veneered plywood, as a composite, or as a nonveneered panel. For special engineered applications, veneered panels conforming to PS 1 may be required. EXPOSURE DURABILITY CLASSIFICATIONS: Exterior, Exposure 1, Exposure 2. COMMON THICKNESSES: 5/16, 3/8, 7/16, 1/2, 5/8, 3/4.

APA STRUCTURAL I and II RATED SHEATHING[3]

TYPICAL
TRADEMARK

APA
RATED SHEATHING
STRUCTURAL I
42/20 5/8 INCH
SIZED FOR SPACING
EXTERIOR
000
PS 1-74 C-C   NRB-108

Unsanded all-veneer PS 1 plywood grades for use where strength properties are of maximum importance, such as box beams, gusset plates, stressed-skin panels, containers, pallet bins. Structural I.more commonly available. EXPOSURE DURABILITY CLASSIFICATIONS: Exterior, Exposure 1. COMMON THICKNESSES: 5/16, 3/8, 1/2, 5/8, 3/4.

APA RATED STURD-I-FLOOR

TYPICAL
TRADEMARK

APA
RATED STURD-I-FLOOR
20 OC 19/32 INCH
SIZED FOR SPACING
EXPOSURE 1
000
NRB-108

Specially designed as combination subfloor-underlayment. Provides smooth surface for application of resilient floor coverings and possesses high concentrated and impact load resistance. Can be manufactured as conventional veneered plywood, as a composite, or as a nonveneered panel. Available square edge or tongue-and-groove. EXPOSURE DURABILITY CLASSIFICATIONS: Exterior, Exposure 1, Exposure 2. COMMON THICKNESSES: 19/32, 5/8, 23/32, 3/4.

(1) Specific grades, thicknesses and exposure durability classifications may be in limited supply in some areas. Check with your supplier before specifying.

(2) Specify Performance-Rated Panels by thickness and Span Rating. Span Ratings are based on panel strength and stiffness. Since these properties are a function of panel composition and configuration as well as thickness, the same Span Rating may appear on panels of different thickness. Conversely, panels of the same thickness may be marked with different Span Ratings.

(3) All plies in Structural I panels are special improved grades and limited to Group 1 species. All plies in Structural II panels are special improved grades and limited to Group 1, 2, or 3 species.

APA RATED STURD-I-FLOOR 48 oc (2-4-1)

TYPICAL
TRADEMARK

APA
RATED STURD-I-FLOOR 1-1/8 INCH
48 OC (2-4-1)
SIZED FOR SPACING
EXPOSURE 1
T&G   000   INT/EXT GLUE
NRB-108 FHA-UM-66

For combination subfloor-underlayment on 32-and 48-inch spans and for heavy timber roof construction. Manufactured only as conventional veneered plywood. Available square edge or tongue-and-groove. EXPOSURE DURABILITY CLASSIFICATIONS: Exposure 1. THICKNESS: 1-1/8.

Figure 10-37. Guide to APA performance-rated panels. (Courtesy American Plywood Association)

APA A-A

TYPICAL TRADEMARK

```
A-A · G-1 · INT-APA · PS1-74 · 000
```

Use where appearance of both sides is important for interior applications such as built-ins, cabinets, furniture, partitions; and exterior applications such as fences, signs, boats, shipping containers, tanks, ducts, etc. Smooth surfaces suitable for painting. TYPES: Interior, Exterior. COMMON THICKNESSES: 1/4, 3/8, 1/2, 5/8, 3/4.

APA A-B

TYPICAL TRADEMARK

```
A-B · G-1 · INT-APA · PS1-74 · 000
```

For use where appearance of one side is less important but where two solid surfaces are necessary. TYPES: Interior, Exterior. COMMON THICKNESSES: 1/4, 3/8, 1/2, 5/8, 3/4.

APA B-D

TYPICAL TRADEMARK

```
APA
B-D    GROUP 2
INTERIOR
000
PS 1-74 EXTERIOR GLUE
```

Utility panel for backing, sides of built-ins, industry shelving, slip sheets, separator boards, bins and other interior or protected applications. TYPE: Interior. COMMON THICKNESSES: 1/4, 3/8, 1/2, 5/8, 3/4.

APA UNDERLAYMENT

TYPICAL TRADEMARK

```
APA
UNDERLAYMENT
GROUP 1
INTERIOR
000
PS 1-74 EXTERIOR GLUE
```

For application over structural subfloor. Provides smooth surface for application of resilient floor coverings and possesses high concentrated and impact load resistance. TYPE: Interior. COMMON THICKNESSES: 3/8, 1/2, 19/32, 5/8, 23/32, 3/4.

(1) Specific grades and thicknesses may be in limited supply in some areas. Check with your supplier before specifying.

(2) Can also be manufactured in Structural I (all plies limited to Group 1 species) and Structural II (all plies limited to Group 1, 2 or 3 species).

Figure 10-37. (Continued)

169

APA RATED STURD-I-FLOOR

TYPICAL TRADEMARK

```
════ APA ════
RATED STURD-I-FLOOR
20 OC  19/32 INCH
SIZED FOR SPACING
EXTERIOR
──── 000 ────
NRB-108
```

Specially designed as combination subfloor-underlayment. Provides smooth surface for application of resilient floor coverings and possesses high concentrated and impact load resistance. Can be manufactured as conventional veneered plywood, as a composite, or as a nonveneered panel. Available square edge or tongue-and-groove. EXPOSURE DURABILITY CLASSIFICATIONS: Exterior, Exposure 1, Exposure 2. COMMON THICKNESSES: 19/32, 5/8, 23/32, 3/4.

Figure 10–38. Grade-trademark. (Courtesy American Plywood Association)

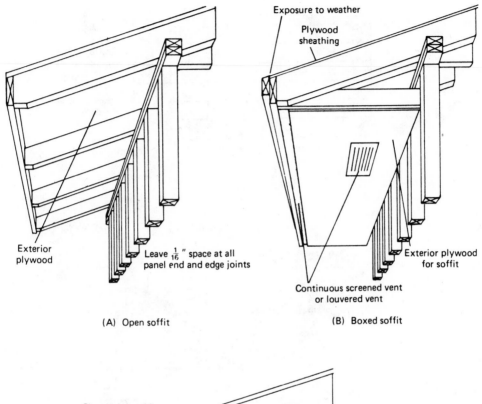

Exposure to weather

Plywood sheathing

Exterior plywood

Leave $\frac{1}{16}$" space at all panel end and edge joints

Exterior plywood for soffit

Continuous screened vent or louvered vent

(A) Open soffit

(B) Boxed soffit

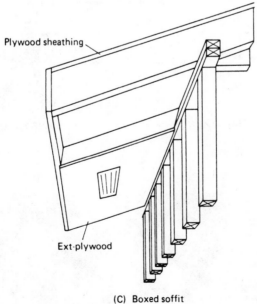

Plywood sheathing

Ext-plywood

(C) Boxed soffit

Figure 10-39. Uses of exterior plywood.

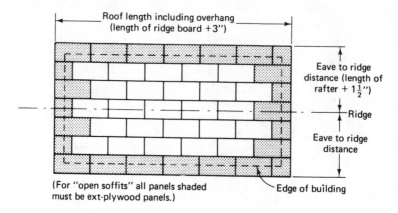

Figure 10-40. Roof sheathing plywood layout.

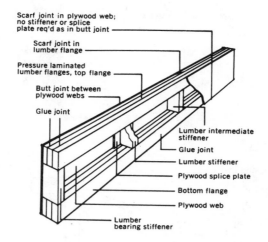

Figure 10-41. Plywood box beam. (Courtesy American Plywood Association)

Panel Grade	Thickness (in.)	No. of Plies[a]	Span Rating	Maximum Span (in.)
APA STRUCTURAL I RATED SHEATHING	3/8	3	24/0	12
	1/2	4	32/16	24
	1/2	5	32/16	24
	5/8	5	42/20	24
	3/4	5&6	48/24	24
	1-1/8	7	—	48

Figure 10-42. Structural I Identification Index. (Courtesy American Plywood Association)

APA Plywood Underlayment(e)

Plywood Grades(a) and Species Group	Application	Minimum Plywood Thickness (in.)	Fastener Size and Type	Fastener Spacing (in.)(b) Panel Edges	Intermediate
Groups 1, 2, 3, 4, 5 APA UNDERLAYMENT INT (with interior or exterior glue) APA UNDERLAYMENT EXT APA C-C Plugged EXT	Over smooth subfloor	1/4	18 ga. staples or 3d ring-shank nails(c)(d)	3	6 each way
	Over lumber subfloor or other uneven surfaces.	11/32	16 ga. staples(c)	3	6 each way
			3d ring-shank nails (d)	6	8 each way
Same grades as above, but Group 1 only.	Over lumber floor up to 4" wide. Face grain must be perpendicular to boards.	1/4	18 ga. staples or 3d ring-shank nails(c)(d)	3	6 each way

(a) When thicker underlayment is desired, APA RATED STURD-I-FLOOR may be specified.
(b) If green framing is used, space fasteners so they do not penetrate framing.
(c) Use 16 ga. staples for 11/32 inch and thicker plywood. Crown width 3/8 inch for 16 ga., 3/16 inch for 18 ga. staples, length sufficient to penetrate completely through, or at least 5/8 inch into. subflooring.
(d) Use 3d ring-shank nail also for 1/2 inch panels and 4d ring-shank nail for 5/8 inch or 3/4 inch panels.
(e) For underlayment recommendations under ceramic tile, refer to ANSI Standard A108 or contact APA.

Figure 10-43. Plywood underlayment. (Courtesy American Plywood Association)

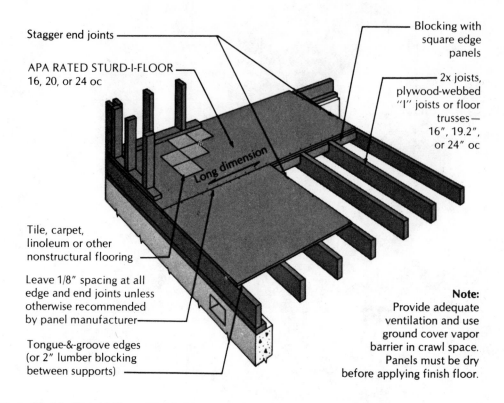

Stagger end joints

APA RATED STURD-I-FLOOR
16, 20, or 24 oc

Long dimension

Blocking with
square edge
panels

2x joists,
plywood-webbed
"I" joists or floor
trusses—
16", 19.2",
or 24" oc

Tile, carpet,
linoleum or other
nonstructural flooring

Leave 1/8" spacing at all
edge and end joints unless
otherwise recommended
by panel manufacturer

Tongue-&-groove edges
(or 2" lumber blocking
between supports)

Note:
Provide adequate
ventilation and use
ground cover vapor
barrier in crawl space.
Panels must be dry
before applying finish floor.

Figure 10-44. Sturd-I-Floor 16, 20, 24 o.c. (Courtesy American Plywood Association)

Chapter 11

Gypsum

Gypsum is known chemically as *hydrous calcium sulfate* and is a fairly common rocklike mineral found in rock formations in several areas of the world, with significant deposits in the United States, Canada, England, France, China, areas of South America, and Russia. Pure gypsum is white, but it is usually found combined with impurities such as clay, limestone, and iron compounds, which change the color to gray or brown. Gypsum deposits may be located near the surface, easily obtained by strip mining,

or they may be buried deeply in the ground and have to be mined out.

11.1
GYPSUM WALLBOARD

This type of wallboard is composed of a gypsum core encased in a heavy manila-finished paper on the face side and a strong liner paper on the back side (Figure 11–1). It is available

ceiling attachments

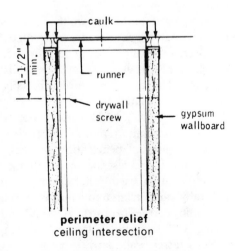

perimeter relief
ceiling intersection

floor attachments & bases

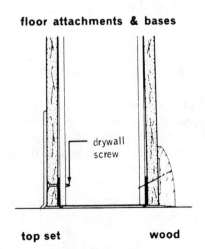

top set   wood

Figure 11-1. Gypsum wallboard.

175

in a 4-foot width, with lengths of 6, 8, 10, or 12 feet, and thicknesses of $\frac{1}{4}$, $\frac{3}{8}$, $\frac{1}{2}$, or $\frac{5}{8}$ inches. Plain gypsum wallboard will require a finish of some type such as painting or wallpaper.

Gypsum wallboard with a rugged vinyl film, factory laminated to the panel, is also available. The vinyl finished panel is generally used in conjunction with adhesive fastenings and matching vinyl covered trim. It is also available in fire-resistant gypsum wallboard where fire-rated construction is desired or required.

Fire-resistant gypsum wallboard is available generally in $\frac{1}{2}$-inch and $\frac{5}{8}$-inch thicknesses. These panels have cores containing special mineral materials, and they can be used in assemblies that provide up to 2-hour ratings in walls and 3-hour ratings in ceilings and columns.

Other commonly used gypsum wallboards include insulating panels (aluminum foil on the back), water-resistant panels (for use in damp areas; they have special green paper and core materials), and backing board, which may be used as a base for multi-ply construction and acoustical tile application and which may be specially formulated for fire-resistant base or acoustical tile application.

11.2
DRYWALL CONSTRUCTION

Drywall construction consists of putting wallboard over supporting construction. The materials used for this type of construction will depend on the requirements of the project with regard to appearance, sound control, fire ratings, strength requirements, and costs.

The supporting construction for partitions generally is made of wood or steel studs, but gypsum wallboards are often applied over concrete and masonry. The supporting construction for ceilings generally is built from wood joists and trusses, steel joists and suspended ceilings, or concrete and masonry, often in conjunction with the various types of furring materials available.

Gypsum drywall partitions are usually broken down into two basic types of construction: single-ply and multi-ply (Figure 11–2). Single-ply construction consists of a single layer of gypsum board on each side of the supporting construction, while multi-ply construction consists of two or more layers of gypsum boards, often using different types of boards for the inner layers. The multi-ply construction may be semisolid or solid or may have various combinations of materials. Various types of demountable and reusable assemblies using various types of gypsum wallboards are also available.

Accessories

Accessories for the application and installation of drywall construction include adhesives, tape and compound for joints, fastener treatment, and trim to protect exposed edges and exterior corners, as well as baseplates and edge moldings.

Nails used to fasten wallboard may be bright, coated, or chemically treated; the shanks may be smooth or annularly threaded with a nailhead that is generally flat or slightly concave. The annularly threaded nails are generally used since they provide more withdrawal resistance, require less penetration, and minimize nailpopping. For a fire rating, 1 inch or more of penetration is usually required, and in this case the smooth shank nails are most often used. The spacing of nails generally varies from 6 to 8 inches on center, depending on the size and type of nail and the type of wallboard being used. Screws may be used to fasten wallboard to both wood and metal supporting construction and furring strips. In commercial work, drywall screws have virtually eliminated the use of nails. Typically, these screws have self-drilling, self-tapping threads with flat recessed Phillips heads for use with a power screwdriver. The drywall screws are usually spaced about 12 inches on center, except that when a fire rating is required the spacing is usually 8 inches on center at vertical joists. There are three types of drywall screws: one for fastening wood, one for sheet metal, and one for gypsum board.

Adhesives. Adhesive may be used to attach single-ply wallboard directly to the framing, concrete, or masonry or to laminate the wallboard to a base layer. The base layer may be gypsum board, sound-deadening board, or rigid-foam insulation. Often the adhesives are used in conjunction with screws and nails that provide either temporary or permanent support. The three classes of adhesives used are stud adhesives, laminating adhesives, and contact adhesives. There are various modifications within each class. Information regarding exact adhesives required should be obtained from the manufacturer.

Trim. A wide variety of trims are available in wood and for use on drywall construction. The trim is generally used to provide maximum protection and neat, finished edges throughout the building. The wood trim is available unfinished and prefinished in an ex-

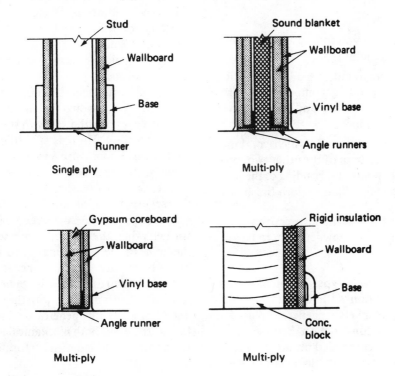

Figure 11-2. Typical drywall assemblies.

tensive selection of sizes, shapes, and costs. The metal trim is available in an almost equal number of sizes and shapes, with finishes ranging from plain steel, galvanized steel, and prefinished painted to permanently bonded finishes that match the wallboard. Even aluminum molding, plain and anodized, is available.

Joint Tape and Compounds.

Joint tape and compounds are employed when a gypsum wallboard is used. It is necessary to reinforce and conceal the joints between wallboard panels and to cover the fastener heads. These tapes and compounds provide a smooth, continuous surface in interior walls and ceilings.

The tape used for joint reinforcement is usually a fiber tape designed with chamfered edges feathered thin and with a cross-fiber design. Joint compounds are classified as follows:

1. Embedding compound, used to embed and bond the joint tape.
2. Topping compound, used for finishing over the embedding compound (it provides final smoothing and leveling over fasteners and joints).
3. All-purpose compound, which combines the features of both of the other two, providing embedding and bonding of the joint tape and giving a final smooth finish.

The compounds are available premixed by the manufacturer or in a powder form to be job-mixed.

Blankets.

Various types of blankets are used in conjunction with the drywall construction. The blankets are generally placed in the center of the construction, between studs or on top of the suspended ceiling assembly. The two basic types of blankets are *heat-insulating* and *sound control*. The heat-insulating blankets are used to help control heat loss (winter) and heat gain (summer), while the sound-control blankets are used to improve Sound Transmission Classification (STC) ratings of the assembly being used (Chapter 17). Both types are available in a variety of thicknesses and widths. Blankets are available with aluminum foil on one or both sides, paper on one or both sides, and various methods of attachment. The most common method of attachment is with staples.

11.3
WETWALL CONSTRUCTION

Wetwall construction consists of supporting construction, lath, and plaster. The exact types and methods of assembly used for this construction will depend on the requirements of the particular job regarding appearance, sound control, fire ratings, strength requirements, and cost.

The supporting construction may be wood, steel, concrete, gypsum tile, masonry, or lath. Certain types of lath used with the plaster are self-supporting. The plaster itself may be two-coat or three-coat, with a variety of materials available for each coat.

Proper use of plasters and bases provides the secure bond necessary to develop the strength required. A mechanical bond is formed when the plaster is pressed through the holes of the lath or mesh and forms keys on the back side. A *suction,* or chemical bond, is formed when the plaster is applied over masonry and gypsum bases, with the tiny needle-like plaster crystals penetrating into the surface of the base. Both mechanical and suction bonds are developed with perforated gypsum lath.

Plaster

Plaster is a material that in its plastic state can be troweled to form and, when set, provides a hard covering for interior surfaces such as walls and ceilings. Plaster is the final step in wetwall construction (although other finishes may be applied over it). Together with the supporting construction and some type of lath, the plaster will complete the assembly. The type and thickness of plaster used will depend on the type of supporting construction, the lath, and the intended use. Plaster is available for one-coat, two-coat, and three-coat work, and plaster is generally classified according to the number of coats required. The last and final coat applied is called the *finish coat,* while the coat or combination of coats applied before the finish coat is referred to as the *basecoat.*

Finish Coats. Finish coats serve as leveling coats and provide either a base for decoration or the required resistance to abrasion. Several types of gypsum finish plasters are available, including those that require the addition of only water and those that blend gypsum, lime, and water or gypsum, sand, and water. The finish coat used must be compatible with the basecoat. Finishing materials may be classified as prepared finishes, smooth trowel finishes, and sand float finishes. Finish coat thicknesses range from $1/16$ to $1/8$ inch.

Specialty finish coats are also available. One such specialty coat is *radiant heat plaster,* for use with electric cable ceilings. It is a high-density plaster that allows a higher operating temperature for the heating system, as it provides more efficient heat transmission and greater resistance to heat deterioration. It is applied in two coats—the first to embed the cable, the second as a finish coat over the top. The total thickness is about $1/4$ to $3/8$ inch. It is usually mill-prepared and requires only the addition of water.

One-coat plaster is a thin-coat interior product used over large sheets of gypsum plaster lath in conjunction with a glass fiber tape to finish the joints. The plaster coat is $1/16$ to $3/32$ inch thick.

Keene cement plaster is used where a greater resistance to moisture and surface abrasion is required. It is available in smooth and sand-float finishes. It is a dead-burned gypsum mixed with lime putty and is difficult to apply unless sand is added to the mixture. However, with sand as an additive it is less resistant to abrasion.

Acoustical Plasters. Acoustical plasters, which absorb sound, are also available. Depending on the type used, they may be troweled or machine-sprayed onto the wall. Trowel applications are usually stippled, or floated to a finish. Some of these plasters may be tinted various colors. Thicknesses range from $3/8$ to $1/2$ inch.

Lath

Lath is used as a base, and the plaster is bonded to the lath. Types of lath include gypsum tile, gypsum plaster board, metal, and wood.

Gypsum Tile. This is a precast, kiln-dried tile (Figure 11–3) used for non-load-bearing construction and fireproofing columns. Thicknesses available are 2 inches (solid), 3 inches (solid or hollow) and 4 and 6 inches (hollow). The 2-inch tile is used only for fireproofing, not for partitions. A face size of 12 x 30 inches (2.5 square feet) is available. Used as a plaster base, it provides excellent fire and sound resistance. Gypsum tile and plaster used as column fireproofing are shown in Figure 11-3.

Gypsum Plaster Lath.   Available in sheet form, this lath provides a rigid base for the application of gypsum plasters (Figure 11-4). Special gypsum cores are faced with multi-layered, laminated paper. The different types of gypsum lath available are plain gypsum, perforated gypsum, fire-resistant, insulating, and radiant heat lath. Depending on the supporting construction, the lath may be nailed, stapled, or glued. The spacing of the attachments depends on the type of construction and the thickness of the lath. Gypsum lath may be attached to the supporting construction by use of nails, screws, staples, or clips.

Plain gypsum lath is available in thicknesses of ³⁄₈ and ¹⁄₂ inch with a face size of 16 × 48 inches. The ³⁄₈ inch thickness is also available in a 16 × 96 inch sheet. (A 16³⁄₁₆ inch width is available in certain areas only.) Gypsum lath 24 × 144 inches is also available in certain areas. When the plaster is applied to this base, a chemical bond holds the base to the gypsum lath.

Perforated gypsum lath is available ³⁄₈ inch thick with a face size of 16 × 48 inches. Holes ³⁄₄ inch in diameter are punched in the lath, spaced 16 inches on center. The perforated lath permits higher fire ratings since the plaster is held on by mechanical as well as chemical bonding.

Fire-resistant gypsum lath has a specially formulated core of special mineral materials. It has no holes, but provides additional resistance to fire exposure. It is available ³⁄₈ inch thick with a face size of 16 × 48 inches.

Insulating gypsum lath is plain gypsum lath with aluminum foil laminated to the back face. It serves as a plaster base, an insulator against heat and cold, and a vapor barrier. It is available in ³⁄₈ and ¹⁄₂ inch thicknesses, with a face size of 16 × 48 inches.

Radiant heat lath is a large gypsum lath for use with plaster in electric cable ceilings. It improves the heat emission of the electric cables and increases their resistance to heat deterioration. It is available 48 inches wide and in ¹⁄₂-inch and ⁵⁄₈-inch thicknesses and lengths of 8 to 12 feet. This type of lath is used with plaster that is formulated for use with electric cable heating systems.

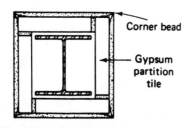

(b) 2″ solid gypsum partition tile, 4 hr.

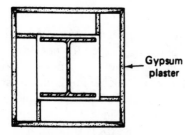

(d) 3″ hollow gypsum partition tile, 4 hr.

Figure 11-3. Gypsum tile. (Courtesy United States Gypsum Co.)

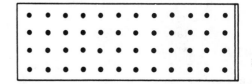

Figure 11-4. Gypsum lath.

Metal Lath. Metal lath is sheet steel that has been slit and expanded to form a multitude of small mesh openings. Ordinary expanded metal lath (such as diamond mesh or flat-rib lath) is used in conjunction with other supporting construction. There are also metal laths such as ⅜-inch rib lath that are self-supporting, requiring no supporting construction (Figure 11–5).

Metal lath is available painted, galvanized, or asphaltum-dipped; sheet sizes are generally 24 x 96 inches (packed 16 square yards per bundle) or 27 x 96 inches (20 square yards per bundle). Basically, the three types of metal lath available for wetwall construction are diamond, flat-rib, and ⅜-inch rib lath. Variations in the designs are available through different manufacturers.

The metal lath should be lapped no less than ½ inch at the sides and 1 inch at the ends. The sheets should be secured to the supports at a maximum of 6 inches on center. The metal lath is secured to the steel studs or channels by use of 18-gauge tie wires about 6 inches on center. For attachment to wood supporting construction, nails with a large head (about ½ inch) should be used.

Diamond lath (Figure 11–5) is an all-purpose lath that is ideal as a plaster base, as a reinforcement for walls and ceilings, and as fireproofing of steel columns and beams. It is easily cut, bent, and formed to curved surfaces. It is available in copper alloy steel, either painted or asphaltum-coated. Galvanized diamond lath is available only in the 3.4 pounds per square yard weight.

Flat-rib lath (Figure 11–5) is a ⅛-inch lath with "flat ribs," which make a stiff type of lath. This increased stiffness generally permits wider spacing between supports than does diamond lath, and the design of the mesh allows for the saving of plaster. The main longitudinal ribs are spaced 1½ inches apart, with the mesh set at an angle to the plane of the sheet. It is available in copper alloy and steel in weights of 2.75 and 3.4 pounds per square yard and in galvanized steel in a weight of 3.4 pounds per square yard. It is used with wood or steel supporting construction on walls and ceilings, and for fireproofing.

Rib lath (Figure 11–5) combines a small mesh with heavy reinforcing ribs. The ribs are ⅜ inch deep, 4½ inches on center. Used as a plaster base it may be employed in studless wall construction and in suspended and attached ceilings. The rib lath permits wider spacing of supports than flat-rib and diamond lath. This type is also used as a combination form and reinforcement for concrete floor and roof slabs. Copper alloy steel lath is available in 3.4 and 4.0 pounds per square yard weights, while the galvanized is available only in the 3.4 pounds per square yard weight.

Wood Lath. Although largely displaced by the other types of laths available, wood lath is used on occasion as a plaster base. The most commonly used size is ⅜ inch thick, 1⅜ inches wide (actual size), and 48 inches long, spaced ⅜ inch apart. The pieces are attached by nails or staples.

Wetwall Accessories

The accessories available for use with wetwall construction include various types of corner beads, control and expansion joints, screeds, partition terminals, casing beads, and a variety of metal trims to provide neat-edged case openings. Metal ceiling and floor runners are also available, as are metal bases. Resilient channels may also be used.

A complete selection of steel clips, nails, staples, and self-drilling screws is available to provide positive attachment of the lath. Special attachment devices are available for each particular wetwall assembly.

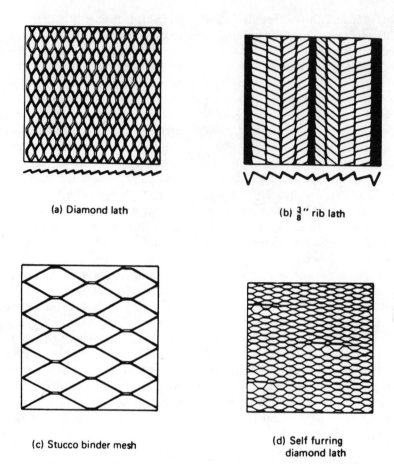

(a) Diamond lath

(b) $\frac{3}{8}''$ rib lath

(c) Stucco binder mesh

(d) Self furring
diamond lath

(e) Flat rib lath

Figure 11-5. Metal lath.

11.4
SUPPORTING CONSTRUCTION

Wallboard and plaster can be applied directly to wood, metal, concrete, or masonry that is capable of supporting the design loads and that provides a firm, level, plumb, and true base.

Concrete and Masonry

These materials often have wallboard or plaster applied to them. When used on exteriors and below grade, furring should be applied over concrete or masonry to protect the wallboard from damage due to moisture in the wall; this is not required for interior walls. Furring may also be required to plumb and align the walls.

Wood and Metal

Wood and metal supporting construction often consists of self-supporting framing members including wall studs, ceiling joists, and roof trusses. Wood and metal furring members such as wood strips and metal channels are used over the supporting construction to plumb and align the framing, concrete, or masonry.

Wood Studs. The most common size used is 2 x 4, but larger sizes may be required on a particular project. The spacing may vary from 12 to 24 inches on center, again depending on job requirements. All openings must be framed around, and backup members should be provided at all corners. The most common method of attaching the wallboard to the wood studs is by nailing, but screws and adhesives are also used.

Wood Joists. When you intend to apply the wallboard directly to joists, the bottom faces of the joists should be aligned in a level plane. Joists with a slight crown should be installed with the crown up, and if slightly crooked or bowed joists are used it may be necessary to straighten and level the surface with the use of nailing stringers or furring strips. The wallboard may be applied by nailing or screwing. The deflection must be limited to the length of the span divided by 360 if plaster is to be applied.

Wood Trusses. When used for the direct application of wallboard, trusses often require cross-furring to provide a level surface for attachment. Stringers attached at third points will also help align the bottom chord of wood trusses, and a built-in camber, on the trusses, is suggested to compensate for future deflection. Since the trusses are made up of relatively small members spanning a large distance, they tend to be more difficult to align and level for the application of wallboard.

Metal Studs. The metal studs (Figure 11-6) generally used are made of 25-gauge, cold-formed steel, electrogalvanized to resist corrosion. Most metal studs have notches at each end and knockouts located about 24 inches on center to facilitate pipe and conduit installation. The size of the knockout, not the size of the stud, determines the maximum size of pipe or other material that can be passed through horizontally.

Often when large pipes, ducts, or other items must pass vertically or horizontally in the walls, double stud walls are used, spaced the required distance apart. Studs are generally available in thicknesses of 1⅝, 2½, 3½, 4, and 6 inches. The metal runners used are also 25-gauge steel and are sized to complement the studs.

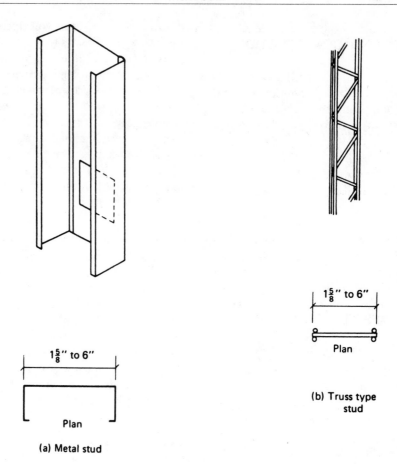

Plan

(a) Metal stud

$1\frac{5}{8}''$ to 6"

Plan

(b) Truss type stud

$1\frac{5}{8}''$ to 6"

Figure 11-6. Metal studs.

A variety of systems have been developed by the manufacturers to meet various requirements of attachment, sound control, and fire resistance. Many of the systems have been designed for ease in erection, yet they are still demountable for revising room arrangements. The wallboard may be attached by use of nails, screws, and in certain applications an adhesive.

Metal Furring.   Metal furring is used with all types of supporting construction. It is particularly advantageous where sound control

or noncombustible assemblies are required. Various types of channels are available (Figure 11-7). Cold-rolled channels, used for drywall or wetwall construction, are made of 16-gauge steel, ¾ to 2 inches wide and available in lengths of up to 20 feet. These channels must be wire-tied to the supporting construction, and they are used primarily as a supporting grid for the lighter drywall channels to which the wallboard may be screw-attached or lath and plaster-applied.

Drywall channels are 25-gauge, electrogalvanized steel and are designed for the

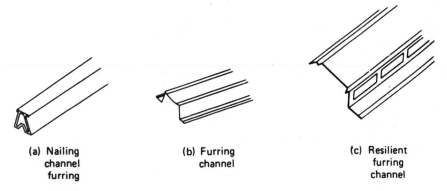

(a) Nailing
channel
furring

(b) Furring
channel

(c) Resilient
furring
channel

Figure 11-7. Metal furring.

screw attachment of wallboard; nailable channels are also available. The channels may be used in conjunction with the cold-rolled channels or installed over the wood, steel, masonry, or concrete supporting the construction. These drywall channels may be plain or resilient. The resilient channels are often used over wood and metal framing to improve sound isolation and to help isolate the wallboard from structural movement.

Wood Furring. Strips are often used with wood frame, masonry, and concrete to provide a suitable plumb, true, or properly spaced supporting construction. These furring strips may be 1 x 2 inches, spaced 16 inches on center, or 2 x 2 inches spaced 24 inches on center. Occasionally larger strips are used to meet special requirements. They may be attached to masonry and concrete with cut nails, threaded concrete nails, and powder or air actuated fasteners.

When the spacing of the framing is too great for the intended wallboard thickness, furring applied perpendicular to the framing members is referred to as cross-furring. If the wallboard is to be nailed to the cross-furring, the wood furring should be a minimum of 2 x 2 inches in order to provide sufficient stiff-

ness to eliminate excessive hammer rebound. Thinner furring (1 x 2 or 1 x 3 inches) is often used for screw and adhesive attached wallboard. The furring is attached by nailing, with the spacing of the nails 16 or 24 inches on center.

Suspended Ceiling Systems

When the plaster, wallboard, or tiles cannot be placed directly on the supporting construction, the wallboard is suspended below the structural system. This may be required if the supporting construction is not properly aligned and true or if lower ceiling heights are required.

There are a large variety of systems available for use in drywall construction, but basically they can be broken down into two classes: exposed grid system and concealed grid system (Figure 11-8). Within each group many different shapes of pieces are used to secure the plaster wallboard or tile, but basically the systems consist of hangers, main tees (runners), cross tees (hangers), or furring channels (Figure 11-9). No matter which type is used, there are accessories such as wall molding, splines, and angles that must be con-

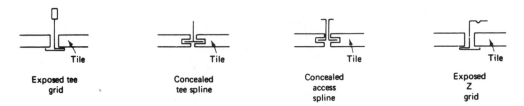

| Exposed tee grid | Concealed tee spline | Concealed access spline | Exposed Z grid |

Figure 11-8. Typical grid systems.

Figure 11-9. Typical suspended ceiling. (Courtesy United States Gypsum Co.)

sidered. For wetwall construction, a lath of some type is attached to the grid system to receive the plaster.

The suspension system and wallboard may also be used to provide recessed lighting, acoustical control (by varying the type of wallboard and panel), fire ratings, and air distribution (special tile and suspension system).

The suspension system is available in steel with an electrozinc coating, as well as pre-painted, and in aluminum with plain, anodized, or baked enamel finishes. Special shapes, such as metal shaped like a wood beam and exposed in the room, are also available.

The suspension may be hung from the supporting construction with 9- or 10-gauge hanger wire spaced about 48 inches on center, or it may be attached by use of furring strips and clips.

Chapter 12

Roofing

Roofing includes all the materials that actually cover the roof deck. Residential roofing materials commonly include asphalt, asbestos, and wood shingles and built-up roofing. Roofing also includes base sheets of felt, vapor barriers, flashing, blocking, nailers, cant strips, curbs, expansion joints, insulation, and any other items necessary for a complete roofing system.

12.1
SHINGLES

Shingles may be asphalt, asbestos, wood, or fiberglass-reinforced. They are rated for fire and wind resistance by the Underwriters Laboratories, Inc. which places them in Class A, B, or C or gives them no rating if they fail to qualify.

The fire rating class is determined by testing a sample application of the shingles in accordance with ASTM specifications, which test their resistance to intermittent flame exposure, spread of flame, and the burning of special wood blocks (called brands) on the shingles. A Class A rating is the highest, indicating about twice the resistance of Class B and about 10 times the resistance of Class C.

Shingles are also tested for wind resistance by the Underwriters Laboratories. They are marked "wind-resistant" if they pass the test.

Many asphalt and asbestos shingles have sealing tabs that hold them firm in high winds and rain.

For added weather protection the first course shingles, at the eaves, should always be doubled (Figure 12–1). Galvanized, large-headed nails ⅞ to 1¾ inches long are used to secure the shingles to the deck. Shingles 36 inches long require 4 nails per shingle.

The incline of the roof often plays a large part in determining what roofing materials and methods of application are required. The incline of a roof is often referred to as the *pitch* or *slope*. The pitch is equal to the vertical rise of the roof over the span of the roof. The slope is the vertical rise of the roof divided by the run (run is one-half the span). The slope is given in inches per foot. Both slope and pitch are illustrated in Figure 12–2.

Asphalt and asbestos shingles are available in a variety of colors. The colors are determined by the color of the stone granules embedded in the shingle, and they vary among the manufacturers. The colors most commonly available are shades of brown, gray, green, and black. These shingles are available in a variety of patterns and layouts that will greatly affect the appearance of the project (Figure 12–3).

Asphalt and asbestos shingles may be made of predominantly one material or may be a combination of materials built up in layers. A

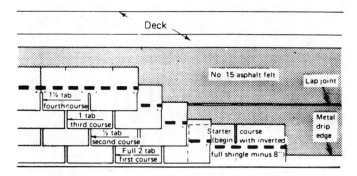

Figure 12-1. Shingle starter course.

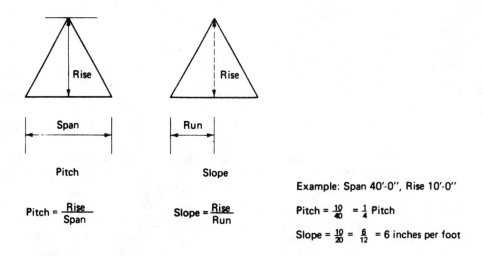

Pitch = $\dfrac{\text{Rise}}{\text{Span}}$

Slope = $\dfrac{\text{Rise}}{\text{Run}}$

Example: Span 40'-0", Rise 10'-0"

Pitch = $\dfrac{10}{40}$ = $\dfrac{1}{4}$ Pitch

Slope = $\dfrac{10}{20}$ = $\dfrac{6}{12}$ = 6 inches per foot

Figure 12-2. Pitch and slope.

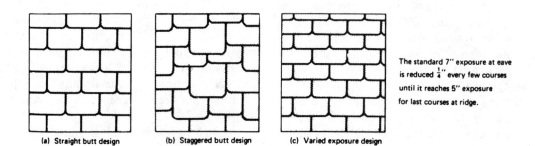

(a) Straight butt design   (b) Staggered butt design   (c) Varied exposure design

The standard 7" exposure at eave is reduced $\frac{1}{4}$" every few courses until it reaches 5" exposure for last courses at ridge.

Figure 12-3. Shingle patterns and layouts (for individual shingles).

typical example of this type of composite construction is shown in Figure 12-4, where six layers of materials are combined to form the shingle. The stone granules placed on the outside surface of the shingles provide protection from the sun's rays drying out the bitumins.

Asphalt Shingles

Strip asphalt shingles are available in a variety of colors, styles, and exposures, and are 12 to 15 inches wide and 36 inches long. They are packed in bundles that contain enough shingles to cover 25, 33, or 50 square feet of roof area. The exposure (amount of shingle exposed to the weather) generally is 4, 4½, or 5 inches. Individual shingles 12 to 16 inches

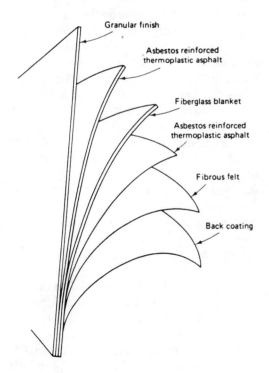

Figure 12-4. Composition shingle.

long are sometimes used. Asphalt shingles may be specified by weight per square (one square equals 100 square feet), which may vary from 180 to 350 pounds. Shingles may be fire rated and wind resistance rated, depending on the type specified.

Asbestos Shingles

These are also referred to as mineral fiber roof shingles. They are available in hexagonal and rectangular shapes as well as in strip shingles of 14 × 30 inches. Bundle sizes range from 20 to 33⅓ square feet, with the weight per square between 250 and 500 pounds. Roof slopes as low as 3 inches per foot (3/12) may be used, provided the proper underlayment is employed. They are also manufactured in individual tiles, about 9⅓ × 16 inches and ¼ inch thick, to resemble slate.

Wood Shingles and Shakes

Wood shingles are sawed and have a relatively smooth surface, while wood shakes are split and have at least one highly textured natural grain surface. They are available in various woods; the most commonly used are cypress, cedar, and redwood. All are resistant to deterioration from weather and insects. These woods may be left in their natural state—in which case the cedar and redwood will eventually weather to a silver or dark gray color—or they may be painted or stained.

The proper weather exposure (the length of shingle exposed to the weather) is very important in wood shingle applications; recommended exposures for shingles and shakes are shown in Figure 12-5.

The minimum recommended roof pitch for shakes is 4 inches per foot, and for shingles it is 3 inches per foot. While satisfactory ap-

Handsplit shakes		Wood shingles	
Minimum roof pitch	$\frac{4}{12}$	Minimum roof pitch	$\frac{5}{12}$
Shake length (in.)	Exposure (in.)	Shingle length (in.)	Exposure (in.)
18	$7\frac{1}{2}$	16	5
24	10	18	$5\frac{1}{2}$
32	13	24	$7\frac{1}{2}$
		Roof pitch $\frac{3}{12}$ to $\frac{5}{12}$	
		16	4
		18	$4\frac{1}{2}$
		24	6

Figure 12-5. Exposure, shingle and shake.

plications have been installed on lesser slopes, proper climatic conditions and extra care in application are required; this should not be attempted without consultations with the shake manufacturer.

Lengths available include 16, 18, 24, and 32 inches, while random widths of 4 to 12 inches are common. Roofing felt is required for shakes, and a double or triple starter course may be required. Valleys will require some type of flashing. Typical installation details are shown in Figure 12-6.

Underlayment

Shingles are often placed over an underlayment of roofing felts. For wood decks with low slopes (2 to 4 inches per foot) the shingle should be applied over two layers of number 15 saturated felt, which should be lapped 19 inches and exposed 17 inches (rolls are 36 inches wide). These layers of felt are referred to as *underlayment* and work with the shingles to keep water out. Wood decks with slopes of 4 inches per foot and greater should have the shingles applied over a single layer of underlayment with all edges lapped a minimum of 2 inches. Underlayment and minimum slopes for nonwood decks vary considerably, and the shingle manufacturer should be contacted to determine the requirements.

12.2
BUILT-UP ROOFING

Built-up roofing consists of layers of overlapping roof felt, with each layer set in a mopping of hot tar or asphalt. Such roofing is usually designated by the number of plies of felt used. For example, a three-ply roof has four coats of bituminous material (tar or asphalt) and three layers of felt. While built-up roofing is used primarily for flat and nearly flat roofs, it can be used on inclines of as great as 9 inches per foot, providing certain special bituminous products are used.

There is no one type of system applicable for all situations. There may be vapor barrier requirements, varying amounts of lapping felt, and different weights and types of felts and bituminous materials used. The deck type, slope, and use the roof is intended to receive are the most important factors in determining the system to be used on a particular project.

Bituminous materials are used to cement the layers of felt into a continuous skin over the entire roof deck. The types of bituminous materials used are coal tar pitch and asphalt. The bituminous material must be applied in moppings, which fully cement each ply of felt to the next in such a way that in no instance

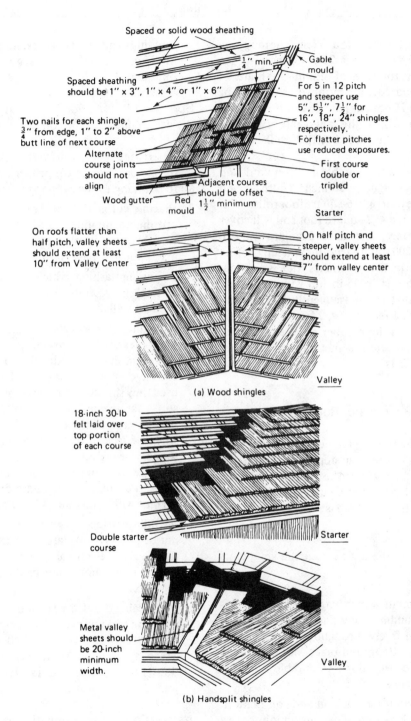

Spaced or solid wood sheathing

¼" min.

Gable mould

Spaced sheathing should be 1" x 3", 1" x 4" or 1" x 6"

For 5 in 12 pitch and steeper use 5", 5½", 7½" for 16", 18", 24" shingles respectively. For flatter pitches use reduced exposures.

Two nails for each shingle, ¾" from edge, 1" to 2" above butt line of next course

Alternate course joints should not align

First course double or tripled

Wood gutter

Red mould

Adjacent courses should be offset 1½" minimum

Starter

On roofs flatter than half pitch, valley sheets should extend at least 10" from Valley Center

On half pitch and steeper, valley sheets should extend at least 7" from valley center

Valley

(a) Wood shingles

18-inch 30-lb felt laid over top portion of each course

Double starter course

Starter

Metal valley sheets should be 20-inch minimum width.

Valley

(b) Handsplit shingles

Figure 12-6. Typical installation details for wood shingles and shakes. (Courtesy Red Cedar Shingle and Handsplit Shake Bureau)

will felt be allowed to touch felt. The mopping between felts averages 25 to 30 pounds per square. The top layer of bituminous material is often poured, not mopped, and may be from 65 to 75 pounds per square. It is this last pour into which the aggregate surfacing materials (if they are specified) will become embedded.

The felts are available in 15-pound and 30-pound weights, 36-inch widths, and 432-square foot or 216-square foot rolls. With a 2-inch lap, a 432-square foot roll will cover 400 square feet, while the 216-square foot roll will cover 200 square feet. In built-up roofing a starter strip 12 to 18 inches wide is applied, over which a strip 36 inches wide is placed. The felts that are subsequently laid overlap the preceding felts by 19, 24⅔, or 33 inches, depending on the number of plies required. Special applications sometimes require other layouts of felts.

Steps to Successful Roofing

A successful roofing job depends on a variety of elements. The first step in roofing is the design of the project; the second step is the creation of proper drawings and specifications; and the third step is proper installation on the job.

Design.   In the design stage the key points to consider include:

1. A minimum ¼ inch per foot slope, to eliminate standing water and to allow the roof to drain freely throughout the life of the building. Poured-in-place roof decks may be poured with slopes to suitable outlets or roof drains.

2. Provision for expansion and contraction. Typical locations of expansion joints are a maximum of 200 feet apart in rectangular

buildings and at all corners, and as shown in Figure 12–7, for irregular shaped buildings.

Specifications.   For best roofing results, the design must also be concerned with:

1. Restricting lateral and vertical movement of the roof deck by the proper fastening of the decking to the supporting members.

2. Placing curbs (Figure 12–8) at all roof openings a minimum of 8 inches above the surface of the finished roof so that proper installation of base flashings is permitted.

3. Providing flashing, which is required at all vertical projections such as curbs and parapet walls.

4. Providing interior roof drains when no other means of roof draining are provided or in conjunction with other methods of drainage. When interior roof drains are used they should be properly located to insure that the water will drain to them. Special care in their location is required when deflection of the decking may reasonably be expected. A sufficient number and size of drains are required so that all water that might accumulate on the roof surface will drain in a 24-hour period.

5. Avoiding surface irregularities, which cannot be properly insulated and roofed. For this reason, electrical conduit, bolts, or similar items should not be placed on the roof deck.

6. Minimizing the effect of vapor pressure by applying two layers of insulation, with the joints of the top layer staggered with those of the layer below. Another effective solution is the taping of all insulation board joints.

Installation.   Job installation is the final step, and cooperation is required between the

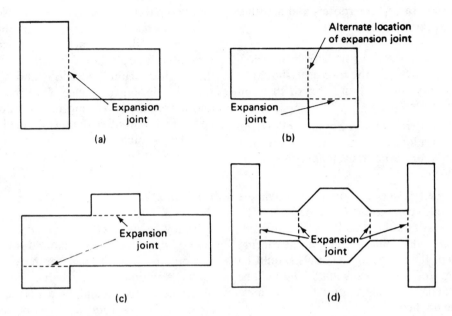

Figure 12-7. Expansion joint location.

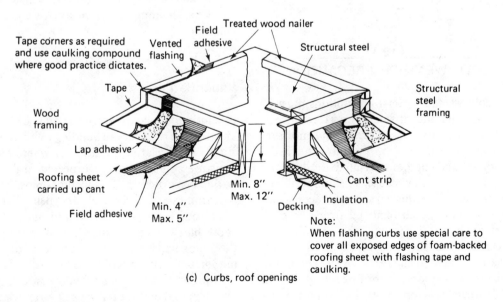

Tape corners as required and use caulking compound where good practice dictates.

Vented flashing

Field adhesive

Treated wood nailer

Structural steel

Structural steel framing

Tape

Wood framing

Lap adhesive

Roofing sheet carried up cant

Field adhesive

Min. 4″ Max. 5″

Min. 8″ Max. 12″

Decking

Insulation

Cant strip

Note:
When flashing curbs use special care to cover all exposed edges of foam-backed roofing sheet with flashing tape and caulking.

(c) Curbs, roof openings

Figure 12-8. Curbs.

193

general contractor, the roofer, and all other trades working on the project.

1. To serve as a satisfactory base for the roofing the surface of the roof deck should be firm, dry, smooth, and free of dirt and loose materials.

2. The roofer should carefully follow the specifications and manufacturer's recommendation in the actual application of the built-up roof.

3. The best results are obtained when the roofing is completed in one operation. The process of putting a building "in the dry," which is the application of all the roofing except the surfacing material, should be avoided. The surfacing material should be applied immediately after the felts have been applied.

4. The roof surface is not a workshop. Other trades should not place scaffolds and other equipment on the roof. Damage to the roofing membranes by other trades is common and often leads to roofing problems.

12.3
COLD WEATHER PRECAUTIONS

In cold weather special precautions must be taken so that the satisfactory performance of the finished roof is assured. Any moisture present at the time of roofing may result in the poor adhesion of the materials to the deck and subsequently may cause blistering of the membrane. For this reason all traces of ice and snow must be removed from the deck before any roofing is applied. Even when the deck is dry asphalt moppings tend to congeal (harden) rapidly, and extra care is required so that the insulation and roofing sets quickly in cold weather. Overheating the bitumin to off-

set the rapid chilling will cause it to lose some of its oil by distillation, resulting in a thin film of mopping that will not be adequate. Instead, the felt roll should be kept no further than 5 feet behind the mop, and all rolls should be broomed at once. The top pouring and any required surfacing material should be applied as soon after the membrane is completed as possible.

Felt Layers

Felt may be laid by hand, spreading a mopping of bituminous material and rolling the felt out, or by stationary or movable felt layers. A stationary felt layer has a roll of felt placed on a spindle, and the felt is pulled out over a drum that rotates in the hot bitumin. The dispenser remains stationary during the operation. This device is particularly useful when sloped surfaces are being roofed and when short runs of felt are being applied. With the movable felt layer, the hot bituminous material is applied to the felt by means of a roller as the equipment is pushed over the insulation.

Vapor Barrier

A continuous vapor barrier should be provided on the roof deck in buildings with a constantly high relative humidity or where high moisture content may be expected to occur during construction. This includes most construction except during hot, dry spells. Sufficient roof insulation is required above the vapor barrier to maintain the temperature at the vapor barrier above the dewpoint of the indoor air during the cold weather. All expansion joints and flashing details must also be designed so that there is a vapor barrier with a warm side up in continuous contact with the vapor barrier on the deck.

Where no vapor barrier is installed moisture will migrate into and through porous insulations, lowering their heat resistance values (Figure 12–9) and reaching the underside of the roofing. With cellular insulation such as urethane and styrofoam the migration takes place at the joints, and no heat resistance is lost. Once the moisture reaches the first felt, usually at the insulation board joint, it works its way into the felt and expands until a ridge is formed. Once the ridge is formed the bitumin tends to flow off it in the summer heat, exposing bare felt. This will either cause a roof blister or a roof leak. On many roofs the outline of every insulation joint is obvious from the ridges raised up.

In lieu of vapor barriers below the cellular insulation the joints may be taped, cutting off the migration of the moisture, or a coated base sheet that will not permit the passage of moisture into the roofing system may be used as the first ply of roofing. A coated base sheet

is felt that has been coated with bituminous materials at the factory.

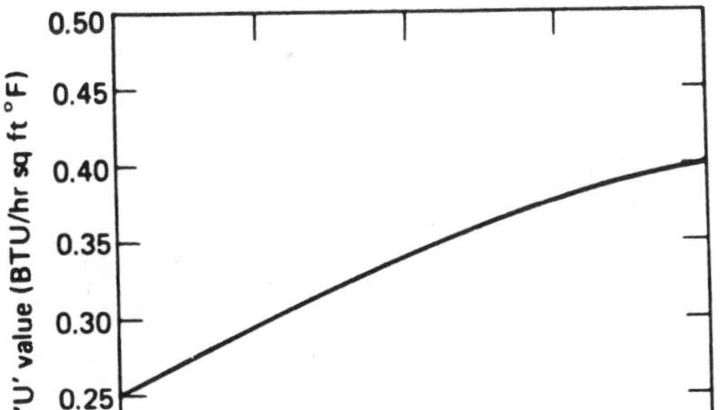

Aggregate Surfaces

An aggregate surface such as slag or gravel is often embedded in the extra-heavy top pouring on a built-up roof. This aggregate acts to protect the membrane against hail, sleet, snow, and driving rain. It also provides weight against the lifting action of high winds. To insure embedment in the top pouring, the aggregate should be applied while the bituminous material is still hot. The amount of aggregate commonly used ranges from 250 to 400 pounds per square.

Almost any type of hard, durable stone or gravel may be used as aggregate surfacing including white marble chips, which may be used to reflect the sun's heat. The stone and gravel used may be crushed or water worn. The slag used is air-cooled blast furnace slag, which is a lightweight aggregate.

Approximate fibrous insulation heating value loss over 20-year period based on 10% water pickup.

Figure 12-9. Fibrous insulation.

12.4
TRAFFIC DECKS

Traffic-bearing roof deck surfacing is applied in a series of trowel applications and provides a flexible neoprene composition traffic surface. A fiberglass mat is included in the thickness to bridge hairline cracks in the deck. Best results are obtained when it is applied over solid decking such as concrete slabs and plywood. The applied deck is about $3/16$ inch thick, weighs about 2 pounds per square foot, and is available in a range of colors including red, green, gray, maroon, blue, and tan. Many manufacturers require that installation of the traffic deck be by an approved contractor, who usually has been trained in the application by their company.

12.5
FLASHING

Flashing is required at all intersecting surfaces of the roof with any other surfaces or materials. This is because of material shrinkage and building movement. The vulnerable points include junctions of walls, skylights, chimneys, and any other roof projections. Since movement is expected at these points, the flashing should be made of elastic material or designed so that there is room for movement. Flashings most commonly used are metal such as copper or aluminum or fabric such as vinyl, plastic, and bituminous impregnated cloth.

Fabric flashings are installed over a coat of adhesive (special adhesives are required for vinyl and plastic flashing), and the flashing is tightly adhered to the wall. The flexibility of this type of flashing is an added advantage. The flashings are usually installed up the wall and into a joint or through the wall. Metal flashing may be copper or aluminum and may be formed to the shape required in the shop or on the job. Metal flashing is installed part way up the wall (base flashing) and then must be capped with metal flashing installed above it covering the top of the base flashing. Some typical installations are shown in Figure 12–10.

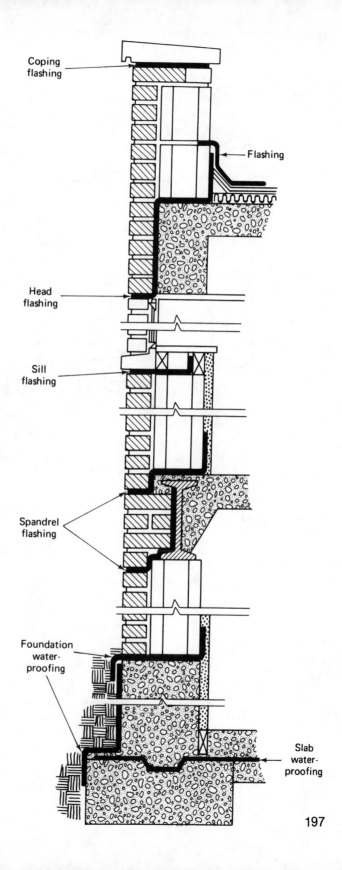

Coping flashing

Flashing

Head flashing

Sill flashing

Spandrel flashing

Foundation water-proofing

Slab water-proofing

Figure 12-10. Typical flashing installations.

197

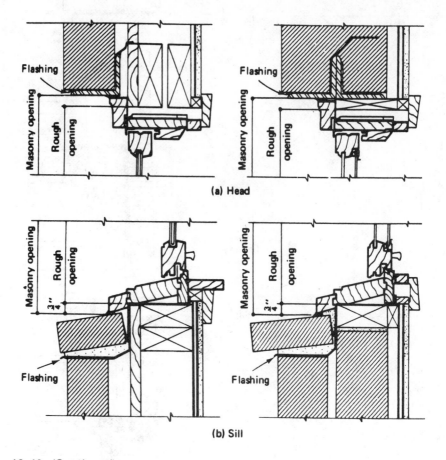

Figure 12-10. (Continued)

Chapter 13

Thermal Insulation

In all construction the flow of heat must be controlled to achieve human comfort inside the building. Heat flows from the area of high temperature to the area of low temperature, and to achieve comfort this heat flow must be controlled. In the winter, the object is to keep the heat inside the building and reduce the *heat loss.* In the summer, for buildings that are cooled, the object is to keep the heat outside the building from entering the building; this reduces the *heat gain.*

13.1
HEAT FLOW

Heat is transmitted by three basic methods: *conduction, radiation,* and *convection.*

Conduction

Heat will flow through a material to the cold side of the construction. All material allows heat to flow through, or *conducts* heat. Some materials such as metals are excellent heat conductors, while others such as cork and glass fibers are relatively poor heat conductors. The amount of heat loss by conduction may be limited by using materials that have low heat conductance values. To help in the selection of materials that limit heat loss by conduction, materials are tested for their heat flow characteristics and are given heat flow values. The values most commonly used are C, k, and R, all of which are interrelated to indicate a material's heat flow characteristics.

The C (conductance) value of a material measures the material's ability to conduct heat through it (Figure 13-1). The C value is expressed in Btu (British thermal units) per square foot per hour for the thickness being used (Figure 13-2). (One Btu is the amount of heat required to heat one pound of water one degree Fahrenheit.) The lower the C value, the less heat that will flow through the material.

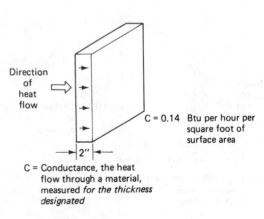

C = Conductance, the heat
flow through a material,
measured *for the thickness
designated*

Figure 13-1. Conductance.

1 lb. water	+ 1 Btu =	1 lb. water
70°F		71°F

Figure 13-2. BTU.

Brick, common	5.0
face	9.0
Cellular glass	8.5
Polystyrene,	
molded beads	0.28
extruded	0.20
Glass fiber	0.33
Maple, oak	1.10
Fir, pine	0.80
Concrete	
lightweight	3.6
sand and gravel	5.0

Figure 13-4. K values.

Example: C value, 2 inch cork = 0.14. This means that the cork that is 2 inches thick will transmit 0.14 Btu per square foot per hour.

The k value of a material (Figure 13-3) measures the material's ability to conduct heat through it, just as the C value does, but it is expressed as Btu per square foot per hour *per inch thickness* of the material.

Example: k value, cork = 0.28. This means that a 1 inch thick piece of cork will transmit 0.28 Btu per square foot per hour. To determine the conductance (C) value for any other thickness (t), use the formula C = k/t. So, a 2 inch thick piece of cork has a value of C = 0.28/2 = 0.14, and ½ inch (0.5 inch) thick cork has a C value of 0.28/0.5 = 0.56.

The lower the k value the less heat that will flow through the material. A variety of several k values are given in Figure 13-4.

The R value of a material is an indication of its *resistance* to the flow of heat through it (Figure 13-5). This is just the opposite of the material's conductance. The higher the R value, the greater the resistance of the material to heat flowing through it. The R value is determined by the formula R = 1/C.

Example: Using the 2 inch thickness of cork, with a C value of 0.14, it has an R value of R = 1/C = 1/0.14 = 7.14. R values may also be determined *per inch* of thickness. In this case R per inch = 1/k, and the cork has an R value of 3.57 per inch.

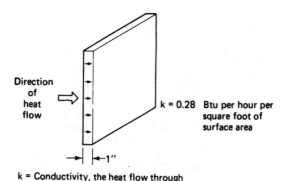

Direction of heat flow →

k = 0.28 Btu per hour per square foot of surface area

←1"→

k = Conductivity, the heat flow through a material, measured for a *one inch thickness* of material

Figure 13-3. Conductivity.

Radiation

The process by which heat (radiant energy) is transmitted from one surface or body to another surface or body by electromagnetic waves that travel through air (or a vacuum) is called *radiation*. Heat passes through the materials of construction by conduction, but it crosses any air space in the material's assembly by radiation, conduction, and con-

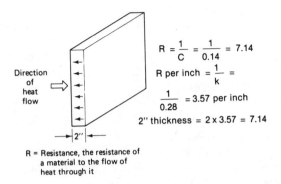

$$R = \frac{1}{C} = \frac{1}{0.14} = 7.14$$

$$R \text{ per inch} = \frac{1}{k} =$$

$$\frac{1}{0.28} = 3.57 \text{ per inch}$$

$$2'' \text{ thickness} = 2 \times 3.57 = 7.14$$

R = Resistance, the resistance of
a material to the flow of
heat through it

Figure 13-5. Resistance.

vection. The use of a reflective material in this space will act to radiate heat back in the direction it is being transmitted from. Metal foils such as aluminum are the most effective reflective materials used to reduce heat flow by radiation. The effectiveness of the reflective material is indicated by an R value and is influenced by the following factors:

1. Thickness of air space.
2. Reflectivity of the material.
3. Position of air space (horizontal, vertical).
4. Direction of heat flow.
5. Mean temperature of air space.

Convection

Heat is moved throughout a building by air circulation through the building. As air heats it expands and becomes lighter, moving in an upward direction, forcing cooler air down, and setting up a flow of air. This is known as *convection*. Typically, the cooler (low) air is warmed by the heating system, causing it to rise. It gives off heat as it passes over cooler surfaces (usually exterior walls, windows, and ceilings), causing it to cool and move in a downward direction, where it is reheated. The heat passed to exterior walls, windows, and ceilings may then be transmitted to the exterior by conduction. Convection can only be controlled to a limited extent, for example, by designing the space into as many small, enclosed spaces as possible, thus restricting the air flow.

13.2
TYPES OF THERMAL INSULATION

Materials whose primary use is to reduce the flow of heat (heat transfer) through the construction are known as thermal insulations. These insulating materials may be divided into two broad groups according to the way they reduce heat transfer. The first group reduces heat transfer by *conduction* and the second by *reducing radiation*.

Conduction

Heat transfer is best reduced by materials that are low density, porous, or fibrous, all of which have low conductivity, which results in a high resistance to the flow of heat through them. These materials' effectiveness is measured by a k value, which indicates the conductivity of the material in Btu per square foot/hour. This type of insulation makes use of air as the basic insulator, since *still air* is one of the best insulators known. However, keeping the air still is a problem since each temperature change causes air to expand or contract, and as it changes temperatures, the lighter, warmer air rises over the heavier, cooler air, thus keeping it from staying still.

The most effective insulators are those that can most effectively keep the air from moving by use of voids (spaces) in the material or microscopic cells that hold the air. These insulators are available in four basic forms: *flexible* (or quilt, blanket, batt), *rigid* (board, slab insulation), *fill*, and *sprayed-on*. The materials most commonly used as rigid, flexible and fill insulators are glass fibers, mineral wool, wood, cane, expanded perlite (mica), foamed glass, and plastic. Other materials sometimes used are cork, hair felt, vegetable fibers, and other light, granular materials.

Flexible Insulation.
Flexible insulation is available in long rolls (blankets) and short lengths (batts or bats). It is also available faced with aluminum foil or vapor barrier (usually Kraft paper) on one side or unfaced. This type of insulation is most commonly used in wood and metal frame construction.

Unfaced Insulation.
This plain, glass fiber insulation can be used in walls and ceilings. Unfaced insulation usually requires the installation of a separate vapor barrier.

Walls. When installed in walls, the insulation will fit snugly between studs spaced 16 or 24 inches on center, and it should butt snugly against the top and bottom plates. All spaces should be completely filled with pieces of in-

sulation. The vapor barrier is applied next. Polyethelyne (2 mil or thicker) is the most commonly used vapor barrier. The polyethelyne is placed on the inside of the insulation and, as installed, covers the entire wall area including doors and windows. Once it is stapled securely to the top and bottom studs and to all door and window frames, the door and window openings may be cut out.

Ceilings. When installed in the ceilings, the unfaced glass fiber insulation is applied by pushing it up between joists. Care must be taken that the insulation will slightly overlap the wall plate of the stud wall but not cover any eave vents (Figure 13–6). A vapor barrier is not required when the attic space is ventilated to meet FHA requirements, but its use does help maintain controlled humidity levels within the building.

Floors. Unfaced insulation should not be used in the floors.

Faced Insulation.
Glass fiber insulation is also available with a vapor barrier on one side. The vapor barrier may be Kraft paper (Kraft-faced) or aluminum foil (foil-faced). The insulation must be installed with the vapor barrier toward the warm-in-winter side

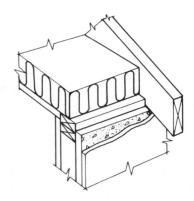

Figure 13-6. Ceiling insulation.

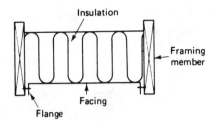

Figure 13-7. Flange stapled to side.

of the construction. The vapor barrier is slightly wider than the insulation. This extra width is referred to as a *flange* and the insulation is secured in place by stapling the flange to the wood frame members. The flange may be stapled to the sides of the framing member, referred to as *inset stapling* (Figure 13-7), or stapled to the face of the framing member, referred to as *face stapling* (Figure 13-8).

Walls. The insulation is placed between the studs, and the flanges are stapled snugly to the sides of the studs. The insulation is secured against the top and bottom plates by peeling back about 1 inch of the facing from the insulation, then fitting insulation snugly against the plate and stapling the facing to the side of the plate.

Ceilings. The insulation is placed between the framing members, and the flanges are either inset or face stapled to the framing members. For best insulating results, the insulation should extend over the top wall plate.

This end is secured in place by peeling the facing back and stapling it to the edge of the top plate (Figure 13-9). The insulation must not extend out so far that it will block ventilation if there are eave vents.

Floors. The insulation is installed with the vapor barrier facing up (toward the warm-in-winter side). The insulation is placed between the framing members, and when the standard faced insulation is used, it may be supported by heavy gauge wire, chicken wire nailed to the bottom of the floor joists, or wire laced between nails that are at intervals along the joists. It is important that the insulation be supported, or portions of it will eventually drop out from between the framing. Some manufacturers have a faced insulation available that has a special breathing paper on the side opposite the facing. This breathing paper has a flange that allows the insulation to be inset or face stapled.

Rigid Insulation.   Rigid insulation is available as boards or planks of insulation of various sizes, which may be used as insulation, as structural boards in roof decks, as ex-

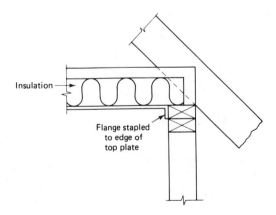

Figure 13-9. Ceiling insulation, flange stapled to face.

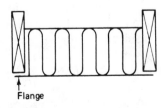

Figure 13-8. Flange stapled to face.

terior sheathing, and as formboards for concrete and gypsum.

Rigid insulation installation will vary with the type of insulation used and the type of structure it is being attached to, as follows:

Wood framing, walls. The rigid insulation will probably be nailed to the wood frame as an exterior sheathing (Figure 13–10).

Wood framing, roof. The rigid insulation will probably be nailed to a sloped wood deck of plywood or solid wood or hot mopped with coal tar onto a flat wood deck.

Flat roofs. The rigid insulation is usually hot mopped into place, but certain types of rigid insulation may have clips that attach it to the supporting construction.

Certain types of rigid insulation (especially polystyrene, polyurethane, and cellular glass) are often used as perimeter insulation around the foundation and concrete slabs (Figure 13–11).

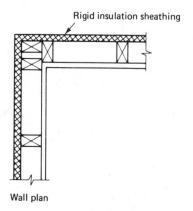

Rigid insulation sheathing

Wall plan

Figure 13–10. Wall plan.

Fill Insulation. Fill insulations most commonly used may be divided into three main types: *fibrous* (floor, ceiling, walls), *granular* (floor, ceiling, walls), and *foamed-in-place* (walls).

Fibrous fill insulations are usually glass fibers or mineral wood. They are available as loose-fill insulation, which can be hand placed into the open areas of construction or blown through large tubes into the areas to be insulated.

Granular fill insulation, usually expanded minerals such as vermiculite and perlite, is available as loose-fill insulation, which may be blown into the spaces but is often dumped into the space by hand. In ceiling areas, the insulation is then spread by hand. In walls, this type of insulation is often used in the cores of the block. It is seldom used as a floor insulation.

Foamed-in-place insulation may be polystyrene, polyurethane, or urea-formaldehyde. They are made from synthetic liquid resins, and as the resins and foaming solution are mixed they are sprayed into the wall area to be insulated. These products have sufficient viscosity to remain in place in the vertical wall, but the urea-formaldehyde will not bond to the surrounding surfaces as the others do, and it will flow into any holes, hollow spaces, or cracks before it sets up. While polystyrene and polyurethane foams are closed-cell and do not absorb water, urea-formaldehyde has open cells and can absorb up to 30 percent by water volume. Also, many people are concerned about possible health hazards from formaldehyde fumes, and its use and effects are being monitored. All of these foams are excellent insulators; polyurethane has a k value of 0.14, polystyrene 0.26, and urea-formaldehyde 0.20.

Sprayed-on Insulation. Sprayed-on insulation is usually a cellular or fibrous material that is mixed with an adhesive and sprayed on

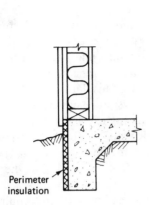
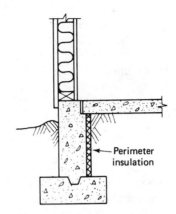

Figure 13-11. Perimeter insulation.

the wall or ceiling area to be insulated. It is often used in industrial facilities, particularly where condensation control inside the building may be a design factor. Use of a sprayed, water-resistant insulator will protect the interior surfaces from condensation being formed on them (especially important on any exposed steel, which would be susceptible to rust). The k value will vary with the base type material used.

Insulation Materials

Glass Fiber. Very fine filaments of glass are made into lightweight insulating blankets used for floor, wall, and ceiling insulation. They are available in various thicknesses and lengths and are most commonly used in wood frame construction. Glass fiber insulation is available with aluminum facing on one side, with a vapor barrier (usually Kraft paper) on one side, and unfaced. This insulation is always installed with the vapor barrier toward the warm-in-winter side of the construction. When unfaced insulation is used, covering it with a 2-mil polyethylene sheet will provide the vapor barrier required.

Glass fiber insulation is available in 15-inch and 23-inch widths to fit 16-inch and 24-inch on center framing. They are available in lengths of 48 inches and 96 inches, referred to as *batts,* and rolls of various lengths depending on insulation thickness, width, and manufacturer. The most common thicknesses and their respective R values are:

		R VALUE
Thickness	Walls	Ceilings & Floors
3½ inches	11	13
6 inches		19

Mineral Fiber. Mineral wood fibers are mixed with bituminous binders and formed into insulation boards used on roof decks. These insulation boards may be applied with asphalt to relatively flat or low-pitched roof decks or nailed to wood roof decks with greater pitches. Mineral fiber insulation has a k value of 0.36 when used as an insulation board.

Mineral wool fibers are also formed into flexible insulation in the form of blankets and batts and loose insulation to be blown or placed into the construction. Care must be taken in installing the mineral wool insulation, since any moisture or condensation will reduce its insulating value considerably. In blanket, batt, or loose form, it has a k value of 0.27.

Wood Fiber. Wood fibers are asphalt-impregnated and are formed into insulation boards used on roof decks. This insulation may be nailed to wood decks or applied with hot asphalt to metal decks. It has a k value of 0.36 and is also available plain (unimpregnated) when desired.

Polystyrene and Polyurethane. Polystyrene and polyurethane are thermosetting, rigid foam plastics that are also known as cellular plastics. Both of these plastics are used extensively as insulation in construction. They are both available as plain insulation boards, with aluminum foil facing, and as insulating core in sandwich panels, which have a structural backing and may also have a finished surface that may be exposed to view. Both of these lightweight plastics have high insulating values.

The principle behind these insulations is that the plastic is used to form enclosed cells that contain air or gas. The basic insulating effectiveness of the board is derived from the air or gas enclosed in the cells.

Polystyrene boards are available in two types, expanded beadboard and expanded, extruded polystyrene. Expanded beadboard is manufactured by first mixing an expanding agent with polystyrene granules or powder and applying heat. As the heat melts the polystyrene a gas is formed, and the melted polystyrene expands into a foam. This foam is cooled, placed in a closed mold, and reheated until it is fused together. Expanded, extruded polystyrenes are manufactured by feeding polystyrene granules or powder and foaming agents into one end of a molding machine. The polystyrene liquefies and is extruded as blocks of insulation.

Polystyrene has either air or pentane as its insulating gas and has a k factor of 0.25 and an R value of about 4 per inch of thickness, based on a density of about 2 pounds per cubic foot. It is available in other densities with varying insulating values.

Foamed polystyrene requires some special precautions. It has only limited resistance to heat; only specially formulated adhesives may be used to attach it; it is brittle; and care must be taken not to have solvents come in contact with it since they will attack it, destroying its effectiveness. Its advantage over polyurethane is its lower initial cost.

Polystyrene boards are often used as exterior sheathing in wood frame construction, insulating boards in masonry construction, form liners for poured concrete, roof insulation placed below built-up roofing, and perimeter insulation. For roof insulation a higher density foam is preferred, and it is faced with an asphalt-impregnated Kraft paper to protect it from any possible damage from the hot asphalt.

Polyurethane is manufactured by mixing the appropriate resin with an expanding agent and a curing agent which is heated in a mold. Large blocks of polyurethane (referred to as buns) are formed and then sawed to the desired thickness. The insulating gas enclosed in the polyurethane cells is either carbon dioxide or freon gas, both of which have higher insulating values than the air entrapped in polystyrenes. As a result, the k value of polyurethane is lower, as low as 0.11 for freon-foamed polyurethane and 0.14 for the carbon dioxide, both significantly more effective than the other types of insulation. However, exposure to air will cause the k factor to rise to 0.16 since air will eventually pass through the cell walls and replace the enclosed gas. This will not happen if the insulation is installed or enclosed in the construction in such a manner that air will not have access to it.

Polyurethane is less susceptible to attack from adhesives and solvents and is generally stronger than polystyrene. However, it is more expensive.

Polyurethane boards are often used as exterior sheathing in wood frame construction, as insulating boards in masonry construction, and as roof insulation below built-up roofing. A heavier density (4.0 pounds per cubic foot) board is used for roof insulation.

Cellular Glass.
Cellular glass is lightweight, rigid insulation composed of completely sealed glass cells—each cell enclosing an insulating air space—which is formed into boards or blocks of insulation. Cellular glass insulation is strong, solvent-resistant, impervious to water, and noncombustible.

It is used as insulation in masonry walls, sandwich panels, and roof decks. Its k value of about 0.40 makes it much less effective per inch of thickness than either polystyrene or polyurethane.

Perlite.
Perlite is a volcanic material obtained through mining. Because of its volcanic origin, each perlite particle has moisture chemically locked within it. The perlite is mined and then heated to about 1700° F, during which time the perlite expands to about 12 times its original size. Each tiny bead of perlite is a honeycomb of air cells that give the material its insulating qualities. This lightweight material is then mixed with mineral binders and fibers to form insulation boards that are used primarily as roof deck insulation. The resulting product is fire and water resistant. It is placed on the roof deck and then covered by built-up roofing.

As an insulation, it has a k value of about 0.36, which is slightly better per inch than cellular glass but is not nearly as effective as polyurethane or polystyrene.

Reflective Insulation.
In reflective insulation thin metal foil is used to reflect the heat that radiates to it. The metals commonly used are tin, copper, and aluminum. The metal may be mounted on rigid insulation board, gypsum board, flexible insulation, or reinforced Kraft paper.

Proper installation is the key to obtaining the best results from the reflective insulation, and the following guides should be followed:

1. An enclosed air space should be provided on at least one side of the reflective material.

2. The air space should range from ¾ to 2 inches for best results.

3. When reflective insulation is placed in the middle of an air space, two air spaces will be formed, increasing its effectiveness. Metal foil on reinforced Kraft paper would most likely be used in this installation.

NOTE: If there is no other vapor barrier in the construction assembly of materials, condensation can be controlled by using two reflective surfaces, separated by an air space.

Care must be taken in the assembly and installation of materials since aluminum foil should not be in contact with other metals or with the alkalies in wet plaster or cement, which may cause the aluminum to corrode.

To reduce heat gain of a building in the summer (thus reducing the amount of cooling required), reflective insulation with one air space should face the exterior of the building. When the insulation is installed in this manner a vapor barrier should be installed on the interior (warm-in-winter) side of the assembly to prevent condensation.

Chapter 14

Flooring

Flooring materials include materials that are applied over the subfloor or underlayment. Finish flooring materials include wood, resilient tile, clay tiles, terrazzo, and carpet. The material selected will vary with:

1. Type of occupancy.

2. Type of floor construction.

3. Floor usage.

4. Specific requirements.

5. Cost.

14.1
WOOD FLOORING

The basic wood flooring types are strip, plank, and block (Figure 14–1). The most widely used wood for flooring is oak; other popular species are maple, southern pine, and Douglas fir, with beech, ash, cherry, cedar, mahogany, walnut, and teak also available. Flooring is available unfinished or factory finished.

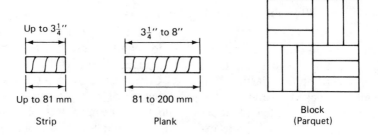

Up to $3\frac{1}{4}''$ $3\frac{1}{4}''$ to 8''

Up to 81 mm   81 to 200 mm

Strip   Plank   Block (Parquet)

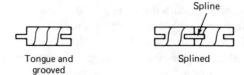

Spline

Tongue and grooved   Splined

Figure 14–1. Basic wood flooring types.

Strip Flooring

Strip flooring is up to 3 ¼ inches (81mm) wide and comes in various lengths. Plank flooring is from 3 ¼ to 8 inches (81 to 200mm) wide and comes in various thicknesses and lengths. The most common thickness is ²⁵/₃₂ inch (20mm), but other thicknesses are available. The strips may be tongue-and-grooved, square-edged, or splined. Flooring is usually installed with nails, screws, or mastic. When mastic is used, the flooring used should have a mastic recess so that excess mastic will not be forced to the face of the flooring. Nailed wood flooring should be blind nailed (concealed), which is accomplished by nailing just above the tongue with the nail at a 45° angle, as shown in Figure 14–2.

Block Flooring

Block flooring is available as parquet (pattern) floors, which consist of individual strips or larger units of wood installed in decorative patterns. Block sizes range from 6¾ x 6¾ inches to 30 x 30 inches (170 x 170mm to 750 x 750mm), with thicknesses from ⁵/₁₆ to ¾ inch (8 to 19mm). They are available tongue-and-grooved or square-edged. Construction of the blocks varies considerably: they may be pieces of strip flooring held with metal or wood splines in the lower surface; laminated blocks; cross-laminated plies of wood; or slat blocks, which are slats of hardwood assembled in basic squares and factory-assembled into various designs.

The wood flooring may be unfinished or factory finished. Unfinished floors must be sanded with a sanding machine on the job and then finished with a penetrating sealer, which leaves virtually no film on the surface, or with a heavy solid finish, which provides a high luster and protective film. The penetrating sealer will usually require a coat of wax also. Sanding the floors will require from three to five passes with the machine. On especially fine work, hand sanding may be required. Factory finished wood flooring requires no finishing on the job, but care must be taken during and after installation to avoid damaging the finish.

Various types of supporting systems may be used for the wood floor. Among the more common are treated wood sleepers, a combination of ⅜-inch hot asphalt fill and treated sleepers, steel splines, and cork underlayment (Figure 14–3).

14.2
RESILIENT FLOORING

Resilient flooring tiles may be made of asphalt, vinyl-asbestos, vinyl, rubber, and cork. Resilient sheets are available in vinyl. The flooring may be placed over wood or concrete subfloors, using the appropriate adhesive. The location of the subfloor—below grade, on grade, or suspended above grade—will affect the selection of resilient flooring since moisture adversely affects some floors. All types may be used on suspended wood subfloors and on concrete, as long as they are sufficiently cured. Where moisture is present below grade and on grade, all of the mater-

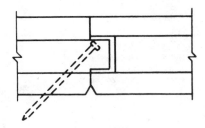

Figure 14–2. Nailing details.

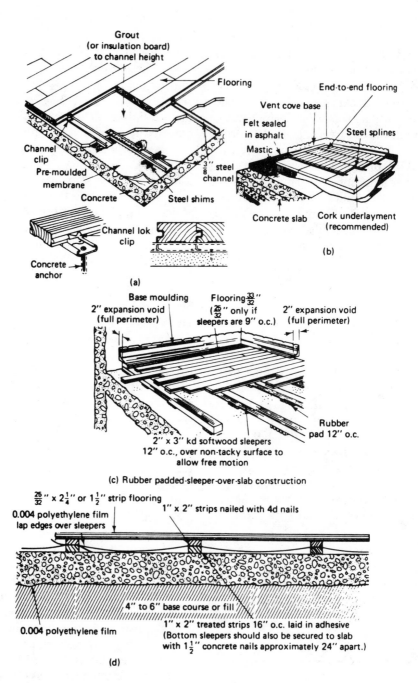

Figure 14-3. Wood support systems.

ials may be used except cork and rag felt-backed vinyl. Tile sizes range from 9 × 9 to 12 × 12 inches (225 × 225mm to 300 × 300mm). Thicknesses range from 0.050 to ⅛ inch (2 to 3.8mm) except for rubber and cork tiles, which are available in greater thicknesses. The most common sheet sizes range from 6 to 12 feet wide with a 4′-6″ width also available.

A subfloor may require an underlayment on it to provide a smooth, level, hard, clean surface for the placement of the tile or sheets. Underlayments of plywood, hardboard, or particle board panel may be used over wood subfloors, while concrete subfloors generally receive an underlayment of mastic. The panel underlayments may be nailed or stapled to the subfloor. The mastic underlayments may be latex, asphalt, polyvinyl-acetate resins, or portland and gypsum cements.

Adhesives used for the installation of resilient flooring may be troweled on with a notched trowel or brushed on. Since there are so many types of adhesives, it is important that the proper adhesive be selected for each application. An adhesive selection chart is shown in Figure 14-4. The wide range of colors and design variations available is in part responsible for the wide use of resilient flooring.

Flooring	Subfloor and location		
	Wood (suspended)	Concrete (suspended)	Concrete (on or below grade)
Vinyl (solid)	Latex Waterproof resin	Latex Waterproof resin	Latex Epoxy
Vinyl (asbestos backing)	Linoleum paste Latex	Linoleum paste Latex	Latex
Vinyl (rag felt backing)	Waterproof resin Linoleum paste	Waterproof resin Linoleum paste	Not recommended
Vinyl-asbestos	Asphalt emulsion Asphalt cutback Asphalt rubber	Asphalt emulsion Asphalt cutback Asphalt rubber	Asphalt emulsion Asphalt cutback Asphalt rubber
Asphalt	Asphalt emulsion Asphalt cutback Asphalt rubber	Asphalt emulsion Asphalt cutback Asphalt rubber	Asphalt emulsion Asphalt cutback Asphalt rubber
Rubber	Linoleum paste Waterproof resin	Linoleum paste Waterproof resin	Epoxy Latex
Linoleum	Linoleum paste Waterproof resin	Linoleum paste Waterproof resin	Not recommended
Cork	Linoleum paste Waterproof resin	Linoleum paste Waterproof resin	Not recommended

Figure 14-4. Adhesive selection.

Accessories available include wall bases, stair treads, stair nosings, thresholds, feature strips, and reducing strips. The color and design variations are more limited in the accessories than in the flooring materials. Wall base (also referred to as cove base) is available in vinyl and rubber, with height of 2½, 4, and 6 inches and in rolls with lengths of 42, 50, and 100 feet. Corners are available preformed.

14.3
CERAMIC TILE

Ceramic tile is used extensively where sanitation, stain resistance, easy cleanability, and low maintenance are desired. Areas where ceramic tiles are commonly used for walls and floors include bathrooms, laundry rooms, showers, kitchens, laboratories, swimming pools, and locker rooms. The tremendous range of colors, patterns, and designs available in ceramic tile even includes three-dimensional sculptured tiles. Extensive use has been made of ceramic tile for decorative effects throughout buildings, both inside and on the exterior. Many tile companies have design departments to assist in developing new and different patterns and color schemes as well as designing mosaic patterns where special effects are desired.

Tile is usually classified by exposure (interior or exterior) and location (walls or floors), although many tiles may be used in all locations. Since exterior tile must be frostproof, the tiles are fired in the kiln to a point where they have a very low absorption. Tiles vary considerably in quality among manufacturers. This may affect the use of them in various exposures and locations. The tiles used should be in conformance with industry standards as set up by the Tile Council of America (TCA).

Tile is generally available in the following square sizes: $4¼ \times 4¼$, 6×6, 3×3, and $1⅜ \times 1⅜$ inches. Rectangular sizes available include: $8½ \times 4¼$, $6 \times 4¼$, and $1⅜ \times 4¼$ inches. Tile often comes mounted into sheets (usually between 1 and 2 square feet) with some type of backing on the sheet or between the tiles to hold them together. Tiles with less than 6 square inches of face area and about ¼ inch thick are called *ceramic mosaics*. Ceramic mosaic tile sizes range from $⅜ \times ⅜$ inch to about 2×2 inches, and they are available from the manufacturers in both sheet and roll form. Often, large tile is scored by the manufacturer to resemble small tiles.

Tile finishes include glazed, unglazed, textured (matte) glaze, and abrasive finishes. Glazed and matte glazed finishes may be used for light duty floors but should not be used in areas of heavy traffic where the glazed surface may be worn away. Glazed ceramic wall tiles usually have a natural clay body (nonvitreous, 7 to 9 percent absorption), and a vitreous glaze is fused to the face of the tile. This type of tile is not recommended for exterior use. Unglazed ceramic mosaics have dense, nonvitreous bodies uniformly distributed through the tile. Certain glazed mosaics are suggested for interior use only or for wall use only. Porcelain tiles have a smoother surface than mosaics and are denser, with an impervious body of less than ½ of 1 percent absorption. This type of tile may be used throughout the interior and exterior of a building. Glazed tile should never be cleaned with acid, which will mar the finish; instead only soap and water should be used. An abrasive finish is available as an aggregate embedded in the surface or an irregular rippled surface texture.

Tiles are available with square edges, cushioned edges (slightly rounded), and self-spacing lugs (Figure 14–5). The lugs assure easy setting and uniform joints. The edges available vary with the size of the tile and the manufacturer.

A complete line of ceramic trim shapes is also available from the manufacturers. Other accessories include towel bars, shelf supports, paper holders, grab rails, soap holders, tumbler holders, and combination tooth brush and tumbler holders, to list a few of the more popular units.

Installation

Ceramic tile's resistance to traffic depends primarily on base and bonding material rigidity, grout strength, hardness, and the accurate leveling and smoothness of the individual tiles in the installation. The Tile Council of America comparison on installation methods is shown in Figure 14–6. The four basic installation methods are: cement mortar, the only thick bed method; and dry-set mortar, epoxy mortar, and organic adhesives (mastic), the thin setting-bed methods.

Cement Mortar.   This traditional material for setting ceramic tiles is composed of a mixture of portland cement and sand. The mix proportion for floors is about 1 to 6 by volume. For walls a portland cement, sand, and hydrated lime mix of 1 to 5½ to 1 is used. The mortar is placed on the surface—¾ to 1 inch thick on walls and ¾ to 1¼ inches thick on floors—and a neat cement bond coat is ap-

plied over it while the cement mortar is fresh and plastic. After soaking in water, the tile is installed over the neat cement bond coat. This type of installation, with its thick mortar bed, permits wall and floor surfaces to be leveled accurately and also allows floor surfaces to be sloped. This installation provides a bond strength of 100 to 200 pounds per square inch. A waterproof backing is sometimes required, and the mortar must be damp-cured.

Dry-Set Mortar.   This technique uses a thin-bed mortar of premixed portland cement, sand, and admixtures that control the setting (hardening) time of the mortar. It may be used over concrete, block, brick, cellular foamed glass, gypsum wallboard, and unpainted dry cement plaster, as well as other surfaces. A sealer coat is often required when the base is gypsum plaster. It is not recommended for use over wood or wood products. It may be applied in one layer 3/32 inch thick, and it provides a bond strength of 500 pounds per square inch. This method has excellent water and impact resistance and may be used on exteriors. The tile does not have to be presoaked, but the mortar must be damp-cured.

Epoxy Mortar.   A bed as thin as ⅛ inch is possible with a two-part epoxy resin compound. When the epoxy resin and hardener are mixed on the job the resulting mixture will harden into an extremely strong, dense setting bed. Pot life, once the parts are mixed, is

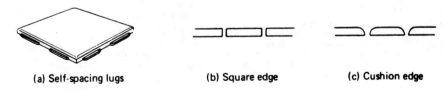

(a) Self-spacing lugs   (b) Square edge   (c) Cushion edge

Figure 14–5. Ceramic tile edges.

FLOORS
INSTALLATION PERFORMANCE LEVELS
Performance-Level Requirement Guide

Use Guide to find Performance Level required. Then consult Selection Table at right and choose installation which meets or exceeds it. For example: Method F113, rated Heavy, can also be used in any area requiring lower Performance Level.

GENERAL AREA DESCRIPTIONS		RECOMMENDED PERFORMANCE-LEVEL RATING
Office Space Commercial Reception Areas	a) General	Light
Public Space in Restaurants and Stores, Corridors, Shopping Malls	a) General	Moderate
Kitchens	a) Residential b) Commercial c) Institutional	Residential or light Heavy Extra Heavy
Toilets, Bathrooms	a) Residential b) Commercial c) Institutional	Residential Light or Moderate Moderate or Heavy
Hospitals	a) General b) Kitchens c) Operating Rooms	Moderate Extra Heavy Heavy—use Method F122
Food Plants, Bottling Plants, Breweries, Dairies	a) General	Extra Heavy
Exterior Decks	a) Roof Decks b) Walkways and Decks on Grade	Extra Heavy—use Method F103 Heavy, Extra Heavy—use Method F101 or F102
Light Work Areas, Laboratories, Light Receiving and Shipping, etc.	a) General	Moderate or Heavy

Selection Table

Maximum Performance Level

RESIDENTIAL:
Normal residential foot traffic and occasional 300 pound loads on soft (70 or less Shore A Durometer) rubber wheels. (Equivalent to passing test cycles 1 thru 3 of ASTM Test Method C 627.)

LIGHT:
Light commercial and better residential use, 200 pound loads on hard (100 or less Shore A Durometer) rubber wheels. (Equivalent to passing test cycles 1 thru 6 of ASTM Test Method C 627.)

MODERATE:
Normal commercial and light institutional use, 300 pound loads on rubber wheels and occasional 100 pound loads on steel wheels. (Equivalent to passing test cycles 1 thru 10 of ASTM Test Method C 627.)

HEAVY:
Heavy commercial use, 200 pound loads on steel wheels, 300 pound loads on rubber wheels. (Equivalent to passing test cycles 1 thru 12 of ASTM Test Method C 627.)

EXTRA HEAVY:
Extra heavy commercial use, high impact service; meat packing areas, institutional kitchen, industrial work areas, 300 pound loads on steel wheels. (Equivalent to passing test cycles 1 thru 14 of ASTM Test Method C 627.)

Notes:
Consideration must also be given to (1) wear properties of surface of tile selected, (2) fire resistance properties of installation and backing, (3) slip-resistance.

Tile used in installation tests listed in Selection Table were unglazed unless otherwise noted. Unglazed Standard Grade tile will give satisfactory wear, or abrasion resistance, in installations listed. Decorative glazed tile or an especially soft body decorative unglazed tile should have the manufacturer's approval for intended use. Color, pattern, surface texture and glaze hardness must be considered in determining tile acceptability on a particular floor.

For waterproof floors (to prevent seepage to substrate or story below), refer to Method F121 and also specify setting method desired.

Selection Table Notes:
Tests to determine Performance Levels utilized representative products meeting recognized industry standards: Dry-Set mortar — TCA Formula 759; epoxy mortar and grout — TCA Formula AAR-II; and epoxy adhesive — TCA Formula C-150.

a. Data in Selection Table based on tests conducted by Tile Council of America, except data for F144 and RF Methods which are based on test results from an independent laboratory through Ceramic Tile Institute.

b. Floor covered after grouting with polyethylene sheeting. Water added to entire surface on second day and sheeting replaced.

c. Rates "Heavy" if Dry-Set is wet cured for three days before grouting.

d. Floor may show surface wear under constant steel wheel traffic.

Figure 14-6. Tile installations. (Courtesy Tile Council of America, Inc., from *1982 Handbook for Ceramic Tile Installation.* Current Handbooks should be reviewed.)

Specification must include Handbook Method Number, grout, setting method and tile description as tabulated to achieve the intended performance level.

Handbook Method Number	Case	Page	Description	Grout	Comments On Use
F116		13	Organic adhesive on concrete Ceramic mosaic or glazed floor tile	Wet cured[b] 1 pc: 1 sand	Dry-Set or Latex-portland cement mortar preferred
F142		16	Organic adhesive on wood Ceramic mosaic or quarry tile	Latex-portland cement	Residential, low cost, bathroom, foyer
F143		16	Epoxy mortar on wood Ceramic mosaic tile	Wet cured[b] 1pc: 1 sand	High bond strength in residential use
TR711		29	Epoxy adhesive over existing resilient tile Ceramic mosaic or quarry tile	Latex-portland cement	Residential renovation
F141		16	Portland cement mortar on wood Ceramic mosaic tile	1 pc: 1 sand	Depressed wood subfloor in residence
F143		16	Epoxy mortar on wood Ceramic mosaic tile	ANSI A118.3 epoxy	Best for wood subfloors
F144[a]		17	Latex-Portland cement mortar on glass mesh mortar unit. Ceramic mosaic or quarry tile	Latex-portland cement	Light weight installation over wood subfloor
RF511[a] RF511	3,4 6	24 24	Glass mesh mortar unit — plywood Glass mesh mortar unit — matting	Commercial portland cement	Wood or concrete subfloor Concrete subfloor
F112		12	Dry-Set mortar on cured mortar bed Ceramic mosaic tile	Wet cured[b] 1 pc: 1 sand	Economy for smooth surface
F113		12	Dry-Set mortar on concrete[c] Ceramic mosaic tile	Latex-portland cement	Economy
F113 F114		12 13	Dry-Set mortar on concrete[c] Ceramic mosaic tile	ANSI A118.3 epoxy	Mild chemical resistance
F122		14	Conductive Dry-Set mortar Conductive tile	ANSI A118.3 epoxy	Hospital operating rooms, other special
RF511[a] RF511	5 8	24 24	Glass mesh mortar unit — matting Portland cement mortar — matting	Commercial portland cement	Wood subfloor Concrete subfloor
F111 F112		12 12	Portland cement mortar Ceramic mosaic tile	1 pc: 1 sand	Smoothest floor surface
F112		12	Dry-Set mortar on cured mortar bed Quarry Tile	Wet cured[b] 1 pc: 2 sand	Economy for smooth surface
F113		12	Dry-Set mortar on concrete Ceramic mosaic tile	Wet cured[b] 1 pc: 1 sand	Best general thin-set method
F122		14	Conductive Dry-Set mortar Conductive tile	Wet cured[b] 1 pc: 1 sand	Hospital operating rooms, other special
RF511[a] RF511	1,2 7	24 24	Portland cement mortar — fiberglass bound Portland cement mortar — matting	Commercial portland cement	Wood or concrete subfloor Wood subfloor
F111 F112 F101		12 12 11	Portland cement mortar Quarry tile or Packing house tile	1 pc: 2 sand	Smooth, hard service best ceramic tile floor
F113 F102		12 11	Dry-Set mortar on concrete Quarry tile or packing house tile	Wet cured[b] 1 pc: 2 sand	Best general thin-set method
F113 F114 F115		12 13 13	Dry-Set mortar on concrete Quarry tile or Packing house tile	ANSI A118.3 epoxy	General, on concrete, for mild chemical resistance
F143		16	Epoxy mortar on wood Quarry tile or packing house tile	ANSI A118.3 epoxy	Hard service on wood subfloor, chemical resistance
F131 F132		14 15	Epoxy mortar on concrete Quarry tile or packing house tile	ANSI A118.3 epoxy	Chemical resistance
F134		15	Chemical resistant mortar on acid resistant membrane, packing house tile[d]	Furan or ANSI A118.3 epoxy	For continuous or severe chemical resistance

Figure 14-6. (Continued)

about 1 hour if the temperature is 82°F. This mortar has excellent resistance to the corrosive conditions often encountered in industrial and commercial installations. It may be applied over bases of wood, plywood, concrete, or masonry. This type of mortar is nonshrinking and nonporous. A bond strength of over 1,000 pounds per square inch is obtained with this installation method.

Organic Adhesives (Mastic). These are applied in a thin layer with a notched trowel. They are a solvent-base, rubber material. Porous materials should be primed before mastic is applied to prevent some of the plasticizers and oils from soaking into the backing. Suitable surfaces include wood, concrete, masonry, gypsum wallboard, and plaster. The bond strength available varies considerably among manufacturers, but the average is about 100 pounds per square inch.

Grouts

The joints between the tiles must be filled with a grout selected to meet the tile requirements and exposure. Tile grouts may be portland cement base, epoxy base, or furans; the five different types available are cement, drywall, epoxy, furan resin, and latex.

Cement Grout. A grout consisting of portland cement and admixtures—which are better in terms of waterproofing, uniform color, whiteness, shrink resistance, and fine texture than a plain cement—should be selected. It may be colored and used in all areas subject to ordinary use. When the grout is placed the tile should be wet, and moisture is required for proper curing.

Drywall Grout. This grout has the same characteristics as dry-set mortar and is suitable for areas of ordinary use. It does not

require wetting of tile except when very dry conditions prevail.

Epoxy Grout. This grout consists of an epoxy resin and hardener. It produces a joint that is stainproof, resistant to chemicals, hard, smooth, impermeable, and easy to clean. It is used extensively in counters that must be kept sanitary for foods and chemicals. It has the same basic characteristics as epoxy mortars.

Furan Resin Grout. In industrial areas that require high resistance to acids and weak alkalis, a cold-atalyzed, thermosetting furan resin grout may be used. Special installation techniques are required with this type of grouting.

Latex Grout. This is used for a more flexible and less permeable finish than cement grout. It is made by introducing a latex additive into the portland cement grout mix.

_____ 14.4 _____

QUARRY TILE

Quarry tile is made from shale and fire clays. This type of tile is easy to clean, fade-proof, and scratch-proof. It is vitreous and acid resistant and requires no waxing or special cleaning agents. It is available in thicknesses of ½, ¾, and 1 inch, face sizes of 6 × 6, 4 × 4, 2¾ × 2¾, 3⅞ × 8, and 9 × 9 inches, and various shapes, some of which are shown in Figure 14–7.

Quarry tile is usually installed on a thick bed of cement mortar to allow easy leveling and any sloping of floors required. Reinforcing mesh is required in mortar beds on flat top structural steel decking and wood construction. The grouts used are the same as for ceramic tile.

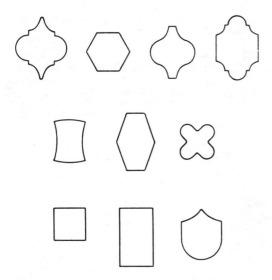

Figure 14-7. Quarry tile shapes.

14.5
TERRAZZO

Terrazzo consists of a binder material—referred to as *matrix*—decorative chips, and color pigments. The binder materials used are portland cement and resinous materials such as polyester, latex, and epoxies. The decorative chips may be any clean, hard material; marble and granite stone chips are most commonly used. The color pigments may be almost any color and may blend with the chips or contrast with them to show them off.

The unbonded method for applying terrazzo (Figure 14-8) uses portland cement as the binder. It provides the greatest control against cracking of the terrazzo. Wherever structural movement in the building is anticipated the unbonded method is suggested, since the isolation between subfloor and underbed permits independent movement. The separation may be accomplished by an isolation membrane such as 15-pound felt or 4-mil polyethylene, or a membrane may be used in conjunction with a sand cushion.

Whenever a sand cushion is used the installation is referred to as *floating*.

The bonded methods of terrazzo installation are shown in Figure 14-9. The resinous binders are used in thin-set terrazzo floors only; hydraulic cement also can be used. Thin-set terrazzo should be applied over firm subfloors such as wood, metal, or concrete. It should range in thickness from ¼ to ½ inch. Portland cement is used as the binder in the monolithic, chemically bonded, and bonded underbed systems. The subfloor surface is wire broomed or roughed to remove any laitance (powdery concrete) and to assure a good mechanical bond with the portland cement terrazzo placed over it. The surface of the slab must be protected from damage and dirt until the terrazzo is placed. Chemically bonded installations have a bonding agent of epoxy or polyvinyl chloride, and the cement terrazzo is placed on the bonding agent. The subfloor must be clean and free from any coatings such as curing compounds, oil, wax, and paint, which would interfere with the bond. A bonded-underbed installation has a mortar underbed bonded to the rough finished slab, with the portland cement terrazzo placed on the underbed.

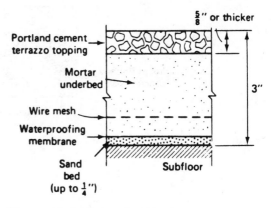

Figure 14-8. Unbonded terrazzo.

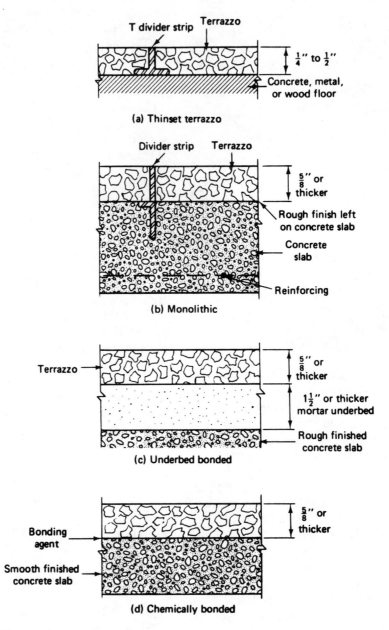

Figure 14-9. Bonded terrazzo.

Cracking of the terrazzo topping must be considered, particularly when portland cement is used as a binder. Cracking is a result of drying shrinkage, temperature variations, and minor structural movements. Divider and expansion strips are placed in the finish to control cracking. The strips are also used for decorative purposes by placing different colors in each section; various designs are available. The strips also assure a constant topping thickness and convenient leveling of the topping. These strips are usually made of zinc, brass, or plastic. Various strip shapes are shown in Figure 14–10. Expansion joints are required wherever substantial structural movement is anticipated, for example, over beams, bearing walls, and expansion joints or control joints in the underlying floor.

Terrazzo topping is mixed on the job, spread on the subfloor, and troweled even with the tops of the divider strips, which act as screeds to level the surface. The top surface has additional wet chips seeded and troweled in. The area is then rolled with a roller, bringing excess water to the surface. The surface is then troweled level with the top of the divider strips, smoothed and leveled, and cured. Portland cement binders must be damp-cured for 3 to 7 days. Once partially cured, the surface is covered with water and ground twice with a grinding machine; then the surface is thoroughly rinsed with water. The topping is allowed to dry, and then all dust and loose particles are removed. The voids are then filled in with a grout and cured. The final finishing steps include grinding off the skim coat of grout, cleaning the floor with a liquid cleaner, and applying a sealing compound.

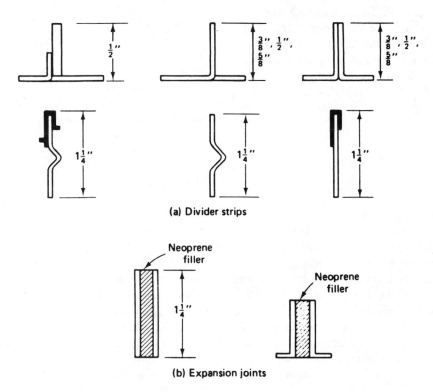

(a) Divider strips

(b) Expansion joints

Figure 14–10. Strip shapes.

Chapter 15

Doors, Windows, and Glass

15.1
DOORS

Doors are generally classified as interior or exterior doors, although exterior doors are often used in interior spaces. The list of materials of which residential doors are commonly made includes wood, aluminum, steel, glass, and plastics. Doors are also grouped according to the mode of their operation. Some different types of operation are illustrated in Figure 15–1. Accessories for doors include glazing, grilles and louvers, weatherstripping (for sound, light, and weather), molding, trim, mullions, transoms, and more.

Wood Doors

Wood doors are available in two basic types, solid core and hollow core (Figure 15–2). Solid-core doors have a core of wood blocks or composition materials such as inorganic minerals, wood, or other particles in a suitable binder. They are generally available in thicknesses of 1⅜ inches and 1¾ inches for the fiber core units and 1¾ through 2½ inches for the wood block core units. Widths to 5′-0″ are available in both types with a height of 12′-0″ as the maximum for wood

block core; the fiber core door is usually available in 6′-8″, 7′-0″, and 8′-0″ heights.

A hollow core door is any door that is not solid. The door may have interlocking wood strips, ribs, struts, or corrugated honeycomb core construction. Thicknesses range from 1⅜ to 1¾ inches with widths of 1′-0″ to 4′-0″ and heights to 8′-0″.

Doors may be flush or paneled as shown in Figure 15–3. Another number used in the description of door construction is the number of plies—3-, 5-, or 7-ply—which denotes the total number of plies including both face panels and the core.

The face veneers may be softwood, hardwood, hardboard, composition board, or plastic laminates. A relative cost guide for some of the face veneers available is shown in Figure 15–4. The doors may be prefinished at the factory or job-finished. Doors that are prefinished at the factory may require some touch-up on the job, and damaged doors may have to be replaced.

Solid core doors are generally used where heavy service is expected and also to provide some sound insulation, fire resistance, and dimensional stability, qualities generally lacking in hollow core doors. Exterior wood doors use a fully waterproof adhesive, called Type 1, while interior doors use Type 11 adhesives, which will delaminate (unglue) when exposed to high-moisture conditions.

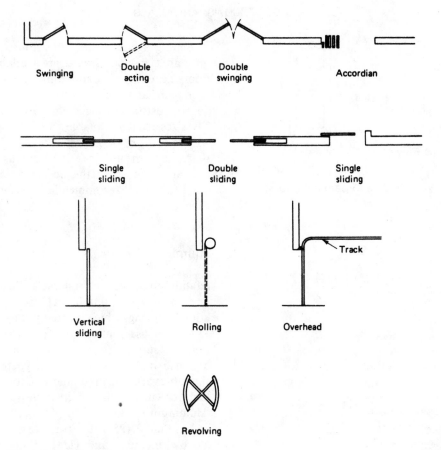

Figure 15-1. Door operation.

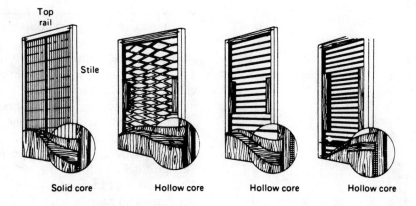

Figure 15-2. Solid and hollow core doors.

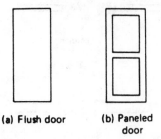

(a) Flush door   (b) Paneled
                             door

Figure 15-3. Flush and paneled doors.

	Index value
Rotary sound ph. mahogany	72
Rotary sound natural birch	92
Rotary good natural birch	98
Rotary premium grade birch	100
Rotary premium red oak	100
Rotary premium select red birch	116
Plain sliced premium red oak	133
Plain sliced premium african mahogany	150
Plain sliced premium natural birch	165
Plain sliced premium cherry	170
Plain sliced premium walnut	180
Sliced quartered premium walnut	250
Plain sliced premium teak	280
Laminated plastic faces (premachined)	250

Figure 15-4. Veneer cost.

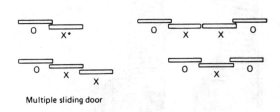

Multiple sliding door

* X denotes doors which are operable (open).

Figure 15-5. Sliding door arrangements.

Sliding wood doors consisting of large glass areas in wood frames are available. The doors generally slide horizontally, and common panel arrangements are shown in Figure 15-5. Custom arrangements are available. The sizes usually available are 6'-8" high with widths of 6, 8, or 12 feet. The openings required to fit the unit in (called the rough opening) are larger than the door size, and the manufacturers' recommendations should be checked.

Aluminum Doors

Aluminum doors are available as flush doors or doors in which aluminum is the frame that holds a glazing, which is usually glass (Figure 15-6). The use of aluminum for the stiles and rails that support the glazing is the most common use of aluminum in doors. There is considerable variation in standards among manufacturers in the size of the stiles and rails. Aluminum thicknesses for doors should not be less than 0.125 inch. Stiles are available in narrow, medium, and wide, but no industry standard for size exists.

Glass is most commonly used to glaze the doors, but materials such as plastics, plywood, and hardboard may also be used. Door widths range to 3'-6" wide and 8'-0" high, but larger sizes can be readily made to order.

Aluminum-faced flush doors are available with cores of honeycomb resin impregnated fiberboard and foamed-in-place insulation. The borders (rails and stiles) are extruded aluminum, and the one-piece aluminum facing is laminated to backing sheets (shock plates) of hardboard. The finishes available are standard mill finish, embossed, anodized, painted (enamel), and wood grained vinyl. Standard door widths are 2'-6", 3'-0", and 3'-6", with a standard height of 7'-0"; special sizes may also be ordered.

Sliding aluminum doors are also available. Usually these doors have aluminum frames and are glazed with glass. They consist of glass panels in a frame designed to allow horizontal movement of some or all of the panels in the frame. Common panel arrangements are shown in Figure 15-5, but custom sliding doors are available where required.

Miscellaneous Doors

Overhead Doors. Overhead doors are available in all sizes. They are most commonly made of wood, plastic, and steel, although aluminum is sometimes used. The overhead doors are designated by the type of operation of the door: rolling, sectional, or canopy. Other types are available.

Folding Doors and Partitions. Folding (accordian) doors and partitions offer the advantage of increased flexibility and more efficient use of the floor space in a building. They are available in fiberglass, vinyl, wood, and combinations of these materials. They can be made to form a radius and to have concealed pockets and overhead tracks; a variety of hardware is also available. Depending on the type of construction, the maximum opening height ranges from 8 to 21 feet and the width from 8 to 30 feet. The doors may be of steel construction with a covering of rigid fiberboard panels, laminated wood, or solid wood panels. Specially constructed dividers are also available with high sound-control ratings.

--------- 15.2 ---------
DOOR SWINGS

The swing of the door (also called the hand of the door) is important to the factory and job coordination of door, frame, and hardware. The door swing (Figure 15-7) is determined by the direction of swing and the location of the butts (hinges) on the door. The swing affects the frame location of the hinges (reinforcement and lock strike reinforcement) (Figure 15-8). It also affects proper location on the door of the beveled edge, and (if the door is

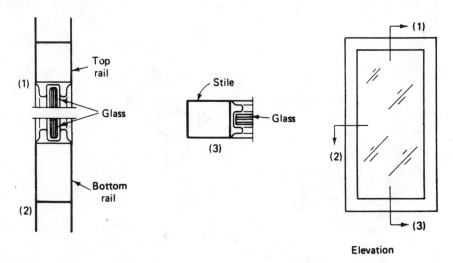

Elevation

Figure 15-6. Aluminum doors.

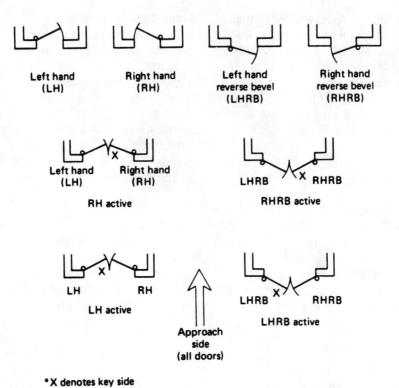

Left hand
(LH)

Right hand
(RH)

Left hand
reverse bevel
(LHRB)

Right hand
reverse bevel
(RHRB)

Left hand
(LH)

Right hand
(RH)

RH active

LHRB   RHRB

RHRB active

LH   RH

LH active

Approach
side
(all doors)

LHRB   RHRB

LHRB active

*X denotes key side

Figure 15-7. Door hand and swing.

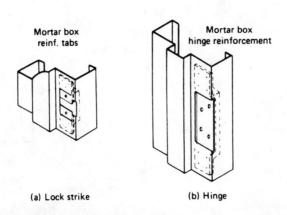

Mortar box
reinf. tabs

Mortar box
hinge reinforcement

(a) Lock strike

(b) Hinge

Figure 15-8. Hinge and lock strike reinforcement.

premachined) the machining of the door to receive the lockset and butts. This coordination is required no matter what type of door, frame, and hinge is used.

15.3
PREMACHINING AND PREFINISHING

Prefitting and Machining (Premachining)

Doors may be machined at the factory to receive various types of hardware. Factory machining can prepare the door to receive cylindrical, tubular, mortise, unit, and rim locks. Other hardware such as finger pulls, door closers, flush catchers, and hinges (butts) are also provided for. Bevels are put on the doors, and any special requirements are taken care of.

Premachining is popular since it cuts job labor costs to a minimum, but coordination is important since the work is done at the factory from the hardware manufacturers' templates. This means that approved shop drawings, hardware, and door schedules, as well as the hardware manufacturer's templates must be supplied to the door manufacturer.

Prefinishing

Prefinishing of doors is the process of applying the desired finish on the door at the factory instead of finishing the work on the job. Doors that are premachined are often prefinished as well. Various coatings are available, including varnishes, lacquers, vinyls, and polyester films (for wood doors). Pigments and tints are sometimes added to achieve the desired visual effect. Metal doors may be prefinished with baked-on enamel or vinyl-clad finishes. Prefinishing can save con-siderable job-finishing time and generally yields a better result than job-finishing. Doors that are prefinished should also be premachined so that they will not have to be "worked on" on the job. The prefinished door must be handled carefully and installed at the end of the job so that the finish will not be damaged; it is often difficult to repair a damaged finish. Care during handling and storage is also necessary.

15.4
DOOR FRAMES

Door frames are available in the same materials as doors. The sides of the door frame are called the *jambs,* and the horizontal piece(s) at the top are called the *heads* (Figure 15-9).

15.5
WOOD FRAMES

Wood door frames may be made of any wood species; often the wood for the frame is selected and finished to match the door selected. Ponderosa pine and fir are two wood species commonly used for door frames and trim and are often in stock at the local supply houses.

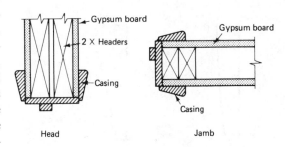

Figure 15-9. Jambs and head(s).

Wood frames are available preassembled (with the door already hung in the frame), knocked down (KD) and ready to be assembled on the job site, or may be constructed at the job site. Door frames built on the job often make use of ¾ inch (18mm) lumber for the head and side jambs and a stock trim for the casing (Figure 15-10).

15.6
ACCESSORIES

Items that may be required to complete the job include weatherstripping, sound control devices, light controllers, and saddles.

Weatherstripping for the jambs and head may be metal springs, interlocking shapes, felt or sponge, neoprene rubber in a metal frame, or woven pile. At the bottom of the door (sill), the weatherstripping may be part of the saddle, attached to the door, or both. It is available in the same basic types as used for jambs and heads, but the method of attachment may be different. Metals used for weatherstripping are aluminum, bronze, and stainless steel.

Various shapes, heights, and widths of sills and saddles may be specified. Sound control and light control usually employ felt or sponge neoprene in an aluminum, stainless steel, or bronze housing. The sill protection usually is automatic closing at the sill, while at

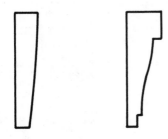

Figure 15-10. Wood frame casing.

the jambs and head it is usually adjustable for best sound and light control results. Some typical installations are shown in Figure 15-11. Exterior doorways may be provided with a sill at the bottom of the opening. Sills are usually masonry, reinforced concrete, stone, or wood (Figure 15-12).

Doors must be high enough off the floor to allow enough clearance so that when they are opened they will freely swing above the floor and above any floor covering materials such as carpets. Since exterior doors must be reasonably weathertight when closed, a threshold is placed above the sill and below the door. Thresholds may be wood or aluminum, and some common shapes are shown in Figure 15-13.

Interior doors may have thresholds under them, particularly if the flooring material on each side of the door is different. In this case the threshold is used as a transition from one material to the other.

In all installations it is desirable to keep air infiltration into the building to a minimum. To protect against the passage of air around the door (cold air in winter and hot in summer) a seal should be provided at the bottom of the door and between the door and frame. The seals for the bottom of the door may be metal spring strips (Figure 15-13) or metal with a vinyl or rubber insert. Seals may be attached to the door or built into the threshold. Those attached to the door are less likely to be damaged and should allow for adjustments up and down for changing conditions.

Astragals (meeting stiles) are used to seal the space between a pair of swinging doors to reduce air infiltration. They may be attached to the closing stile of one of the doors, in which case the closing stile of the other door closes against it, or there may be astragals in both of the doors. The astragals are usually metal with inserts of neoprene, vinyl, or felt, or they may simply be spring metal. Many dif-

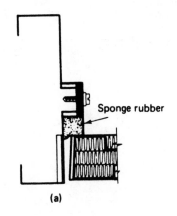

(a)

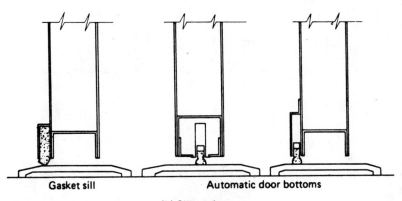

Gasket sill   Automatic door bottoms

(b) Sill sections

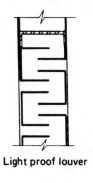

Light proof louver

(c)

Figure 15-11. Light and sound control.

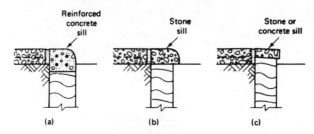

Figure 15-12. Exterior sills.

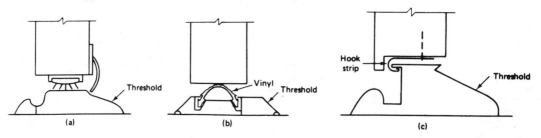

Figure 15-13. Thresholds.

ferent types are offered by the various manufacturers. Some of the adjustable astragals may also be used for weatherstripping at the threshold and frame.

Automatic door bottoms drop down to create a seal between door and threshold when the door is closed. When the door is opened it raises up. They may be built into the bottom of the door or they may be attached to the bottom surface of the door (Figure 15-13). The maximum drop of these seals is about ⅝ inch. Drop inserts may be felt or sponge neoprene. The felt drop may be shaped to seal uneven surfaces by cutting or sanding and is often preferred for sound deadening or lightproofing. However, felt is subject to damage by moths, mildew, and long exposure to water.

15.7
HARDWARE

The hardware required on a project is broken down into two categories: rough and finished. Rough hardware comprises the bolts, screws, nails, small anchors, and any other miscellaneous fasteners. This type of hardware is not included in the hardware schedule, but it is often required for installation of the doors and frames. Finished hardware is the hardware that is exposed in the finished building and includes items such as hinges (butts), hinge pins, door-closing devices, locks, latches, locking bolts, kickplates, and other miscellaneous articles.

The hardware may be made of brass,

bronze, aluminum, steel, or plastic. Finishes available include polished, satin, dull, oxidized, and painted. Each finish and material has its own abbreviation. Often surface finishes may be plated to steel or iron, and if a solid metal is preferred then "solid" must be specified. Often dual finishes are specified; that is, one finish on one side of the door and a different finish on the other side of the door. Not all hardware is available in all finishes. The finishes available for each piece of hardware specified must be checked with the manufacturer.

Hinges

A device that allows the door to turn on the frame is called a *hinge*. The hinge is made of two metal plates (leaves) joined together with a pin passing through the knuckle joints (barrel). Hinges are mounted on the surface of the door. Hinges are usually applied to the edge of a door and to the frame and are referred to as butts. They are usually sold in pairs; 1½ pairs consist of three hinges. Butt hinges range from 2 to 8 inches long, and up to 2 inches wide. Wider butt hinges are also available, and the length and width are often the same.

Butts with ball bearings are used for large, heavy doors that receive high-frequency use. This type of butt is also used where quiet operation is necessary.

Rising butt hinges are available that allow a door to rise above thick carpet as it is opened, yet close as snugly as possible to reduce the passage of sound.

The pins used in hinges may be ordinary loose pins, nonrising loose pins, nonremovable loose pins, or fast (tight) pins. The ordinary *loose pins* are the type that may be removed from the barrel of the hinge to separate the leaves and remove the door. These

pins have a tendency to "climb" out of the barrel by the twisting action of opening and closing the door. Loose pins are found only in the lowest priced hinges. *Nonrising loose pins* may be removed, but are designed and manufactured to eliminate the "climbing" of the pin. *Nonremovable loose pins* are used primarily where security requirements necessitate a pin that is not removable from the outside when the door is closed. A setscrew is set in the barrel and tightens against a groove in the pin, preventing its removal. When the door is closed the setscrew is inaccessible, but when the door is opened the setscrew can be loosened and the pin removed. *Fast pins* are set permanently in the barrel during manufacture and cannot be removed without damaging the hinge.

Special hinges available include friction and door-control hinges, which cause the door to remain in any position it is placed in, and hinges with a spring assembly, which keep a door shut. Exterior door butts should be nonferrous metal or chromium plated steel, as should interior door butts subjected to high moisture such as steam. Other interior butts may be ferrous metal (hardened, cold-rolled steel). Butts may be plated to match other hardware finishes.

15.8
WINDOW FRAMES

Window frames may be made of wood, steel, aluminum, stainless steel, plastic, or bronze. Windows are classified by their method of operation (Figure 15–14) and are available in stock sizes or are custom made. Large quantities of windows may be custom made to almost any size or shape at a cost just slightly above that of stock units. A typical window unit with each of the pieces labeled is shown in Figure 15–15.

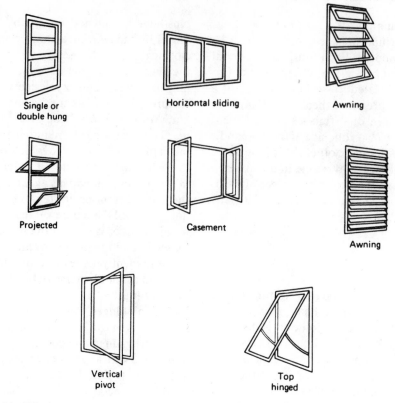

Figure 15-14. Window types.

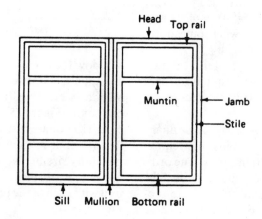

Figure 15-15. Window parts.

Selection

The selection of a window type should depend on the amount of ventilation required, the screening required, air infiltration, cleaning, and repair. Questions that should be answered include whether a window that opens inward, outward, sideways, or vertically is most desirable and whether or not it will interfere with drapes, shades, exterior sun screening devices, and the like.

Wood Window Frames

More than 60 percent of all stock window frames are made from Ponderosa pine. Other commonly used pieces are southern pine and Douglas fir. Custom-made frames may be of these species or of more exotic woods such as redwood and walnut. Frame finishes may be plain, primed, preservative-treated, or shielded with vinyl. Wood frames do not conduct heat as rapidly as aluminum, and aluminum frames are much colder to the touch in winter. Wood windows require periodic painting or staining unless they are vinyl-shielded. The frames should be treated with a water-repellant preservative for protection from moisture, decay, and termites. Some typical installation details are shown in Figure 15–16.

Aluminum Window Frames

The frame shapes are made by either extrusion or roll forming. The greatest advantage of aluminum windows is that they will not corrode except in extreme conditions and they require no painting. Finishes available include mill or natural finish (natural silvery finish); etching, which produces a satin finish;

buffing to a bright finish; an electrolytic finish, which provides a thick protective coating; and colored or opaque coatings. Anodized finishes are available in mill, satin, or bright finishes.

Care on the job is important since exposed aluminum surfaces may be subject to splashing with plaster, mortar, or concrete masonry cleaning solutions that will cause corrosion of the aluminum. Exposed aluminum should be protected during construction with either a clear lacquer or a strippable coating. Concealed aluminum that is in contact with concrete, masonry, or absorbent material such as wood or insulation (and thus may become wet intermittently) must be protected permanently by coating either the aluminum or the adjacent materials. Coatings commonly used include bituminous paint and zinc-chromate primer.

15.9
ACCESSORIES

Screens

Screening mesh may be painted or galvanized steel, plastic-coated fiber, aluminum, or bronze. If the screens are not supplied with the window it must be made certain that the screens and frames are compatible and that a method of attaching the screen to the frame has been determined and is available.

Mullions

Mullions are the vertical or horizontal strips that help connect adjoining sections of frames. The mullions may be of the same material as the frames, or of a different material, color, or finish. Mullions may be small "T"-shaped sections that are barely

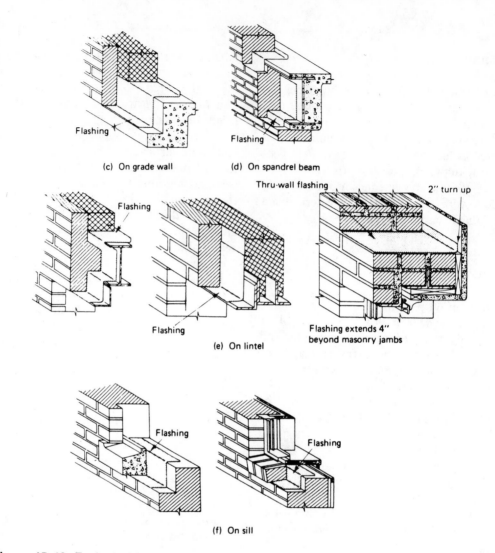

(c) On grade wall   (d) On spandrel beam

(e) On lintel

(f) On sill

Figure 15-16. Typical window installation.

noticeable or large elaborately designed shapes used to accent and decorate. Some typical details are shown in Figure 15–17.

Sills

The sill is the bottom member of the frame. The member on the exterior of the building, just below the bottom member of the frame, is also a sill. The exterior sill serves to direct water away from the window itself. These exterior sills may be made of stone, brick, precast concrete, tile, metal, or wood. Basically, there are two types of sills, the slip sill and the engaged sill. The slip sill is slightly smaller than the opening for the frame and can be slipped in place after the construction of the walls is complete, either just before or just after the frames have been installed (depending on the exact design). The engaged sill must be installed as the walls go up since it is wider than the opening and extends into the wall construction.

Stools

The interior member at the bottom of the frame (sill) is called the stool. The stool may be made of stone, brick, precast concrete, terrazzo, tile, metal, or wood. It may be of the slip or engaged variety.

Flashing

Flashing is most often required at the head or sill of the frame. In masonry walls the flashing is in the head detail so that any moisture that has penetrated the wall is expelled from the building. In most other cases, the flashing is used to keep moisture from entering the construction. Details of the flashing are shown in Figure 15–16.

15.10
GLASS

Glass is the most common material glazed into frames for windows and doors. The most commonly used types, in construction, are sheet, plate, laminated, and insulating. The type of glass used will vary with the project requirements.

Sheet

This material is commonly referred to as clear window glass; it is available in single strength (about 0.086-inch thick), double strength (about 0.120-inch thick), and heavy sheet ($\frac{3}{16}$, $\frac{7}{32}$, or $\frac{1}{4}$-inch thick). The qualities of sheet glass used in construction are AA, A, and B. AA quality is used for specially selected highest grade work; A quality is glass of superior quality; and B quality is used for general work. The designation SSA would mean single strength A quality, DSB would be double strength B quality. Clear window glass has a characteristic surface wave which is more apparent in the larger sizes.

Plate

Plate glass differs from sheet glass in that it has parallel, distortion-free surfaces. This surface is obtained by a grinding or floating process. Grinding consists of the glass being mechanically ground and polished after the glass is rolled. Floated plate glass is manufactured by placing the continuous ribbon of molten glass, which is cast from a furnace, onto a bath of molten metal where it floats, resulting in a perfectly flat configuration, free from distortion, with no need for grinding or polishing. Thickness ranges from $\frac{1}{8}$ to $\frac{1}{2}$ and the qualities available, from highest standard to lowest, are silvering quality, mirror glazing

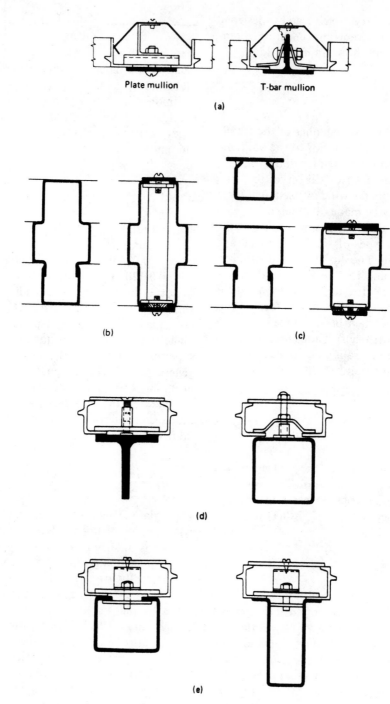

Figure 15-17. Typical mullion shapes.

234

quality, select for mirror quality, glazing, and commercial. The glazing and commercial are commonly used in average applications. The silvering quality is free from defects and is of the highest standards.

Plate glass is also available in heat-absorbing, glare-reducing glass of gray, bluish-green, or bronze colors. These types of glass reduce the solar radiation admitted to the space by about one-half. Comparisons of solar heat radiation values are shown in Figure 15-18.

Laminated Glass

Laminated glass is made by sandwiching a layer of polyvinyl butyral between two or more lights (sheets) of plate or sheet glass. By tinting the plastic inner layer a reduction in heat and glare is achieved. By increasing the number of laminations the glass can be made bullet resistant, and by increasing the thickness of the plastic layer it can be made more sound resistant, especially for sounds about the frequency of airplane noise. Should breakage occur, the glass fragments adhere to the plastic layer, reducing the hazard from flying glass. Thicknesses range from $7/32$ to $7/8$ inch, with sound-control glass ranging from $1/4$ to $5/16$ inch and bullet-resistant glass from $3/4$ to $2\frac{1}{2}$ inches thick.

Insulating Glass

Insulating glass consists of two panes of glass with an air space between; insulating glass can reduce heat transmission values for summer heat and winter cold by 30 to 60 percent. (Comparative values for glass are shown in Figure 15-19.) There are two basic types of insulating glass: sealed and fused. The sealed insulating glass has 2 panes of glass with a protective aluminum frame and a seal keeping the blanket of air dry and clean; this unit is not vacuum sealed (Figure 15-20). The fused unit is made of 2 panes of glass with their edges sealed together and a $3/16$-inch dehydrated air space between them. The round edges are formed by fusing together the two glass panes. The insulating glass may have one pane of heat absorbing glass or may be any of the sheet, plate, tempered or patterned glass.

Glass Size

The sheet sizes available will vary considerably among manufacturers. Often the maximum sheet size will be listed as 84 x 120 inches, although another form of size notation is common, the "united in." United inches means simply the sum of the width and the length of the glass.

Solar radiation: heat transmitted through the glass

$\frac{1}{4}''$ plate glass	88%
$\frac{1}{4}''$ heat absorbing glass	66%
$\frac{1}{4}''$ heat absorbing glass, air space, $\frac{1}{4}''$ plate glass	50%

Figure 15-18. Solar radiation comparison.

		Air space			
		$\frac{3}{16}''$	$\frac{1}{4}''$	$\frac{1}{2}''$	1'' to 4''
Double thermopane		0.69	0.65	0.58	
Triple thermoplane			0.47	0.36	
Single glass	1.13				
Single glass with storm sash					0.56

U'values in BTU per hour per sq. ft. per deg. F

Figure 15-19. Heat loss values of glass.

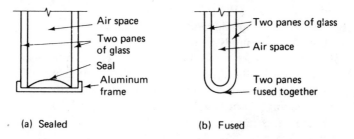

(a) Sealed   (b) Fused

Figure 15-20. Sealed insulating glass.

Chapter 16

Sealants, Fillers, and Finishes

16.1
SEALANTS

Sealants are the materials used to fill the joint that occurs where two different materials meet and expansion and control joints. To be effective a sealant must have several characteristics:

1. Good adherence to surfaces.
2. Workability at a wide range of temperatures.
3. High elasticity and compressibility.
4. Good recovery after being stretched or compressed.
5. Water resistance.
6. Long life.

Polysulfide Synthetic Rubber

This sealant is used to provide a positive, durable seal for construction joints that have relatively constant movement. It cures to a long-lasting synthetic rubber that moves with the joint and offers excellent adhesion, cohesion, and elasticity, while being very resistant to deterioration.

It is recommended both for new construction work and on existing buildings that must be resealed. It is commonly used in wall expansion joints, panel wall connections, and at door, window, and curtainwall connections to the surrounding materials. It bonds to any material with the appropriate use of primers.

Elastomeric Butyl Synthetic Rubber

This is a sealant composed of polymerized butyl rubber, inert pigments, and a mild solvent. Once applied, the outer layer hardens to form a tight seal against the weather. While it has excellent elongation properties, it will not recover more than 10 percent of any elongation that does occur. This makes it most effective in narrow joints where little movement is expected. It is commonly used to glaze glass, door, and window frames, copings, panel walls, skylights, and curtainwalls. It adheres tightly to most building materials such as glass, masonry, concrete, and metals, but not to plastics. It is not recommended for joints that exceed $3/8$ inch in width or where the depth of the joint is greater than its width.

This type of sealant has excellent adhesion properties, exceptional resistance to weathering, and three times the life expectancy of conventional sealants. It will not corrode or stain adjacent surfaces, and it may be painted.

Acrylic Latex

This type of sealant has a latex base and is used as an interior and exterior sealant before painting. It adheres well to clean, dry surfaces with or without a primer. It is recommended for filling joints such as those around door and window frames and may be painted the same day with any type paint.

Acrylic Polymeric

This sealant is used for sealing expansion joints in curtainwalls and masonry joints, glazing wood and metal sash, and filling control joints. For best results a maximum joint size of ¾ inch wide and ⅜ inch deep is recommended. It has an excellent temperature range and if distorted or torn will reseal itself when compressed. It has a life expectancy of up to 20 years.

Oil-Based

Oil-based sealants are formulated from polymerized vegetable oils, nonhardening plasticizers, filler, and color pigments. This type of sealant is commonly used as a crack and joint sealant where expansion and contraction stresses are nominal. It is less expensive than most others and has a lower rate of expansion and recovery, as well as a much shorter life. No oily or asphaltic materials should be used as back-up material with oil-based sealants.

Asphalt

Bituminous products are very water resistant and are often used for weatherproofing. Asphalt sealant compounds combine asphalt with inert fillers, and they are used extensively on flashings, roof ventilators, and skylights. They are available only in black but are nonstaining, and no priming is required. Asphalt sealants tend to sag in extreme heat, and their life span is relatively short.

Silicone

Silicone sealants are composed of silicone and elastomers, and with the use of appropriate primers may be used on almost all materials. These sealants have wide workable temperature ranges; excellent elongation, compression, and recovery characteristics; and are highly water resistant, with a very long life. They are used as sealants for glazing, masonry, metals, flashing, concrete, curtainwall, and for expansion and control joints.

Elastomeric

An elastomeric sealant is any sealant that is formulated with an elastomeric polymer as part of its composition. Use of elastomeric polymers adds significantly to the sealant's characteristics in terms of its adherence, elasticity, compressibility, recovery characteristics, life, and temperature range. It is used with many other materials such as butyl rubber, asphalt, and silicone.

One-Part and Two-Part Sealants

A one-part sealant is a sealant that comes in a container premixed and ready to use. By contrast, the two-part sealant must be mixed on the job before it can be used. Its limited potlife—the length of time after it is mixed and before it must be used up—is a possible disadvantage with this type of sealant.

16.2
FILLERS

Many times the joint to be sealed requires a filler or backup material. The filler may be used to control the depth of the sealant used, prevent the sealant from adhering to a material at the back of the joint, and act as a temporary joint closure until the sealant is applied.

Closed cell polyethylene and urethane filler rods and rods of butyl, neoprene, and elastomeric tubing may be used to control the depth of sealant and act as temporary weather sealers until the sealant is applied. Resilient and nonresilient tapes are used in glazing installations since they do not absorb water, have good recovery, and are flexible even at very low temperatures. Fillers such as corkboard, resin-impregnated fiberboards, elastomeric tubing, and high density foams are commonly used in joints for sidewalks, patios, highways, and other horizontal areas.

16.3
PAINT

Paints are used to finish, protect, and decorate the surfaces of buildings. An evaluation of the various painting systems will provide a basis for the proper selection of paint for each use. Paints are made from vehicles and pigments (or resins). The *vehicle* is the substance used to spread the pigment (or resin) over the surface. Typical vehicles include oil emulsions, alkyd-oil, and polyvinyl formulations. Following are lists of the properties, advantages, and disadvantages of each type of paint currently available commercially.

Alkyd Base

1. One of the most versatile of the synthetic resins, it is made by combining synthetic materials with vegetable oils such as linseed, soya, or tung oil. It produces clear, hard resins.

2. It may be made for interior or exterior use and is formulated in a full range of sheens, from flat to gloss.

3. The formulation for interior use contains an oil that tends to darken with age and is not as easily touched up or patched as latex paints.

4. In flat sheen, alkyd base paints are capable of higher pigment concentrations than most of the latex vehicles, thus providing better one-coat covering.

5. The high pigment concentration of flat alkyd paints results in a coating that has a lower resistance to dirt penetration and marring and has limited washability.

6. Alkyd paints have self-healing properties and can seal surfaces of many materials without requiring a special primer-sealer or undercoating.

Linseed Oil Base

1. It provides excellent penetration and adhesion to wood, forming a flexible exterior wood coating.

2. A primer is required when it is used on bare wood to prevent the absorption of oils from the finish into the porous surface.

3. Self-cleaning and nonchalking types are available.

4. Some color change may occur from exposure to ultraviolet rays of sun.

Alkyd, Chlorinated Paraffin-Linseed Oil Base

This type of paint combines the positive features of alkyd and linseed oil formulations for improved flexibility; resistance to dirt collection, blistering, and fumes; and high color retention.

Latex (Water-Thinned)

General comments on latex painting systems are listed first, and then the three major types of latex paint in use are compared.

1. Latex paint is an oil-free emulsion that can produce either satin or flat sheens. Since it is oil-free, it shows less darkening with age and may be touched up or patched at a later date.
2. It has excellent washability and resistance to stains, grease, and dirt and is not affected by the free alkali in masonry.
3. Special primers are required only on bare metal surfaces, due to the self-sealing qualities of the formulation.

Polyvinyl Acetate (Vinyl Latex, PVA).

1. This type of latex paint provides excellent adhesion to interior and exterior (non-chalky) masonry and cement asbestos surfaces.
2. It does not chalk excessively on exterior exposures.
3. When used on gypsum wallboard, it does not raise the grain of fibers in the wallboard covering.
4. Its fast drying permits the application of two coats of paint in 1 day.

Acrylic (Polyvinyl Chloride, PVC).

1. Formulated for use on exterior wood and masonry, it is available in a flat sheen.
2. As it allows moisture to pass through the paint film, it is a breather-type paint that is blister resistant.
3. It dries in about 20 minutes; two finish coats are easily applied in one day.
4. When it is applied over new wood surfaces, a primer is required.

Acrylic (Metal Finishing).

1. Acrylic paint may be both a primer and finish coat system for metal structures and tanks.
2. It provides excellent gloss and color retention as well as good resistance to chemicals.
3. This nonflammable, fast drying coating may be recoated in 2 to 3 hours.
4. The primer coat has a rust inhibitive formulation.

Chlorinated Rubber Base (Synthetic Rubber, Solvent-Thinned).

1. This paint has good chemical, acid, and alkali resistance.
2. It resists passage of moisture and vapor.
3. It has poor resistance to solvents, animal fats, and vegetable oils.
4. It is used extensively on basement floors and for wall coatings in shower and laundry areas, sewerage disposal plants, and industrial plants.

Epoxy Base (Pigmented).

1. This paint provides excellent resistance to oils, alkali, solvents, water, and abrasion.

2. It has excellent adhesion to all types of surfaces and may even be applied over damp surfaces.

3. This paint withstands temperatures of 250°F on dry surfaces and 150°F on constantly wet surfaces.

4. The paint film loses gloss on exterior applications, but the integrity of the film is not affected.

5. It is made of two components that must be mixed prior to application. Pot-life of the paint is subject to variations due to individual formulations used.

6. Epoxy resin systems may be formulated for masonry wall coatings, caulking compounds, and concrete floor patching.

Epoxy Base (Polyester).

1. This paint provides a tile-like finish and can be applied to any exterior or interior surface in semi-flat to full-gloss sheens.

2. The two components are mixed prior to application and have a pot-life of 24 hours or more.

3. The film has excellent gloss and color retention and high satin resistance and is impervious to moisture.

4. A flame-spreading rating of zero is obtained when it is applied over noncombustible surfaces.

Polyurethane Base (Two-Component).

1. This has outstanding abrasion resistance and is generally superior to epoxy in acid and impact resistance. Epoxy is superior in alkali resistance.

2. It must be applied over an absolutely dry surface.

3. The two components are mixed prior to application.

4. It is used extensively as a concrete floor finish where heavy traffic is expected.

5. Prolonged exterior exposure causes loss of color and gloss, but the integrity of the film is not affected.

Polyurethane Base (Oil Modified).

1. This paint is made of a formulation of polyurethane, alkyd, and soya or linseed oil.

2. It is available as a clear coating for interior or exterior wood or pigmented floor enamel. It surpasses ordinary varnishes and paints in color retention, abrasion resistance, durability, and gloss.

3. Extra care is required in preparing the surface to be coated and in recoating applications.

Silicone Base.

1. This is a water-repellent material for new brick, masonry, mortar, sandstone, and poured concrete.

2. This colorless film preserves a fresh, clean appearance and is effective up to 10 years.

3. It should not be used on limestone.

4. It should not be used on a surface that might be painted at a later time, since adhesion of the new paint to the silicone surface is poor.

16.4
STAINS

Wood stains are available for wood surfaces. Stains may be oil-based, and a variety of color pigments can be added. Creosote oil stains offer protection against decay and insects while dyeing the wood. They may contain bleaching oils that quickly produce a weathered tone (in 6 to 12 months) similar to the tone that would result from years of exposure to salt air. Stains that result in a soft, weathered gray finish and still provide the protection of creosote are also available. Specially formulated stains are available for covering red cedar and redwood. When a finish is to be stained it should be nailed with galvanized or aluminum nails. If regular wire nails are used they should be set in and the holes wood-puttied.

Wood deck stains have an alkyd resin base and are impervious to alcohol, oil, gasoline, salt water, soap solutions, and most acids. They are very resistant to cracking, peeling, and blistering, and they waterproof the surface.

16.5
SURFACE PREPARATION

To assure the best results the surface to be painted must be prepared to receive the finish. The surface preparation that is generally required is outlined in this section.

Exterior Wood

The surfaces should be dry and free of dirt. Any cracks or nail holes should be filled with putty after the first coat is applied, and all knots and sap streaks should be sealed. To avoid rust showing from nail heads, the nails should be countersunk and filled with putty after the first coat is applied.

When painting over previously painted surfaces, all loose, scaling, or peeling paint must be removed. The cause of any paint peeling on previously painted surfaces should be determined and corrected before repainting.

Some woods such as redwood and cedar tend to stain paint films. The use of a primer formulated for this use will control the staining.

Exterior Masonry and Concrete

The surface must be free of all oil, grease, or other foreign matter since adhesion of the paint film to the wall surface will be impaired. All cracks should be cleaned out and filled with mortar that is uniformly textured and similar to the original surface. New brick, stucco, and masonry surfaces should be allowed to cure 60 to 90 days before paint films are applied.

Any evidence of efflorescence must be removed and neutralized with a solution of 10 percent muriatic acid and water, which is followed by a thorough rinsing with clean water and thorough drying.

When repainting previously painted surfaces, all loose paint must be removed. Surfaces previously painted with water emulsion or water-thinned surfaces should be thoroughly wire brushed. Any defective previous coatings must be removed by scraping or sandblasting.

Metal

All surfaces must be clean and dry and free from wax, oil, or grease. Rusting metals should be wire brushed, and all rust flakes and loose particles should be removed. Steel must be coated with a primer; if the primer was applied at the shop it should be touched up with a rust inhibiting primer. Galvanized surfaces must be either treated with a chemical wash or primed with a special primer before painting. Aluminum surfaces should be etched with a special preparation before painting.

Interior Wood, Plaster, and Wallboard

All dirt, grease, oil, and loose foreign matter should be removed. All cracks and open joints should be cut out and properly filled. All nail holes and other defects should be spackled. New plaster should be cured thoroughly before painting since damp walls cause loss of adhesion. Primers are required for some types of paint. A primer is also required over plywood, cedar, and redwood before painting.

Interior Masonry

All surfaces must be thoroughly dry and clean before paint is applied. A stiff brush or broom is used to remove loose sand particles. Filler coats are often used to fill the porous texture of the block.

Interior Concrete

The concrete should be cured at least 30 days before painting. Any form-oil deposits should be thoroughly removed from all poured concrete surfaces before painting, since they reduce the adhesion between the paint and the concrete surface. A primer sealer is used unless the paints selected are self-priming.

Chapter 17

Sound Control/ Materials and Assemblies

This chapter deals with controlling sound in buildings. Sound within a space may be controlled by good acoustical design throughout the building and the use of appropriate materials on walls, floors, and ceilings.

17.1
SOUND CONTROL

Sound Transmission

The goal of sound control is to keep the desired sounds in and the undesirable sounds out. Sound is not selective. It passes (is transmitted) through most building materials and any openings no matter how small. As sound passes through building materials (walls, floors, ceilings) its intensity is reduced, the amount of reduction depending on the type of materials and construction assembly used. Sound passes more readily through lighter, more porous materials than through heavy, dense, massive materials and assemblies of materials. To determine what materials will provide the desired results, it is first important to identify the type of sound control required.

Frequency

Sound is generated by an object vibrating and causing the surrounding air to move. The sound moves in a series of pressure waves that radiate out from the point of generation. As these pressure waves radiate out they vary above and below the normal atmospheric pressure. A sound wave is defined as one complete cycle of pressure variation (Figure 17–1). The number of complete cycles occurring in one second is referred to as the frequency of the wave. For example, if there are 2,000 cycles in one second (cps) the frequency is 2,000 cps. The standard nomenclature for cps is the Hertz (Hz), so the frequency would be 2,000 Hz. The frequency range of human hearing ranges from about 20 Hz to 20,000 Hz.

Decibels

A decibel (dB) is a standard unit of measure of the intensity of sound levels. It is used to measure the sound at the frequency ranges of human hearing. A chart showing relative sound levels (in decibels) is shown in Figure 17–2. A level of 25 to 40 decibels is reasonably

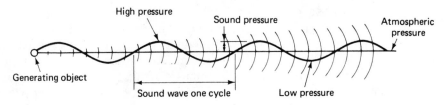

Figure 17-1. Sound wave cycle.

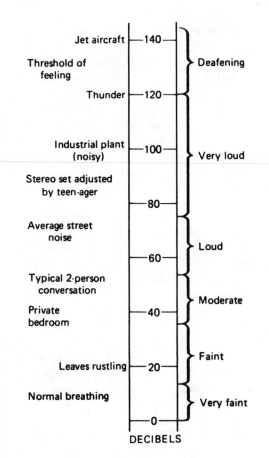

Figure 17-2. Decibels.

Another important point to remember in dealing with decibels is that the apparent change in loudness varies greatly with the change in sound level. A change of 3 decibels is barely noticeable to the listener, while a change of 5 decibels is easily noticeable and a change of 10 decibels makes a significant difference. For example, a change from 50 to 60 decibels would seem twice as loud, while a change from 50 to 40 decibels would seem half as loud.

Sound Control Ratings

Materials and assemblies are rated by a variety of different methods with various meanings. The most common ratings are the Sound Transmission Class (STC), the Average Transmission Loss (TL), the Noise Reduction Coefficient (NRC), the Sound Absorption Coefficient at One Pitch (SAC), the Impact Noise Rating (INR), and the Impact Insulation Class (IIC).

Sound Transmission Class (STC). This rating represents the transmission loss performance of the construction over the broad range of frequencies. The results are plotted, and a standard curve is used to determine the rating given. The higher the rating the better the construction in terms of reducing sound transmission. The curve is based on human response at various frequencies, and this rating is considered one of the most dependable in measuring overall performance.

quiet, while levels of 60 decibels and up become increasingly noticeable. The object of sound control is to reduce the noise to a reasonable level, not to eliminate it.

Average Transmission Loss (TL). The performance of the construction is tested over a range of different frequencies. The results of these tests are averaged together for a transmission loss rating. This rating does not take into account the fact that some people are more sensitive to noise in some of the frequency ranges than in others. Unless the test data are carefully checked, it is possible that the construction will not provide the transmission loss desired at the proper frequencies.

Noise Reduction Coefficient (NRC). This is a rating used to measure the ability of materials such as acoustical tile to absorb sound. Part of this absorbed sound is "killed" within the material, while some of it is passed through to the other side of the material. The NRC is a rating that averages the absorption characteristics of the middle frequencies, rounded off to the nearest 5 percent. Theoretically, these ratings can range from 0 to 1; a perfectly reflective material would have a rating of 0 and a perfectly absorbing material would have a rating of 1. An NRC rating of 0.25 indicates that an average of 25 percent of the sound that strikes the material would be absorbed and 75 percent would be reflected. A difference of 0.10 in the NRC is seldom detectable in a completed installation.

Sound Absorption Coefficient (SAC). This rating for sound absorption ability represents the peak performance of the material. It is not representative of the overall sound absorbing characteristics of the material, but simply represents its highest rating and may be at any of the frequencies tested. The SAC rating ranges from 0 to 1, similar to the NRC.

Impact Noise Rating (INR). This is used to rate the floor-ceiling assembly in terms of its ability to resist the transmission of impact sound. Ratings are based on a plus (+) and minus (−) from a standard curve. Plus ratings indicate a better than standard performance. An INR rating of +5 indicates the assembly averages 5 decibels better than the standard. The standard is based on average background noises that might exist in typical moderately quiet suburban apartments. Where higher background noises are found, as in urban areas, a minus INR rating (to about −5) may be used without any detrimental effects. In areas that are quieter, an INR rating of +5 to +10 would be used since not much background noise is available to mask the sound. The disadvantages of this rating system are its lack of relationship to the STC ratings and the negative values; these tend to cause confusion in use of the INR.

Impact Insulation Class (IIC). This rating is also used to rate the floor-ceiling assembly's ability to resist the transmission of impact sound. This rating is expressed in decibels and more closely relates to the STC ratings for air-borne sound transmission than does the INR rating. The IIC rating of any given assembly will be about 51 decibels higher than its INR rating. Thus, an assembly with an IIC rating of 61 would be comparable to an INR rating of +10. It should be noted that deviations in this can occur and individual test data should be checked.

Sound Isolation

The transmission of air-borne sound between rooms depends on the materials and methods used in the construction. Sound transmission through walls, floors, and ceilings depends on the mass (or unit weight) of the materials and their inelasticity.

Sound Absorption

Noise levels within a space may become objectionable because the sound waves reflect off hard surfaces, which can cause the air-borne sound to build up. The reflection of the sound waves can be controlled by using materials that absorb the sound energy. Highly absorptive materials include carpet, drapes, acoustical ceiling tile, and sprayed on materials; they are often used to control sound reflection. A material's ability to absorb sound will vary with the frequency of the sound. Some materials are more absorbent at lower frequencies while others are more absorbent at higher frequencies. Some typical absorption coefficients for various materials are shown in Figure 17–3. The exact composition of each material being tested will have an effect on that material's coefficients. The figures shown are averages. Information on any one particular product can usually be obtained from the manufacturer.

Knowing the absorption coefficient at each frequency is important to the designer. For example, plywood paneling is quite effective in absorption of low frequency sounds but not as effective in absorption of higher frequency sounds. Therefore, if absorption of higher frequency sounds is desired, another material should be selected.

The most critical frequencies in controlling speech sounds are in the range of 600 to 4,000 Hz, and absorption materials selected should have a high coefficient in these ranges. For other noise problems, the frequency of the noise must be determined.

MATERIAL	500 Hz	1000 Hz	2000 Hz	4000 Hz
Brick, unglazed	.03	.04	.05	.07
unglazed, painted	.02	.02	.02	.03
Carpet (heavy)				
on concrete slab	.14	.37	.60	.65
on felt or foam pad	.57	.69	.71	.73
Concrete block				
porous	.31	.29	.39	.25
painted	.06	.07	.09	.08
Fabrics				
Light weight	.11	.17	.24	.35
Medium weight	.49	.75	.70	.60
Floors				
Concrete	.02	.02	.02	.02
Resilient	.03	.03	.03	.02
Wood	.10	.07	.06	.07
Gypsum Board	.05	.04	.07	.09
Plywood paneling				
⅜" thick	.17	.09	.10	.11

Figure 17-3. Typical absorption coefficients.

17.2
MATERIALS AND ASSEMBLIES

Acoustical Tile

One of the most commonly used absorptive materials is acoustical tile. The effectiveness of the tile varies greatly depending on the material it is made from; its thickness; type of surface (smooth, rough); surface patterns such as drilled, fissured, striated, or plain; and the method used to mount the tile to the supporting construction. Acoustical tile is commonly made from wood or asbestos fibers and is factory painted. It is generally available in thicknesses from $^{3}/_{16}$ to $1\,^{1}/_{2}$ inches (5 mm to 32 mm).

Acoustical tile is generally intended for use as a ceiling tile, placed in a suspended ceiling grid (Figure 17-4) or nailed to furring strips that are attached to the ceiling structure. It is also used on walls and may be applied with adhesives or nailed to furring strips attached to the structure. Some typical ceiling grid systems are shown in Figure 17-5.

Acoustical Units

Various types of acoustical materials are available that are made of a facing over a sound absorbing material. Typically, the facing is a material such as a metal sheet, hardboard, asbestos board, or plastic that is perforated to allow the sound waves to pass through to the sound absorbing materials. The sound absorbing materials commonly used are rock-wool or glass-fiber blankets.

Acoustical units may be used as wall or ceiling finishes, and some are even structural components. Sound absorption coefficients will vary depending on the materials used, thickness of the materials, size and number of face perforations, and type of mounting.

Figure 17-4. Typical suspended ceiling. (Courtesy United States Gypsum Co.)

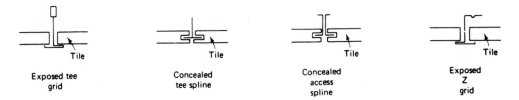

| Exposed tee grid | Concealed tee spline | Concealed access spline | Exposed Z grid |

Figure 17-5. Typical grid systems.

Some typical sound absorption coefficients are shown in Figure 17-6.

Sprayed-On Materials

Sprayed-on acoustical materials are also used to control sound reflection within spaces. Typically, the sprayed-on materials are plaster made with a vermiculite or perlite aggregate or material fiber mixed with an adhesive. These materials may be sprayed on any solid supporting construction such as metal lath, steel, or concrete. Sound absorption coefficients vary with the thickness of materials, aggregates used, and the base on which the materials are applied. Some typical sound absorption coefficients are shown in Figure 17-7.

TILE	THICKNESS	MOUNTING[1]	FREQUENCY, CPS				
			250	500	1000	2000	4000
Fissured	$9/16''$	1	.30	.72	.74	.61	.45
(wood fiber)		7	.39	.53	.72	.63	.51
	$3/4''$	7	.47	.78	.87	.89	.84
Fissured	$5/8''$	7	.32	.59	.85	.83	.85
(Vinyl coated)							
Random	$5/8''$	1	.17	.61	.99	.78	.47
holes		7	.57	.69	.99	.83	.51
(wood fiber)	$3/4''$	1	.21	.70	.99	.77	.52
		7	.69	.80	.99	.82	.56
Metal,	$1 1/4''$	7	.75	.89	.99	.78	.88
perforated							
Brushed	$1/2''$	1	.25	.72	.75	.77	.83
(wood fiber)	$5/8''$	2	.71	.58	.68	.74	.75
Long strand	$1''$	7	.49	.44	.62	.67	.83
wood fiber							
Perforated	$1''$	7	.66	.62	.75	.65	.44
(Asbestos)							

[1]Mounting No. 1—Cemented to gypsum board. Equivalent to cementing to plaster or concrete ceiling
Mounting No. 2—Nailed to 1" × 3" wood furring 12" O.C.
Mounting No. 3—Mechanical mounted to metal supports (16" mounting depth)

Figure 17-6. Typical sound absorption characteristics.

THICKNESS INCHES	MOUNTING	COEFFICIENT OF SOUND ABSORPTION AT CPS						NOISE REDUCTION COEFFICIENT
		125	250	500	1000	2000	4000	
⅜	Solid	.08	.19	.37	.61	.79	.78	.50
½	Solid	.38	.20	.57	.73	.75	.74	.60
¾	Metal lath	.46	.65	.75	.81	.89	.87	.80
¾	Metal lath	.49	.65	.88	.93	.90	.87	.85
1″	Solid	.59	.50	.71	.91	.91	.88	.75

Figure 17-7. Typical absorption coefficients for spray-on acoustical materials.

Massive, thick construction provides excellent sound barriers. Economically there is a point of diminishing return, since as the mass is doubled, the transmission is reduced by only about 5 decibels. Therefore, mass alone is not an economical solution.

Walls

A wall made of 2 x 4 wood studs and ½ inch thick gypsum board has an STC rating of 34 decibels. When a 2-inch thickness of glass fiber or mineral wool unfaced insulation is

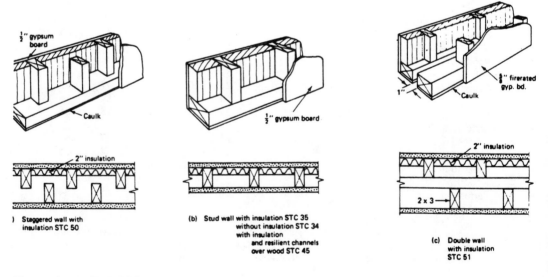

(a) Staggered wall with insulation STC 50

(b) Stud wall with insulation STC 35 without insulation STC 34 with insulation and resilient channels over wood STC 45

(c) Double wall with insulation STC 51

Figure 17-8. Wall STC ratings.

placed between the studs (Figure 17-8), the STC rating increases to 35 decibels, an increase of 1 decibel. This difference of 1 decibel would not even be noticed by the listener. To achieve a higher rating, a possible assembly might include having resilient channels attached to the 2 x 4 studs and then attaching the ½ inch gypsum board to the resilient channels, which will result in an STC rating of 45. By adding 3 inches of fiberglass or mineral wool insulation the rating would be increased to 49. Two other commonly used assemblies for partitions are the staggered wall (STC 50) and the double wall (STC 51).

By comparison, a block wall made of 8-inch normal weight concrete block has an STC rating of 46 to 49 (depending on aggregates used in the block) while 8-inch lightweight block STC ratings range from 34 to 38.

Floors

The typical wood frame floor used on two-story buildings is shown in Figure 17-9. The STC rating for the hardwood flooring ranges from 38 to 42, the same as the assembly with carpeting.

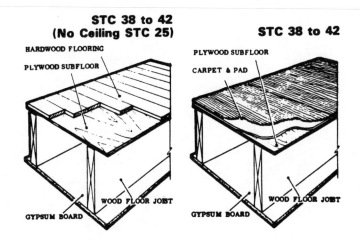

Figure 17-9. Floor STC ratings.

Part III

Mechanical Systems

Chapter 18

Water Supply

18.1
WATER

Basically, water may be *potable* (suitable for human drinking) or *nonpotable* (not suitable for human drinking). While an abundant supply of water is vital to a prosperous economy, on an individual basis a supply of potable water is even more important to survival than food. This potable water must be supplied, or be available, for drinking and cooking. Nonpotable water may be used for flushing water closets (toilets), watering grass and gardens, washing cars and irrigating farms, and for any use other than drinking or cooking.

At this time potable water is commonly used for many activities that could be done with nonpotable water. As potable water becomes scarcer and as the cost of treating nonpotable water to make it potable increases, less and less potable water will be wasted by using it where nonpotable water will adequately serve. In some communities the cost of potable water is so high that many of the residents have shallow wells that provide them with the water to water their lawns and gardens and to wash their cars.

18.2
WATER SOURCES

Rain is the source of most of the water available for our use, and it is classified as *surface water* or *groundwater*. Surface water is the rain that runs off the surface of the ground into streams, rivers and lakes. Groundwater is the water that percolates (seeps) through the soil, building the supply of water below the surface of the earth.

Surface water readily provides much of the water needed by the cities, counties, large industry and others. However, keep in mind that this source is dependent on rain, and during a drought, the flow of water may be significantly reduced. Most surface water will probably have to be treated to provide the potable water required. Where nonpotable water is required, such as for irrigating farms, no treatment of the water may be necessary.

Also classified as surface water is rain which may be collected in a small reservoir or tank (cistern) (Figure 18-1). As it drains from the roof of a building. This water is then pumped into the supply line of the building for use. The need for water is so critical on cer-

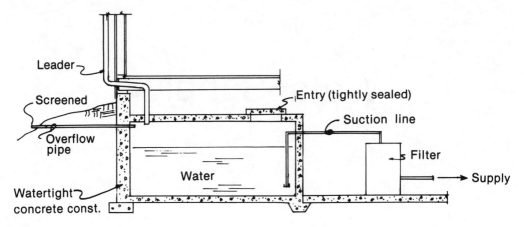

Figure 18-1. Cistern.

tain islands that the government has covered part of the land surface (usually the side of a mountain or hill facing the direction from which the rains usually come) with a plastic so that the rain may be collected and stored for later use.

As groundwater percolates through the soil, it forms a water level below the surface of the earth. This water level is referred to as the *water table*. The distance from the ground surface to the water table (referred to as the *depth* of the water table) varies considerably; generally the more rainfall an area gets, the higher the water table will be. During a dry spell the water table will usually go down, while during a rainy season it will probably rise.

Since the water table is formed by an accumulation of water over an impervious stratum (a layer of earth, usually rock, that the water cannot pass through), the flow of the water follows the irregular path of the stratum, sometimes moving close to the surface while dropping off nearby (Figure 18-2). This underground supply of water flows horizontally, and if it reaches a low spot in the ground surface, it may flow as a spring or seep out creating a swampy area. Or, if the flowing underground water becomes confined between impervious strata, enough pressure

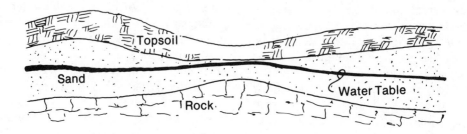

Figure 18-2. Water table.

may be built up in the water that if opened (by drilling through the top stratum or by a natural opening in the stratum), it will create an artesian well.

Groundwater may require treatment to become potable, *but* often it does not. When treatment is required, it is generally less than is required to make surface water potable.

The increased use (and misuse) of our potable water supply has forced the development of additional sources of water. This need for potable water has led to the desalination (removal of salt) of water from the oceans and the purification of waste (sewage) water to be returned to the water system for reuse. To date, these methods involve a great deal of additional cost compared with the use and treatment of surface and groundwater.

18.3
IMPURITIES

All water sources contain some impurities. It is the type and amount of these impurities which may affect the water's suitability for particular use.

As surface water runs over the ground, it may pick up various organic matter such as algae, fungi, bacteria, vegetable matter, animal decay and wastes, garbage wastes, and sewage.

As groundwater percolates down through the soil, it dissolves minerals such as calcium, iron, silica, sulphates, fluorides and nitrates, and it may also entrap gases such as sulfide, sulphur dioxide, and carbon dioxide. It may also pick up contamination from public or private underground garbage and sewage wastes. Generally, as it percolates, it filters out any organic matter which may have been accumulated at the surface or in the ground.

The impurities in the water may be harmful, of no importance, or even beneficial to a person's health. To determine what is in the water it must be tested.

18.4
TESTS

All potable water supplies should be tested before put in use and periodically checked during their use. It is assumed that whatever agency of a city, municipality, etc., controls the supply of water to a community regularly tests its water to be certain it is potable. Private water supplies, such as wells and streams, should always be checked before the water system is put into use and periodically thereafter.

Such tests are usually performed free of charge, or at a very low cost, by the local governmental unit in charge of public health; the governing unit (town hall, city hall, county health department) will put you in touch with the proper authorities, or it will refer you to a private testing laboratory.

The test for potable water provides a chemical analysis of the water, indicating the parts per million (ppm) of each chemical found in the water. A separate test is made for bacteriological quality of the water, providing an estimate of the density of bacteria in the water supply. Of particular concern in this test is the presence of any coliform organisms which indicate that the water supply may be contaminated with human wastes (perhaps seepage from a nearby septic tank field). Since the test reports mean little to most people, a written analysis of the test or a standardized form is included with the test results saying whether the water is potable or not.

Water may have an objectionable odor and taste, even be cloudy and slightly muddied or colored in appearance, and yet the test may show it to be potable. This may not mean you *want* to drink it, but it does mean that it is drinkable. Such problems are often overcome by use of water-conditioning equipment, such as filters. As any traveler can quickly tell you, water varies considerably from place to place, depending on the water source of the area, the chemical and bacteria content of the water and the amount and type of treatment given the water before it is put into the system.

18.5
WATER SYSTEMS

The design of any water supply begins with a check of the water system from which the water will be obtained. Basically, water is available through systems which serve the community or through private systems.

Community Systems

Systems which provide water to a community may be government owned, as in most cities, or privately owned, such as in a housing development where the builder or real estate developer provides and installs a central supply of water to serve the community. The water for these systems may have been obtained from any of the water sources listed in Section 18.2, and quite often it is drawn from more than one source. For example, part of the water may be taken from a river, and it may be supplemented by deep wells.

Before proceeding with the design of the water supply, the following information should be obtained:

1. What is the exact location of the water main (pipe) in relation to the property being built on?

2. If the main is on the other side of the street from the property, what procedures must be followed to get permission (in writing) to cut through the street, set up barricades and patch the street? Also, what permits are required from local authorities, how much do the permits cost and who will inspect the work and when? If available, obtain the specifications (written requirements) concerning the cutting and patching of the street.

3. If the water main does not run past the property, can it be extended from its present location to the property (Figure 18-3), and who will pay for the extension?

4. Is there a charge to connect (tap) onto the community system? Many communities charge just to tap on, and the charge is often hundreds of dollars.

5. What is the pressure in the main at the property? If it is too low for a residence [less than 30 psi (pounds per square inch)], a storage tank and pump may be required to raise the pressure. Such a system is often used on commercial and industrial projects where the pressure may have to be quite high to meet the water demands. Water pressure that is too high (above 60 psi for a residence) will probably require a

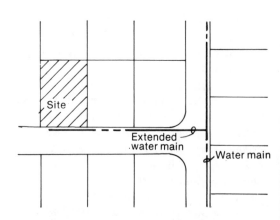

Figure 18-3. Water main location.

pressure-reducing valve in the system to cut the pressure to an acceptable level.

Since plumbing fixtures are manufactured to operate efficiently with water pressures from about 30 to 60 psi, higher pressure may result in poor operation of the fixtures, rapid wearing out of the washers and valves and noises in the piping. Low pressure may cause certain fixtures to operate sluggishly, especially dishwashers, showers, flush valve water closets (toilets), and garden hoses.

6. What is the cost of the water? Typically, a water meter is installed, either out near the road or somewhere in the project, and there is a charge for the water used. After determining what the charges are, a cost analysis may show that it is cheaper to put in a private system. Some areas do not allow private systems for potable water, but quite often it will be desirable to put in a well to provide nonpotable water for sprinkling the lawn and garden and for washing the car. Where costs for potable water are extremely high, it may be feasible to use separate potable and nonpotable water supply systems within the project (especially industrial and commercial projects).

Private Systems

Private systems may also use any of the water sources discussed in Section 18.2. Large industrial and commercial projects may draw all of their supply from one source, or they may draw part of their supply from one source (such as a stream) and supplement the supply with another source (such as a well). Such systems often include treatment plants, water storage towers, and sometimes even lakes or reservoirs to store the water.

Small private systems, such as those used for residences, usually rely on a single source of water to supply potable water through the system. Installing a well is the most commonly used method of obtaining water, and springs may be used when one is available.

Experts, usually consulting mechanical engineers, soil engineers, or water supply and treatment specialists, should be consulted early in the planning for any large project requiring its own private water system. Such specialists can make tests, interpret what the tests mean to the project, and make recommendations as to the quality and amount of water available.

18.6
WELLS

Most private water systems use wells to tap the underground water source. Wells are classified according to their depth and the method used to construct the well.

Depth	Construction Method
Shallow (25 ft or less)	Dug Driven Drilled
Deep (in excess of 25 ft)	Drilled Bored

The depth of the well is determined by the depth of the water table and the amount of water which can be pumped. This flow of water is considered the *yield* or capacity of the well. Where the water table is high, it may not be necessary to go 25 feet deep, but it is not unusual for wells to be 100 feet deep, and in some areas well depths of several hundred feet are required to provide an adequate supply of water.

Dug wells (Figure 18-4) should be 3 to 5 feet in diameter and not more than 20 to 25 feet deep. To minimize the chances of surface contamination, the well should have a watertight top and walls. The top should be either above the ground (Figure 18-5) or sloped so that surface water will run away from it and not over it. The watertight walls should extend at least 10 feet down. The walls may be concrete block, poured concrete, clay tile, precast concrete tile or curved masonry units referred to as *manhole block* (Figure 18-6).

The water will flow into the well through the bottom of the well (and the water in the well will rise to about the level of the water table). Some wells also allow water to seep through the walls by use of porous construction near the bottom of the wall. This porous construction may be concrete block or manhole block placed without mortar (normally used to hold the blocks together and make them watertight).

The placing of washed gravel in the bottom of the well, and on the sides of the well when porous walls are used, will reduce the sand particles or discoloration in the well water. Washed gravel is gravel (stone) that has been put through a wash (water sprayed over the stone) to remove much of the sand or clay from the stone. To further protect the water from possible contamination, tightly seal around the suction line pipe where it passes through the wall. And don't take water out with a bucket or other container since it may

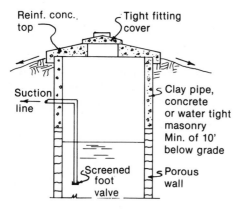

Figure 18-4. Dug well.

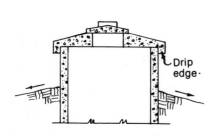

Figure 18-5. Dug well—top above ground.

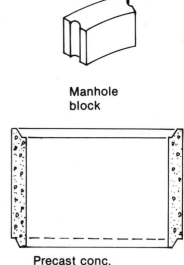

Manhole
block

Precast conc.
tile (pipe)

Figure 18-6. Manhole block.

have contamination on it, and by dipping such a container into the water, the contaminant is transferred to the water supply.

Shallow wells may also be driven. To drive a well, first attach a point to a drive pipe and drive cap (Figure 18-7). Then, by means of an impact loading device such as a small pile driver or even a sledge hammer for very shallow wells in soft, sandy soil, the well point is driven into the ground until it is into the water table. The well point has holes or slots in the side, allowing water to be sucked up to the surface by a shallow well pump (Section 18-7). As the point is driven, additional lengths of pipe may be attached (usually 5-foot lengths are used) by the use of a coupling (Figure 18-7). Driven wells will not pass through rock formations, and the maximum diameter commonly available is 2 inches.

Shallow wells may have to be drilled if it is necessary to pass through rock to get to the water table.

Drilling and boring methods are used for deep wells. A well-digging rig (Figure 18-8) is used to form the well hole. Drilled and bored wells differ in that drilled wells have the holes formed by using rotary bits (Figure 18-9) and spudders, while bored wells have the holes formed by using augers (Figure 18-10). Only the drilling method is most effective in passing through rock.

As the hole is formed, a casing (pipe) is lowered into the ground. This steel or wrought-iron pipe (usually 3 to 6 inches in diameter) protects the hole against cave-ins where unstable soil conditions are encountered and keeps out surface drainage and possible surface or underground contamination. To further protect against surface drainage and contamination, a concrete apron, sloping away from the well, is poured around the casing at the surface (Figure 18-11).

Well location and construction are often controlled by governmental regulations that set minimum distances between the well and any possible ground contaminant. When certain types of construction methods are used, these regulations may even require that licensed well drillers install the well. Various authorities and governmental regulations require different minimum distances, and there is no single set of standards used. The table in Figure 18-12 shows minimum distances required in one locale. It is important that local regulations be checked for each project.

Well contamination from an underground flow of contaminants through rock formations which allow free-flowing groundwater to travel long distances is always possible, especially through strata of eroded limestone. Constant testing of water quality is required wherever there is a possibility of such contamination. For the well contaminated in this manner, three methods used to eliminate the problem are water treatment, relocation of the well and elimination of the source of contamination.

Before planning the well, local conditions should be checked to provide some background information. For example, existing local wells should be checked for depth and

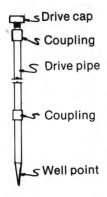

Figure 18-7. Shallow well materials.

Figure 18-8. Well-drilling rig.

Figure 18-9. Rotary bit.

Figure 18-10. Auger.

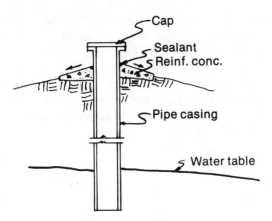

Figure 18-11. Drilled well.

Type of system	Distance from well
Building sewer	50'
Septic tank	50'
Distribution box	50'
Disposal field	100'
Seepage pit	100'
Dry well	50'

Figure 18-12. Well locations.

yield of water. This information can be obtained from local well drillers and governmental agencies and, if possible, verified by testing existing or just completed wells.

Where insufficient information on well yields is available, and especially where large projects will require substantial water supplies, it may be necessary to have test wells made so that the yield can be checked. The well(s) should be tested by the driller to determine the yield, and a sample should be taken so that the quality of the water can be analyzed.

When more than one well is used, they must be spaced so that the use of one well will not lower the water table in the other well. Generally, deep wells must be 500 to 1,000 feet apart, while shallow wells must be 20 to 100 feet apart. Due to soil variables, the minimum distance between wells can be determined only by testing (usually trial and error).

18.7
PUMPS

Pumps used to bring well water to the surface are referred to as *shallow well* and *deep well,* depending on the type of well.

Shallow-well pumps are located above the ground, and a suction line extends into the well below the water table (Figure 18–13). The pump cannot lift or pull the water up more than about 25 feet, so any well with the water table deeper than 25 feet is considered a deep well, and a deep-well pump is used. The shallow-well pumps commonly used are the shallow-well jet, rotary and reciprocating piston pumps.

The deep-well pumps most commonly used are the jet and submersible pumps. Jet pumps are located above ground, either directly over or offset from the well (Figure 18–14). Submersible pumps have a waterproof motor and are placed in the well below the water table.

18.8
PLUMBING FIXTURES

The plumbing fixtures may be selected by the designer of the plumbing system, the architect, the owner or a combination of these people. It is important that the designer of the plumbing system know what fixtures will be used (and even the manufacturer and model number, if possible) in order to do as accurate a job as possible in the design.

The fixtures are the only portion of the plumbing system that the owners or occupants of the building will see regularly since most of the plumbing piping is concealed in walls and floors. All fixtures should be carefully selected since they will be in use for years, perhaps for the life of the building.

The available sizes for each fixture should be carefully checked in relation to the amount of space available. Most manufacturers supply catalogs which show the dimensions of the fixtures they supply.

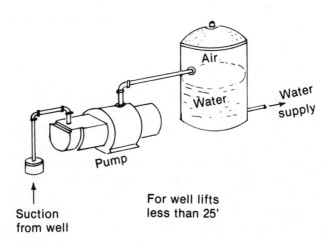

Figure 18-13. Shallow-well pump.

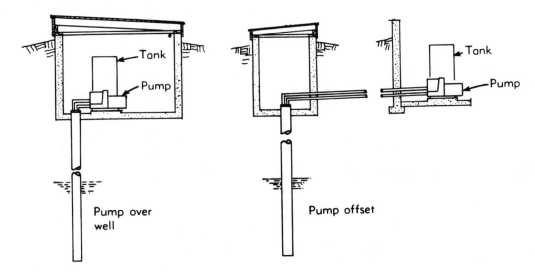

Figure 18-14. Deep-well pump.

Whoever selects the fixtures should check with the local supplier to be certain that they are readily available; if not, they may have to be ordered far in advance of the time they are required for installation. In addition, most of the fixtures are available in white or colors, so the color must also be selected.

Fixtures are grouped according to their use:

Water closets

Bidets

Bathtubs

Showers

Lavatories

Kitchen sinks

Service sinks

Water Closets

Water closets are made of solid vitrified china and are available as flush tank or flush valve fixtures.

A flush tank water closet (Figure 18–15) has a water tank as a part of the fixture. As the handle (or button) is pushed, it lifts the valve in the tank, releasing the water to "flush out" the bowl. Then, when the handle is released, the valve drops and the tank fills through a tube attached to the bottom of the tank. This type of fixture cannot be effectively flushed again until the tank is refilled. There are several types of flushing action available on water closets, as illustrated in Figure 18–16.

Flush tank models range from those having the tank as a separate unit set on the closet bowl to those having a low tank silhouette, with the tank cast as an integral part of the water closet. Generally, this low-slung appearance is preferred by clients, but it is considerably more expensive.

Flush valve water closets (Figure 18–17) have no tank to supply water. Instead, when the handle is pushed, the water to flush the bowl comes directly from the water supply system at a high rate of flow. When used, it is important that the water supply system be designed to supply the high flow required.

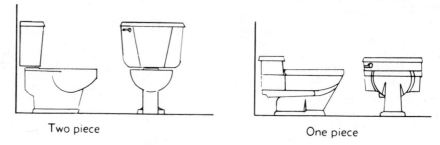

Two piece   One piece

Figure 18-15. Typical flush tank water closets.

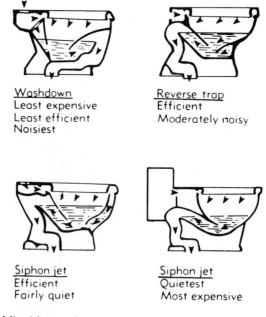

Washdown
Least expensive
Least efficient
Noisiest

Reverse trap
Efficient
Moderately noisy

Siphon jet
Efficient
Fairly quiet

Siphon jet
Quietest
Most expensive

Figure 18-16. Types of flushing action.

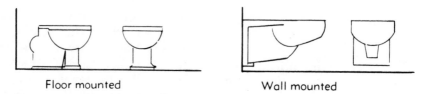

Floor mounted   Wall mounted

Figure 18-17. Flush valve water closets.

While most of the fixtures operate effectively at 20 psi pressure, the manufacturer's specifications should be checked since many times higher pressure is required and must be considered in the design.

Water closets may be floor or wall mounted, as shown in Figures 18–17 and 18–18. The floor-mounted fixture is much less expensive in terms of initial cost, but the wall-mounted fixture allows easier and generally more effective cleaning of the floor. Wall-mounted fixtures are considered desirable for public use, and some codes even require their use in public places. When wall-mounted fixtures are used in wood stud walls, a 2-inch x 6-inch stud will be required instead of the 2-inch x 4-inch stud sometimes used with floor-mounted fixtures.

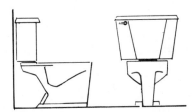

Floor mounted

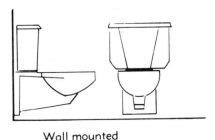

Wall mounted

Figure 18-18. Floor- and wall-mounted water closets.

Bidets

Bidets (Figure 18–19) are designed to wash the perineal area after using the water closet. The bidet is used extensively in Europe and South American and is enjoying increased usage in Canada and the United States. It is designed for use by the entire family and is installed beside the water closet. The user sits on the fixture facing the wall (and the water controls) and is cleansed by a rinsing spray. It is available in vitreous china.

Bathtubs

Bathtubs are available in enameled iron, cast iron or fiberglass. Tubs are available in quite a variety of sizes, the most common being 30 or 32 inches wide; 12, 14 or 16 inches high; and 4 to 6 feet long.

Enameled iron tubs (formed steel with a porcelain enamel finish) are generally available in lengths of 4½ and 5 feet, widths of 30 to 31 inches, and typical depths of 15 to 15½ inches.

Fiberglass bathtubs have been in widespread use since about 1968. The only length commonly available is 5 feet, and it takes 34 to 36 inches of width to install. Generally, the units are cast in a single piece which includes three walls (eliminating the need for ceramic tile around the tub). It is this single-piece feature, with no cracks or sharp corners to

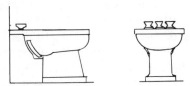

Figure 18-19. Bidets.

clean, which makes the fiberglass tub so popular with the clients. The size of the unit makes it almost impossible to fit it through the standard bathroom door so it must be ordered and delivered early enough to be set in place before walls and doors are finished. In wood frame buildings, they are usually delivered to the job and put in place before the plaster or gypsum board is put on the walls or the doors installed. When selecting fiberglass tubs, be certain to specify only manufacturers which are widely known and respected, with long experience in the plumbing fixture field. Off-brands often give unsatisfactory results in that the fiberglass "gives" as it is stepped on, making a slight noise. In addition, some may be far more susceptible to scratching and damage.

Bathtub fittings may be installed on only one end of a tub, and the tub is designated by the end at which they are placed. As you face the tub, if the fittings are placed on the left, it is called a *left-handed* tub, and, if placed on the right, it is *right-handed*.

Showers

Showers (Figure 18–20) are available in units of porcelain enameled steel or fiberglass. They may be built in with a base (bottom) of tile, marble, cement or molded compositions, and walls may be any of these finishes or porcelain enameled steel. Showers have overhead nozzles which spray water down on the bather. Shower fittings may be placed over bathtubs instead of having a separate shower space; this is commonly done in residences, apartments, and motels. However, it is important that when a shower head is used with a bathtub fixture, the walls be of an impervious material (one that will not absorb water).

Showers of tile, concrete, or marble may be built to any desired size or shape. Preformed shower stall bases are most commonly available in sizes of 30 inches x 30 inches and 30 inches x 36 inches; other sizes may be ordered. Steel showers are usually available in sizes of 30 inches x 30 inches and 30 inches x 36 inches; special sizes may also be ordered. Fiberglass showers are commonly available

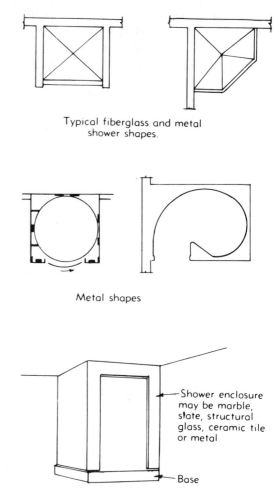

Typical fiberglass and metal shower shapes.

Metal shapes

Shower enclosure may be marble, slate, structural glass, ceramic tile or metal

Base

Base and enclosure

Figure 18-20. Showers.

in sizes of 36 inches x 36 inches and 36 inches x 48 inches. Corner units are also available.

Lavatories

Lavatories (Figure 18–21) are generally available in vitreous china, enameled iron, cast in plastic, or in a plastic compound with the basin an integral part of the countertop. They are available in a large variety of sizes, and the shapes are usually square, rectangular, round or oval (and even shell shaped). The lavatory may be wall hung, set on legs, or on a stand or built into a cabinet (Figure 18–21). Lavatory styles are usually classified as flush-mount, self-rimming, under-the-counter, or integral, or as units which can be wall hung or supported on legs.

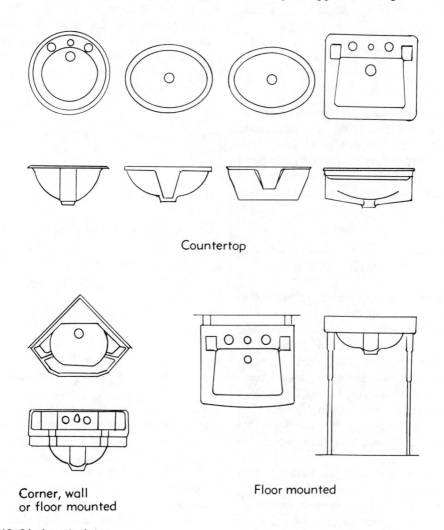

Countertop

Corner, wall
or floor mounted

Floor mounted

Figure 18-21. Lavatories.

Kitchen Sinks

Kitchen sinks are most commonly made of enameled cast iron or stainless steel. Sinks are usually available in a single- or double-bowl arrangement; some even have a third bowl, which is generally much smaller. Quite often a garbage disposal is connected to one of the sinks. Kitchen sinks are generally flush mounted into a plastic laminate (such as For-mica®) or into a composition plastic counter.

Service Sinks

Service sinks are made of enameled cast iron or vitreous china, and they are often called *slop sinks*. Most service sinks have high backs, and there may be two or three bowl compartments. Other sinks commonly used are laundry trays, pantry sinks, and bar sinks.

------------ 18.9 ------------
WATER REUSE SYSTEM

At this time, there are experimental houses being built throughout the world in order to study and analyze the new products and technology available and to examine the possibility of preserving our natural resources by using these new materials. One such project is the Tech House, the NASA Technology Utilization House in Hampton, Virginia. One portion of the design involves the possibility of processing household waste water for reuse.

In this experiment a typical family of four is used with typical household appliances. Water from the bathtub, shower, and washing machine is run into a collection tank in-stead of into the sewer lines. From the collection tank the water is filtered and chlorinated and then reused as water to flush the toilets. The flow of the water is shown in Figure 18–22. This water reuse system cuts water consumption by one-half. Of course, the potable water system is kept completely separate from the reuse portion of the system, and all waste from water closets goes directly into the sewer.

In another experiment all of the household water, except for that from the garbage disposal and water closets, is processed for multiple reuse in the system. This results in savings of up to 70% of overall household water consumption.

Of course, the real significance of this type of reuse system is the reduced amount of water required. The savings from such a system result in:

1. Smaller community or private sewer systems.
2. Smaller community sewage treatment plants.
3. Smaller community treatment plants required to treat supply water (when required).

------------ 18.10 ------------
WATER SUPPLY DESIGN

Plumbing systems are used to perform the two primary functions of water supply and waste disposal. The water supply portion of the system consists of the piping and fittings which supply hot and cold water from the building water supply to the fixtures, such as lavatories, bathtubs, water closets, dishwashers, clothes washers and sinks.

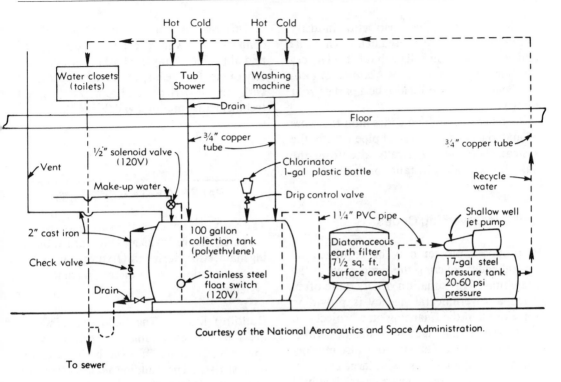

Courtesy of the National Aeronautics and Space Administration.

Figure 18-22. Water reuse system.

_____ 18.11 _____

CODES

Building codes are the regulations which govern the private actions of those who build or modify buildings. They are for the protection of public health, safety and welfare. The codes establish certain minimum requirements for the construction and subsequent occupancy of buildings.

Plumbing codes may be a part of the general building code or, more commonly, a separate code. The code in force in any locale is determined by the municipality involved (each individual area, or governmental unit, selects its own code). The most commonly used codes are the national, regional and state codes which may be used intact (complete) or with changes to meet the local needs and requirements. It should also be noted that the governmental unit also has a right to decide to have no code at all.

Since the code in a given locale is a law, it must be complied with in all of the buildings constructed under its jurisdiction. For this reason, it is important that all of the people involved in the planning, design and construction phases of the project become familiar with the code.

The *National Plumbing Code* has been used as a basis for this text since this is the code most often used, referred to or adapted from. But it is important for anyone using the text and learning about the construction industry to find out what code is used in a given locale and get a copy. You will probably find it very much in agreement with the *National Plumbing Code,* which should also be a part of your library.

In general, plumbing codes limit the types of materials and the sizes of pipe used in the system. The codes generally also form the basis for regulating installation methods.

Multiple Governing Codes

A proposed project may be regulated by several codes covering the same items at the same time. This situation occurs quite often when a governmental agency is providing some or all of the financing on a project and they have certain rules or guidelines which must be followed. This situation also may occur when doing business with large corporations. In such cases, the designers involved will have to become familiar with all of the applicable codes and regulations and, when they are in conflict, use the more stringent requirements.

Administration

Where codes are in force, there will probably be some form of Building Department in the government. The local administration and enforcement of the codes are performed by a building inspector or an engineer who usually reviews the proposed contract documents (drawings and specifications) for compliance with the codes and then checks for compliance during construction. It is important to know just what work must be inspected. Inspections should be scheduled so that work which must be inspected will not be covered before being inspected; otherwise, the inspector may require it to be uncovered so it can be checked.

Such responsibilities require qualified personnel—those who are experienced, informed and objective. Many levels of government offer courses for the inspectors to provide them with great experience and to make the latest technical information available to them.

_____ 18.12 _____
PARTS OF THE SYSTEM

The parts of a typical water supply system are shown in Figure 18–23 and include the building main, riser, horizontal fixture branch, fixture connection, and a meter (in community systems).

Building Main. The building main connects to the community or private water source and extends into the building to the highest riser. The building main is typically run (located) in a basement, crawl space, or below the concrete floor slab.

Riser. The riser extends vertically from the building main in the building to the furthest horizontal fixture branch. (The riser is typically run vertically in the walls).

Horizontal Fixture Branch. The horizontal fixture branch extends horizontally from the riser to the furthest fixture to be connected. It is usually run in the floor or in the wall behind the fixtures.

Fixture Connection. The fixture connection extends from the horizontal fixture branch to the fixture.

Meter. Most community water supply systems require a meter to measure and record the amount of water used. The meter may be placed in a meter box located in the ground, near the street, or inside the building.

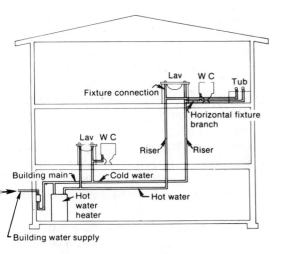

Figure 18-23. A typical water supply system.

18.13
PIPING MATERIALS

Only piping used for water supply is included in this section. Drainage, waste, and vent piping is covered in Section 18.22. Piping most commonly used for water supply includes copper, wrought-iron, steel, plastic, and occasionally brass.

Copper is one of the most popular water supply pipes. The pipe types available are K, L, and M, with K having the thickest walls, and L and M the thinnest. (DWV copper tubing is used for drainage, waste, and vent piping.) The thin walls of copper pipe are usually soldered to the fittings. This allows all of the pipes and fittings to be set into place before joints must be "finished" (in this case, by soldering). This advantage generally allows faster installation of copper pipe. Copper pipe also has the advantage (compared with iron or steel pipe) of not rusting and of being highly resistant to any accumulation of scale (particles) in the pipe.

Type K copper tube is available either rigid (hard temper) or flexible (soft temper). It is used primarily for underground water service in water supply systems. Soft temper tubing 1 inch and smaller is usually available in coils 60 or 100 feet long while 1 ¼- and 1 ½-inch tubing is available in 40- or 60-foot coils. Hard temper is available in 12- and 20-foot straight lengths. Type K copper tubing is color-coded in green for quick visual identification.

Type L copper tube is also available in either hard or soft temper in coils and straight lengths. The soft temper tubing is often used as replacement plumbing because the flexibility of the tube allows easier installation. Hard temper tubing is often used for new installations, particularly in commercial work. Type L copper tubing is color-coded blue. This type of tubing is most popular for use in water supply systems.

Type M copper tube is made in hard temper only and is available in straight lengths of 12 and 20 feet. It has the thinnest wall and is used for branch supplies where water pressure is not too great, but it is not used for risers and mains. It is also used for chilled water systems, exposed lines in hot water heating systems, and drainage piping. Type M copper tubing is color-coded red.

Copper tubing has a lower friction loss per 100 feet than wrought-iron or steel, providing the designer with an additional advantage. Also, the outside dimensions of the fittings are smaller, which makes a neater, better-looking job. With wrought-iron and steel pipe the bigger outside dimensions of the fittings sometimes require that wider walls be used in the building.

Red brass piping, consisting of 85 percent copper and 15 percent zinc, is also sometimes used as water supply piping. The pipe is threaded for fitting connections, but this requires thicker walls to accommodate the threading, making installation and handling more difficult than for copper. In addition, its relatively higher total cost, due to on-the-job installation, limits its usage.

Plastic pipe is also available for water supply systems. Its economy and ease of installation make it increasingly popular, especially on projects where "cost economy" is most important (low-cost housing, apartments). Available in 10-foot lengths, it is lighter than steel or copper and requires no special tools to install. While many plumbing subcontractors, engineers and architects still prefer copper, the use of plastic pipe will continue to increase. It is important to check the plumbing code in force in your locale since some areas still do not allow the use of plastic pipe for water supply systems. This dates back to early concern about possible toxicity (poisoning) resulting from the use of plastic pipe; this concern has long since been proved groundless. Plastic pipe used for water supply should carry the NSF (National Sanitation Foundation) seal. However, not all plastic pipe available should be used for water supply; much of it has been manufactured for use in the drainage portion of the plumbing system. The chart in Figure 18–24 shows plastic pipe materials and their usual use in the water supply system.

Wrought-iron pipe is available in diameters from ⅛ inch to 24 inches. Lightweight wrought-iron pipe, designated standard (or schedule 40), is the type most commonly used for water supply systems. The wrought-iron pipe used is most commonly galvanized to add extra corrosion resistance. Quite often it is used as the service main from the community main to the riser. Wrought-iron pipe is threaded for connection to the fittings, and it can be identified by a red spiral stripe on the pipe. The higher cost of wrought-iron pipe limits its use. Wrought-iron pipe also has a higher friction loss per 100 feet than copper.

Steel pipe is available in diameters from ⅛ inch to 12 inches. Plain steel pipe is usually used only when the water is not corrosive. Galvanized steel pipe is moderately corrosion resistant and suitable for mildly acid water. It is not used extensively in water supply systems; most plumbers and engineers prefer copper tubing because of its superior resistance to corrosion. Steel pipe is connected to its fittings with threaded connections. Steel pipe also has a higher friction loss per 100 feet than copper.

18.14
FITTINGS

A variety of fittings must be used to install the piping in a project and make all the pipe turns, branch lines, joinings on the straight runs, and stops at the end of the runs. Fittings for steel and wrought-iron pipe are threaded and made of malleable iron and cast iron. The fittings for plastic, copper and brass pipe are made of the same materials as the pipe being connected. Typically fittings for the various materials are shown in Figures 18–25 through 18–27.

The 45° and 90° elbows are used to change the direction of the pipe. Unions and couplings are used to join straight runs of pipe. A clamping piece on the coupling allows it to be more easily disengaged for uncoupling of the pipes when future piping revisions are ex-

Type	Cold water	Hot water
Polyethylene (PE)	●	
Polyvinyl Chloride (PVC)	●	
Acrilylonitrile Butadiene Styrene (ABS)	●	●
Polyvinyl Dichloride (PVDC)	●	
Chlorinated Polyvinyl Chloride (CPVC)	●	●

Figure 18-24. Plastic pipe use.

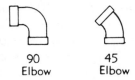

Figure 18-25. Copper tubing and fitting.

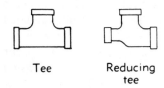

Tee   Reducing
              tee

Figure 18-26. Plastic fittings.

Coupling   Reducer   Union

Figure 18-27. Wrought-iron fittings.

building), branches (horizontal pipe serving the fixtures) and any pipes to individual fixtures or equipment. The proper location of valves simplifies repairs to the system, fixtures or equipment being serviced.

The *globe valve* (Figure 18-28) is a compression-type valve, commonly used where there is occasional or periodic use, such as lavatories (faucets) and hose connections (called hose bibbs). This type of valve usually closes the flow of water and is partially or fully opened only periodically to allow the water to flow. In reviewing Figures 18-28, the handle is turned and a washer on the bottom of the stem is forced against the metal seat which stops the flow of water. To allow the water to flow, the handle is turned and the washer separates from the seat; the more flow desired, the more the valve is opened. The design of the globe valve is such that the water passing through is forced to make two 90° turns which greatly increases the friction loss in this valve when compared with that in a gate valve.

The *angle valve* (Figure 18-29) is similar in operation to the globe valve, utilizing the same principle of compressing a washer against a metal seat to cut the flow of water. It

pected at a given point. Tees are used when branch lines must be made; the reducing tee allows different pipe sizes to be joined together. Adapters are used when threaded pipe is being connected to copper or plastic. Adapters have one end threaded to accommodate the steel pipe.

——————— 18.15 ———————
VALVES

Valves are used to control the flow of the water throughout the system. There are usually valves at risers (vertical pipe serving the

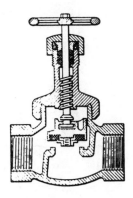

Figure 18-28. Globe valve.

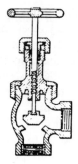

Figure 18-29. Angle valve.

is commonly used for outside hose bibbs. The angle valve has a much higher friction loss than the gate valve and about half the friction loss of the globe valve.

The *gate valve* (Figure 18–30) has a wedge-shaped leaf which, when closed, seals tightly against two metal seats which are set at slight angles. This type of valve is usually used where the flow of the water is left either completely opened or closed for most of the time. Because the flow of water passes straight through the valve, there is very little water pressure lost to friction. The gate valve is usually used to shut off the flow of water to fixtures and equipment when repairs or replacement must be made.

The *check valve* (Figure 18–31) has a hinged leaf which opens to allow the flow of water in the direction desired (indicated by an arrow in the illustration). But the leaf closes if there is any flow of water in the other direction. This eliminates any possible flow of water in a direction other than that desired, or required, by the designer. The check valve works automatically, so there is no need for a handle. This valve is used in such places as the water feed line to a boiler (heating unit), where the water from the boiler might pollute the system if it backed up. The inside of the valve is made accessible for repairs by removing the cover (see Figure 18–31).

Valves referred to as *standard weight* will withstand pressures up to 125 psi; high-pressure valves are also available. Most small valves used have bronze bodies, while larger valves (2 inches and larger) have iron bodies with noncorrosive moving parts and seats which may be replaced. They are available threaded or soldered to match the pipe or tubing used.

_____ 18.16 _____
WATER SHOCK

Water pressure surges from the quick closing of water valves (faucets) may cause the water system to be noisy. This abrupt closing of the

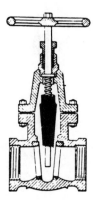

Figure 18-30. Gate valve.

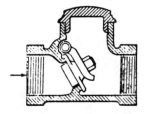

Figure 18-31. Check valve.

valves causes fast-flowing water to stop quickly and make the pipes rattle. A length of pipe, installed above the water connection, will act as a cushion or shock absorber as it controls the pressure surge of the water. Special shock absorbers are also available. Often the noise in the system is referred to as *water hammer*.

18.17
HOT WATER

By piping part of the water in the building main (Figure 18–32) into a heating device, the hot water required on the project can be supplied. The heating device may be a direct or an indirect heater and may operate on oil, gas, electricity, or the sun (solar heat). Direct heaters (Figure 18–33) are designed solely to provide the hot water required. Indirect heaters (Figure 18–34) use some type of boiler (heater) to heat the hot water and also to provide heat or steam to the heating system of the project.

Direct heaters come in a variety of sizes and capacities that allow them to be located in the basement, crawl space, or closet; in a cabinet under the counter; or as units which look similar to clothes washers. Many times for projects such as apartments, cold water is run to a direct heater (usually electric) in each apartment, instead of using one large hot water heating unit. This also allows for each apartment to be on a separate electric meter and for each resident to pay for the electricity he or she uses. Residences commonly use a direct heater, and in large homes two units are sometimes used (one near the kitchen-laundry and one near the bathrooms) to cut down the amount of hot water piping required and to provide the almost instant availability of hot water when a faucet is turned on. (The hot water in a pipe will cool off when not used for awhile.)

Indirect heaters use the same heating unit to provide hot water or steam to the heating system and to heat the hot water required for use at the fixtures. The same water used in the heating system is not used for the fixtures; instead, a separate compartment or coil containing the water is fed through the unit to be heated. Such units have been used in residences as well as commercial projects. This method is more commonly used in colder climates where the heating system is in operation for more months of the year. Often it is not an economical solution because during

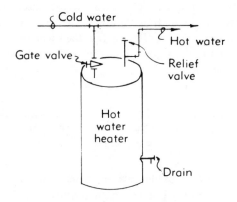

Figure 18-32. Hot water heater.

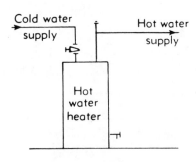

Figure 18-33. Direct heater.

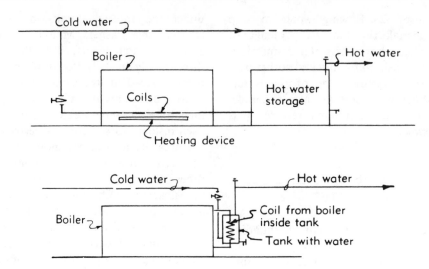

Figure 18-34. Indirect heater.

the warmer months, the heating system (boiler or furnace) will have to go on to provide hot water when no heating is required. Indirect heaters usually have tanks to store the heated water, or they may have a high-capacity coil capable of providing hot water very quickly.

Some projects use a combination of direct and indirect hot water heaters (Figure 18–35) whereby the cold water is piped through the indirect heater and then to the direct heater. When the indirect heater is being used for heating, it will provide fully or partially heated water to the direct heater. This means that at times the direct heater will have little or no additional heating to do.

The cold water supply should have a cutoff valve so that the water supply to the heater can be cut off if necessary. This allows for easier repair or replacement of the heater if required. The hot water pipe exits off the top of the tank and should have a relief valve to allow escape of any excess pressure built up in the system. Heaters which operate on oil or gas will require ventilation to the exterior, usually through a chimney, to get rid of poisonous gases (Figure 18–36). Electric and solar units do not require venting to the exterior.

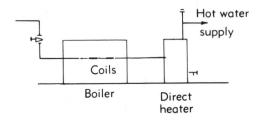

Figure 18-35. Combination heater.

18.18

HOT WATER DISTRIBUTION SYSTEMS

When designing hot water piping systems, the best solution is one which requires the hot water to travel as short a distance as possible

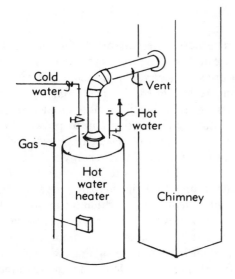

Figure 18-36. Water heater exhaust.

from the heater to the point at which it will be used. The longer the supply pipe, the less efficient the system because as the water stays in the pipe, it quickly loses its heat to the surrounding air, even if the pipe is insulated.

For example, review the situation shown in Figure 18–37, in which the bathroom faucet is about 75 feet from the heater. This means that if the water has had time to cool in the pipe (say, during the night), cool water will come from the faucet until the 75 feet of cool water has flowed through the faucet. Only at that time will hot water come out of the faucet. Then, once the faucet has been shut off, there is 75 feet of hot water in the pipe which begins to lose its heat. Back at the heater, as the hot water is drawn out, cold water begins to enter and be heated, and enough extra hot water must be heated to fill the 75-foot supply pipe again. This process is repeated over and over, several times a day, every day.

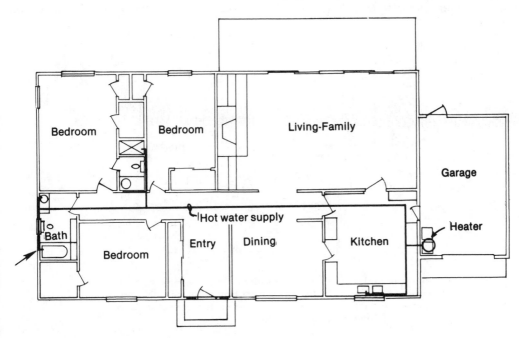

Figure 18-37. Hot water supply length.

In large residences, it is quite common to put in two (or perhaps more) heaters to reduce the distance from heater to source of use. Quite often, one heater will be located near the kitchen, with another in the bedroom area. These units are commonly located in a crawl space, basement, or kitchen or bathroom closet.

_____ 18.19 _____

SOLAR HOT WATER

Solar hot water heating systems are no longer a dream of the future. Development continues on systems which capture the heat from the sun to heat water for use in the home or project. The final chapter cannot yet be written on these systems as experimentation continues and as millions of dollars from government and private enterprise are poured into further research and development. Meanwhile, many dependable and cost-effective hot water heating systems are now available for purchase.

The most successful solar hot water heater, in the author's opinion, is the solar flat-plate absorber/collecter; the principles of operation are shown in Figure 18–38. In many areas, the solar heating unit is supplemented with an electric heating coil to provide hot water in case of prolonged cloudiness, which limits the amount of solar energy available to heat water. This is not to imply that the solar heater cannot be designed to provide all of the hot water required—just that it is more cost-effective to design the solar system to provide 85 percent to 90 percent of the hot water required during a year. Designing the system to definitely provide *all* of the hot water required means the solar system has to be about 25 percent to 33 percent larger.

The cost of the unit *must* be analyzed and the purchase of such a system viewed as an investment of money. Whether it is a good investment or not depends on whether it will

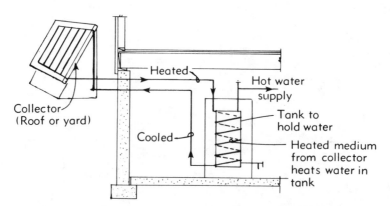

Tank—This tank could be used to just hold water heated by the collector medium or it could be a direct hot water heater used to supplement the collector during times when the collector cannot provide all of the hot water required.

Figure 18-38. Solar hot water system.

give a good return on the investment. Naturally, the economics of the system will vary according to the geographic location, the total amount of hot water required and the current cost of hot water. There is no doubt that an installed solar hot water heating system will cost significantly more than a fuel-burning hot water heating system—in general about 8 to 10 times as much. But a determination on cost is not based on "8 to 10 times as much." It is based on a careful examination of dollars and cents.

18.20
PLUMBING DRAINAGE

Following the flow of the water through the system, the next step will be to dispose of the waste matter, both fluid and organic, which is accumulated. The wastes come from almost all sections of the building—bathrooms, kitchens and laundry areas and, in commercial projects, even the equipment being serviced. Because all of the wastes tend to decompose quickly, one of the primary objectives of the plumbing system is to quickly dispose of the wastes before the decaying wastes cause objectionable odors or become hazardous to health.

Water is used to transport the wastes into the drainage piping and then to the point where they will enter a community sewer line leading either to a community sewage treatment plant or to a private sewage treatment system. The sewage from residences, apartments, motels, office buildings, and other similar types of buildings is referred to as *domestic sewage*. Special sewage from laboratories and many industrial plants requires special handling and treatment and is not discussed here. However, it is important to note that such wastes should not be put into a community sewage system without first getting approval from the community.

The minimum size of a fixture-supply pipe shall be as follows:

Type of Fixture or Device	Pipe Size (Inches)	Type of Fixture or Device	Pipe Size (Inches)
Bath Tubs	½	Sinks Flushing Rim	¾
Combination Sink and Tray	½	Urinal (Flush Tank)	½
Drinking Fountain	⅜	Urinal (Direct Flush Valve)	¾
Dishwasher (Domestic)	½	Water Closet (Tank Type)	⅜
Kitchen Sink, Residential	½	Water Closet (Flush Valve Type)	1
Kitchen Sink, Commercial	¾	Hose Bibbs	¾
Lavatory	⅜	Hose Bibbs— Toilet Rooms	½
Laundry Tray, 1, 2 or 3 Compartments	½	Wall Hydrant	¾
Shower (Single Head)	½	Washing Machines	½
Sinks (Serv., Slop)	½		

For fixtures not listed, the minimum supply branch may be made the same as for a comparable fixture.

Extracted from American Standard National Plumbing Code (ASA A 40.8 - 1955) with permission of the publisher, The American Society of Mechanical Engineers.

Figure 18-39. Minimum fixture supply.

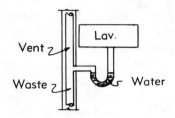

Figure 18-41. Traps.

18.21
PLUMBING CODES

The *National Plumbing Code* is used as the basis for the plumbing in this text. Increasingly, all plumbing codes are being modeled after this code, with only minor revisions to meet the requirements of specific geographic locations. References in this portion of the text relate to the *National Plumbing Code,* and everyone who is learning about plumbing, its design, and its requirements should have a current copy available. In addition, any relevant state or local plumbing codes should be made a part of your library for future reference.

18.22
THE DRAINAGE SYSTEM

In this section the terminology and function of each of the parts of the drainage system are explained. The basic parts of the system are illustrated in Figure 18-40.

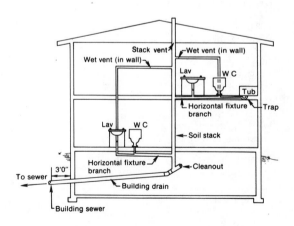

Figure 18-40. Parts of the drainage system.

Traps

A *trap* (Figure 18-41) is a device which catches and holds a quantity of water, thus forming a seal which prevents sewage decomposition gases from entering the building through the pipe. Traps are installed at each fixture as bent pipes unless the fixture is made with the trap as an integral part of it (as in the case of the water closet in Figure 18-42).

Traps may be made of copper, plastic, steel, wrought iron, or brass, with brass most commonly used. Traps in water closets are made of vitreous china and are cast right into the fixture.

The trap is located as close to the fixture as possible, usually within 2 feet of it. Occasionally, more than one fixture is tied to the trap. Quite often a laundry tray and a kitchen sink, a dishwasher and a kitchen sink, or two kitchen sinks may be connected to a single trap, provided all fixtures are close to one

Figure 18-42. Integral trap.

Figure 18-43. Trap cleanout.

another. There should never be more than three closely located fixtures (such as lavatories) on a single trap, or the trap may not operate properly (it may lose its water seal, also called trap seal). Since the trap may occasionally need to be cleaned, either there should be a plug in the bottom which may be removed (Figure 18-43), or the trap should have screwed connections on each end for easy removal (Figure 18-44).

In locations where the fixtures are infrequently used, care must be taken or the water in the traps may evaporate, and once the water seal is gone, gases may back up from the sewer and drainage pipes through the fixture and into the building. Floor drains (Figure

18-45), which are used to take away the water after washing floors or which may be used only in case of equipment malfunctions or repairs, present the most serious possibility of losing the water seal. When these floor drains are connected to the drainage system, the possibility of a serious gas problem exists. The designer of the system can avoid such a situation by *not* tying the floor drain into the drainage system. Instead, the floor drains should be tied into a dry well from which there will be no gases. Many building departments and plumbing codes prohibit the connection of floor drains to the sewage drainage system.

The water seal may also be broken if there is a great deal of air pressure turbulence in the pipes. To reduce the turbulence and equalize the pressure throughout the system, it is opened to the outside at the top, with air supplied throughout the system through *vent pipes*.

Vents

Vent pipes allow gases from the sewage drainage system to discharge to the outside and sufficient air to enter the system to reduce the air turbulence in the system. Also, without a vent, once the water discharges from a fixture, the moving waste tends to siphon the water from other fixture traps as it goes

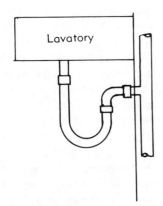

Figure 18-44. Trap connection.

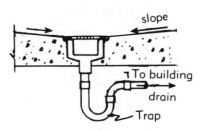

Figure 18-45. Floor drain.

through the pipes. This means that the vent piping must serve the various fixtures, or groups of fixtures, as well as the rest of the sewage drainage system. The vent from a fixture or group of fixtures ties in with the main vent stack (Figure 18–46) or the stack vent (Figure 18–47), which goes to the exterior. Vent piping may be copper, plastic, cast iron, or steel.

A *stack vent* is that portion of the vertical sewage drainage pipe (which may be a soil or waste stack), which extends above the highest horizontal drain that is connected to it. It extends through the roof to the exterior of the building (Figure 18–48).

A *vent stack* is used in multistory buildings where a pipe is required to provide the flow of

Multistory

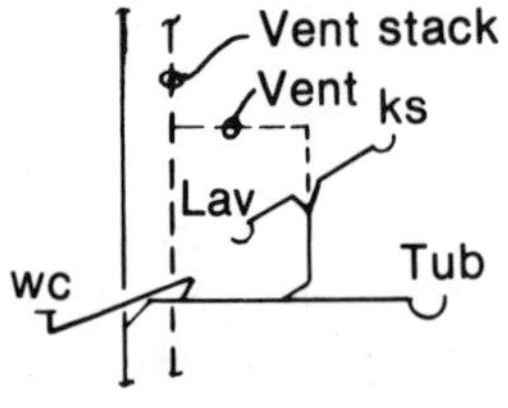

Figure 18–46. Vent to vent stack.

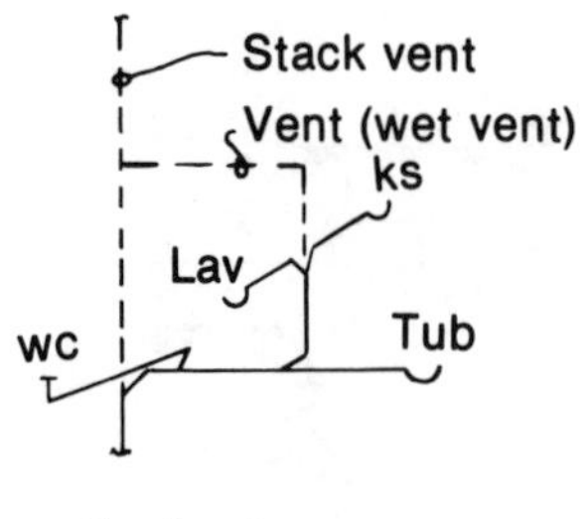

Single story

Figure 18–47. Multistory vent to vent stack.

air throughout the drainage system. The vent stack begins at the soil or waste pipe, just below the lowest horizontal connection, and may go through the roof or connect back into the soil or waste pipe not less than 6 inches above the top of the highest fixture.

Fixture Branches

The fixtures at a floor level are connected horizontally to the stack by a drain called a *fixture branch* (Figure 18–40). Beginning with the fixture farthest from the stack, the branch must slope ⅛ to ½ inch per foot for proper flow of wastes through the branch. Branch piping which serves urinals, water closets, showers, or tubs is usually run in the floor (Figure 18–49). When these fixtures are not on the branch, the piping may be run in the floor or in the wall behind the fixtures (Figure 18–50). Branch piping may be copper, plastic, galvanized steel, or cast iron.

Soil and Waste Stacks

The fixture branches feed into a vertical pipe referred to as a *stack*. When the waste that the stack will carry includes human waste from water closets (or from fixtures which have similar functions), the stack is referred to as a *soil stack*. When the stack will carry all wastes *except* human waste, it is referred to as a *waste stack*. Soil and waste stacks may be copper, plastic, galvanized steel, or cast iron. These stacks service the fixture branches beginning at the top branch and go vertically to the building drain (Figure 18–40).

In larger buildings, the point where the stack ties into the building drain rests on a masonry pier or steel post so that the downward pressure of the wastes will not cause the piping system to sag. In addition, the stack must be supported at 10-foot intervals to limit

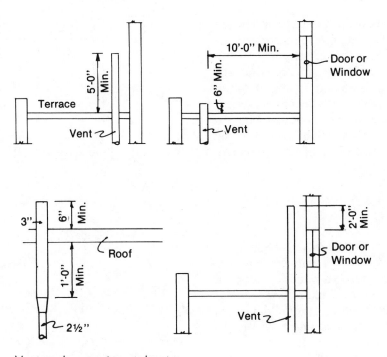

Most codes require at least a 3" vent through the roof beginning at least one foot below the roof.

Figure 18-48. Vent termination.

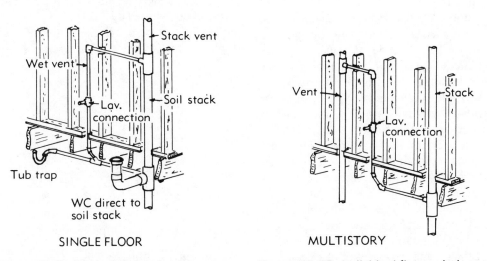

SINGLE FLOOR

Figure 18-49. Typical drainage piping.

MULTISTORY

Figure 18-50. Individual fixture drainage.

movement of the pipe. When a stack length is greater than 80 feet, horizontal offsets are used to reduce free fall velocity and air turbulence. Connections to fixture branches and the building drain should be angled 45° or more to allow the smooth flow of wastes.

Most designers try to lay out plumbing fixtures to line up vertically floor to floor so that a minimum number of stacks will be required.

Building Drains (Also Called House Drains)

The soil or waste stacks feed into a horizontal pipe referred to as the *building drain*. The building drain slopes ⅛ to ¼ inch per foot as it feeds the waste into the building sewer outside the building. By definition, the building drain extends to a point 3 feet *outside* the wall of the building (Figure 18–40).

Provision is made to allow cleaning of the building by putting a *cleanout* at the end of the drain (Figure 18–40). Another cleanout is sometimes placed just inside the building wall in case it is necessary to clean the building drain or sewer line. Cleanouts should also be placed no more than 50 feet apart in long building drains.

Location of the building drain in the building depends primarily on the location (elevation) below grade of the community sewer line. Ideally, all of the plumbing wastes of the building will flow into the sewer (whether it is a community or a private sewer system) by gravity. Typically, the drain is placed below the first floor or below the basement floor (Figure 18–51). If the height of the sewer requires the drain to be placed above the lowest fixtures, it will be necessary for the low fixtures to drain into a sump pit (Figure 18–52). When the level in the sump pit rises to a certain point, an automatic float or control will activate a pump, which raises the waste out of the pit and into the building drain.

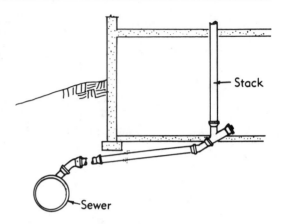

Figure 18–51. Underfloor building drain.

Building Traps (Also Called House Traps) and Fresh Air Inlets

Some codes may require a *building trap* on the building drain near the building wall (Figure 18–53). This trap acts as a seal to keep gases and vermin (rats and mice) from entering the sewage system from the sewer line. The *National Plumbing Code* and most regional and state codes do not feel a building trap is necessary; instead it is felt that this trap will impede the flow of wastes in the system. However, when required by local, state or regional codes, it must be put in the system. When a building trap is used, a *fresh air inlet* (Figure

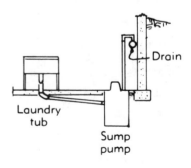

Figure 18–52. Sump pump.

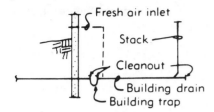

Figure 18-53. Building trap.

18-53) is required to allow fresh air into the system to be certain that the trap seal is not siphoned through. The fresh air inlet must be a minimum of 4 inches or one-half the diameter of the building drain, whichever is larger.

Building drains are usually made of plastic, copper (for above the floor) or extra-heavy cast-iron (for below the floor) pipe.

Building Sewers (Also Called House Sewers)

The *building sewer* (Figure 18–40) begins 3 feet out from the building wall and extends to the community sewer or the private sewage disposal tank. The building sewer slopes $\frac{1}{8}$ to $\frac{1}{4}$ inch per foot and should never be smaller in diameter than the building drain. Building sewers are usually made of cast iron or plastic (not less than 6 inches in diameter). Vitrified clay pipe is occasionally used, but there is the possibility that roots from shrubs or trees may penetrate through the mortar joints and obstruct the flow of the wastes.

18.23
PIPES AND FITTINGS

Drainage lines and vents make use of most of the same types of piping used in the water supply system, except that vitrified clay tile may be used in the building sewer line. Copper tub-

ing, the DWV, is very commonly used in drainage piping but is *not* used in water supply. The *DWV* on the tubing means it can be used for drainage, waste and venting on the job. Type M copper may be used above grade and type L copper below grade. Plastic piping is used extensively in drainage systems because of its low cost and speedy installation. Other piping sometimes used in the building sewer line includes concrete, asbestos—cement, and bituminous pipes. Fittings used in drainage systems are shown in Figure 18–54.

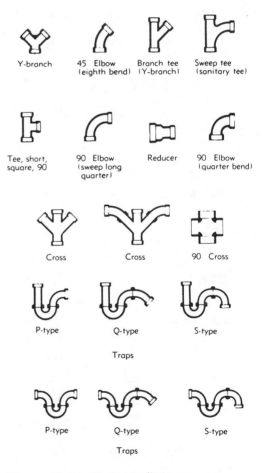

Figure 18-54. Drainage fittings.

_____ 18.24 _____
VENTING

The basics of venting have been outlined, and further explanation and several illustrations to show methods of venting are included in this section.

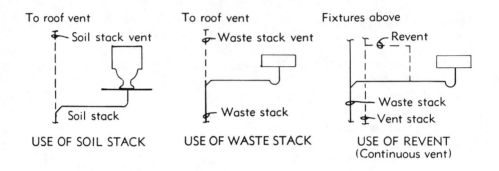

USE OF SOIL STACK   USE OF WASTE STACK   USE OF REVENT
(Continuous vent)

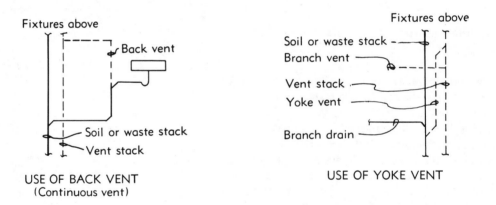

USE OF BACK VENT
(Continuous vent)

USE OF YOKE VENT

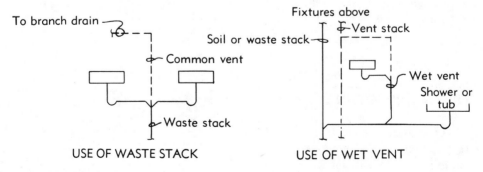

USE OF WASTE STACK   USE OF WET VENT

Figure 18-55. Vents.

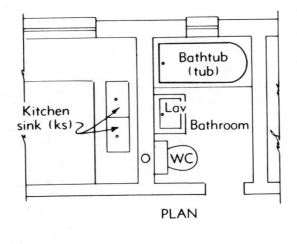

PLAN

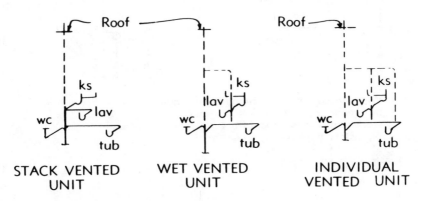

Figure 18-56. Vent for a one-family dwelling.

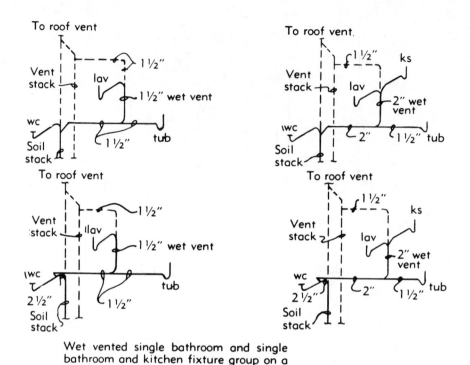

Wet vented single bathroom and single
bathroom and kitchen fixture group on a
stack or at the top floor of a stack serving
multistory bathroom groups.

Figure 18-57. Wet venting—top floor.

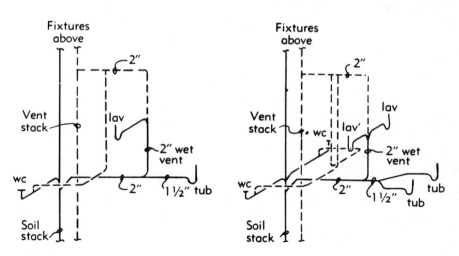

Figure 18-58. Wet venting—below top floor.

WET VENTING

Single Bathroom Groups—A single bathroom group of fixtures may be installed with the drain from a back-vented lavatory, kitchen sink, or combination fixture serving as a wet vent for a bathtub or shower stall and for the water closet, provided that: (1) not more than one fixture unit is drained into a 1½-inch diameter wet vent or not more than four fixture units drain into a 2-inch diameter wet vent, and (2) the horizontal branch connects to the stack at the same level as the water closet drain or below the water-closet drain when installed on the top floor. It may also connect to the water-closet bend.

Multistory Bathroom Group—On the lower floors of a multistory building, the drain from one or two back-vented lavatories may be used as a wet vent for one or two bathtubs or showers provided that: the wet vent and its extension to the vent stack is 2 inches in diameter;

each water closet below the top floor is individually back vented; and the vent stack is sized in accordance with the following table:

SIZE OF VENT STACKS

Number of wet-vented fixtures	Diameter of vent stacks in inches
1 or 2 bathtubs or showers....	2
3 to 5 bathtubs or showers....	2½
6 to 9 bathtubs or showers....	3
10 to 16 bathtubs or showers..	4

In multistory bathroom groups, wet vented in accordance with the paragraph above, water closets below the top floor group need not be individually vented if the 2-inch wet vent connects directly into the water-closet bend at a 45-degree angle to the horizontal portion of the bend and in the direction of flow.

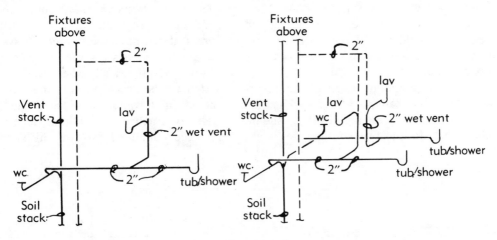

Figure 18-58. Wet venting—below top floor. (Continued)

_____ 18.25 _____
PLUMBING ECONOMY

Economies in plumbing are possible by locating fixtures in clusters, back-to-back, or otherwise grouped to form as few wetwalls (walls in which the plumbing pipes are located) as possible.

In multi-story construction, locating fixtures above each other saves considerable money since a minimum amount of piping and the smallest sizes possible may be used for both supply and disposal.

In residences designed with low cost as a primary objective, it is even possible to use the same wet-wall for a back-to-back bathroom and kitchen (Figure 18–59). For middle-priced and custom-designed residences, the primary concern is the location of fixtures where they will best suit the plan. Most designers find no problem in planning and designing a building so that a certain amount of economy is achieved at no sacrifice to the overall plan.

While the plumbing designer is not an architect (and, typically, he will be working on a layout designed by the architect), he should not hesitate to make suggestions that might save money for the client or provide for more effective use of the space.

The designer should also check the plans to be certain that the fixtures will fit properly in the space and have at least the minimum clearances required.

_____ 18.26 _____
PRIVATE SEWAGE DISPOSAL

Whenever possible, the sanitary drainage system should connect to a community (public) sewer, but when no community sewer is available, a private sewage disposal system must be installed. This is particularly true in suburban and rural areas and is one of the first things the designer should check. The information contained in this chapter is based on the *National Plumbing Code,* but the codes in force in the geographic area of construction should always be checked *before* the system is designed. In some areas, the municipality will not only specify exactly where on the site the system will be placed but will also simply state the size of all equipment required and the installation method. In addition, many areas limit the minimum lot size on which a private

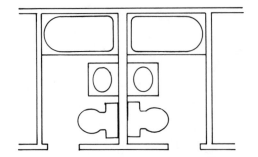

Figure 18-59. Plumbing back to back.

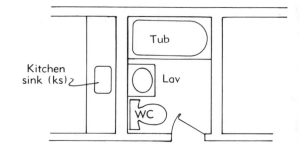

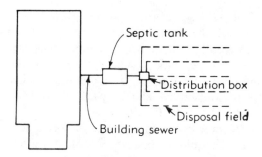

Figure 18-60. Private sewage system.

leach fields. There are variations on this design: as shown in Figure 18–61, the septic tank may feed into a seepage pit(s).

The watertight septic tank is placed underground where it receives the sewage from the building and holds it for about a day, while the suspended solids settle to the bottom and putrefy. The liquids pass out of the tank at the other end into the distribution box. Septic tanks may be concrete, of rectangular shape (Figure 18–62) or asphalt-protected steel, usually round (Figure 18–63).

sewage disposal system might be placed (often one-half acre) and the minimum size if both a sewage disposal system and a well are required (often about one acre). Most of these requirements are established because of the type of soil in the area and because there is always a concern that the potable water supply might be contaminated by a sewage disposal system (either the system serving this project or a neighbor's sewage system).

Private sewage disposal systems (Figure 18–60) usually consist of the building sewer which leads from the project into a septic tank and then the line into a distribution box which feeds the fluid (effluent) into the disposal or

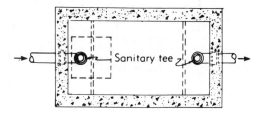

PLAN

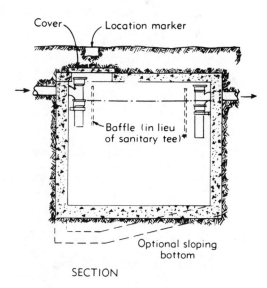

SECTION

Figure 18-62. Rectangular septic tank.

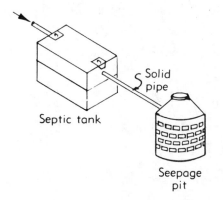

Figure 18-61. Septic tank to seepage pit.

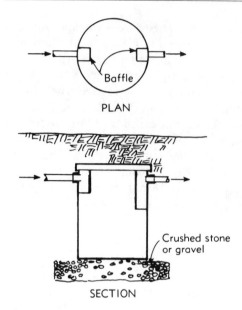

Baffle

PLAN

Crushed stone
or gravel

SECTION

Figure 18-63. Steel septic tank.

It is the size of the septic tank that will be one of the first considerations in private sewage disposal design. For individual residences, the tank design is based on the number of bedrooms, the maximum number of persons served, and on whether a garbage disposal is used (Figure 18-64).

A three-bedroom home housing five people will require a septic tank size of 600 gallons minimum (Figure 18-64). Since a garbage disposal is being installed, the code requires an additional 50 percent to be added to the tank size, so a 900-gallon tank is required. Also listed in Figure 18-64 are *recommended* inside dimensions of the tank. It should be noted, however, that the tank need only be *approximately* the dimensions shown. In many locales the tank sizes available may be limited, and it may be necessary to use a larger tank than the minimum. Using a larger tank will not interfere with the operation of the system.

MINIMUM CAPACITIES FOR SEPTIC TANKS SERVING
AN INDIVIDUAL DWELLING

Number of. Bedrooms	Maximum Number of Persons Served	Nominal Liquid Capacity of Tank	Recommended Inside Dimensions			
			Length	Width	Liquid Depth	Total Depth
	Persons	Gallons	Ft. In.	Ft. In.	Ft. In.	Ft. In.
2 or less	4	500	6 0	3 0	4 0	5 0
3	6	600	7 0	3 0	4 0	5 0
4	8	750	7 6	3 6	4 0	5 0
5	10	900	8 6	3 6	4 6	5 6
6	12	1,100	8 6	4 0	4 6	5 6
7	14	1,300	10 0	4 0	4 6	5 6
8	16	1,500	10 0	4 6	4 6	5 6

NOTE: Liquid capacity is based on number of bedrooms in dwelling. Total volume in cubic feet includes air space above liquid level.

Add 50% to requirements when a garbage disposal is used.

Extracted from American Standard National Plumbing Code (ASA A 40.8 - 1955) with permission of the publisher, The American Society of Mechanical Engineers.

Figure 18-64. Residential septic tank requirements.

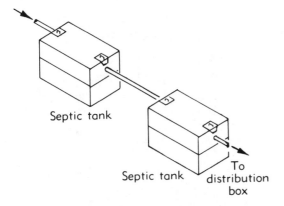

Figure 18-65. Septic tanks in series.

If a larger capacity is required than is available in one tank (for example, if 900 gallons is required and only 600-gallon tanks are available), two 600-gallon tanks may be connected (Figure 18–65). The tanks should be set close enough together that a single length of pipe (with no joints) can be used to join the outlet of the first tank to the inlet of the second.

Many municipalities require tanks larger than those recommended by the *National Plumbing Code*. It is the designer's responsibility to know or find out local requirements *before* designing the system. When the designer is working in an area where he is not familiar with local codes and requirements, he must visit the local authorities and discuss the design, obtain local requirements in writing, or call to determine what codes are in effect in the locale.

The code also sets minimum distances for the location of the various parts of the private sewage disposal system (Figure 18–66); the septic tank must be a minimum of 50 feet from any well or suction line (Figure 18–67), and when possible, the tank is put even farther away. Many local codes require longer distances and therefore must be checked; in general, about 100 feet is the preferred distance, but this is not always feasible. These

	Distance						
Type of System	Well or Suction Line	Water Supply Line (Pressure)	Stream	Dwelling	Property Line	Disposal Field	Seepage Pits
	Feet	Feet	Feet	Feet	Feet	Feet	Feet
Building sewer	50	10	....	....	....	....	....
Septic tank	50	....	....	....	....	....	....
Distribution box	50	....	....	....	....	....	....
Disposal field[1]	100	....	25	10	10	....	....
Seepage pit	100	....	50	20	10	20	20
Dry well	50	....	....	10	....	....	....
Cesspool[2]	150	....	50	20	15	15	15

[1] This separation may be reduced to 50 ft when the well is provided with an outside watertight casing to a depth of 50 ft or more.
[2] Not recommended as a substitute for a septic tank. To be used only when approved by the Administrative Authority.

Extracted from American Standard National Plumbing Code (ASA A 40.8 - 1955) with permission of the publisher, The American Society of Mechanical Engineers.

Figure 18-66. Sewage disposal system distances.

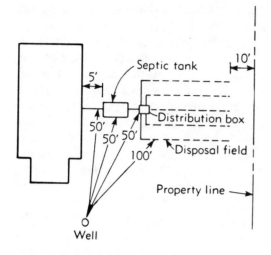

Figure 18-67. Private disposal system distances.

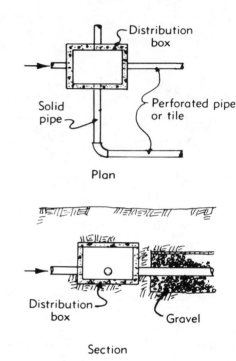

Plan

Section

Figure 18-68. Distribution box.

distances greatly reduce the danger of contaminating drinking water if leaks should occur in the tanks or pipes (lines).

The depth of the tank will be determined by the depth of the sewer line from the building to the tank. It is important that the sewer line be sloped gradually toward the tank (about ⅛ to ¼ inch per foot). Too much slope will cause the sewage to flow into the tank too rapidly and disturb the natural action in the tank. Too little slope and the sewer line may become clogged. The top of the tank is usually located 12 to 36 inches below the ground surface so it can be serviced as required.

The *distribution box* receives the effluent from the septic tank and distributes it equally to each individual line of the disposal field (as illustrated in Figure 18-68). Distribution boxes are not used when one seepage pit is used (Figure 18-69). The box is connected to the septic tank with a tight sewer line.

Disposal or tile fields are the preferred method for distributing the effluent. They consist of rows (called *lines*) of pipe through

which the effluent passes. The lines may be made of clay tile (usually 12 inches long), bituminized pipe, or plastic pipe. The clay tile is laid with about ¼ inch of space between them to allow the effluent to be absorbed into the underlying gravel fill (Figure 18-70). The top portion of the ¼-inch space (Figure 18-70) is covered with a piece of felt so that soil will not fall into the pipe and clog or stop the flow of effluent. Most fields installed today use the bituminized or plastic pipes which have holes in them. These pipes are installed much faster and do not have any open joints to be covered with felt as the tile does. These pipes are installed with the holes down, and as the effluent flows through the pipes, it is absorbed into the gravel.

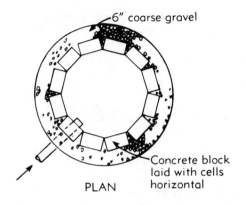

6" coarse gravel

Concrete block
laid with cells
horizontal

PLAN

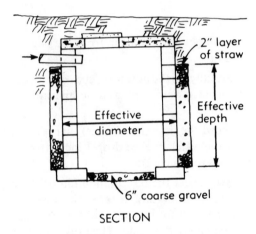

2" layer
of straw

Effective
depth

Effective
diameter

6" coarse gravel

SECTION

Figure 18-69. Typical seepage pit.

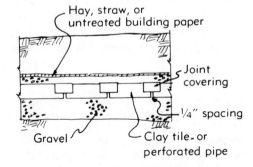

Hay, straw, or
untreated building paper

Joint
covering

¼" spacing

Gravel

Clay tile- or
perforated pipe

Figure 18-70. Clay tile disposal field.

The tile or pipe is set into a trench which varies from 18 to 30 inches in depth and 24 to 36 inches in width (Figure 18-71). Trenches must be sloped in the direction of flow; the code limits the maximum slope to 6 inches in 100 feet so that the effluent will not simply flow to the end of the line and then back up. The bottom of the trench is filled with a layer of filter material not less than 6 inches deep below the pipe line, extending the full width of the trench and a minimum of 2 inches above the pipe, and covered with a layer of straw. The code also limits any individual line to a length of 100 feet and sets the minimum separation between lines at 6 feet. The disposal field may take any number of shapes (Figure 18-72), depending on the contours (slope) of the ground, the size of the lot, and the location of any well or stream on the property. The minimum distance between the field and the building is 10 feet, between field and property line 10 feet, between field and stream 25 feet, and between field and well 100 feet. If the well has an outside watertight casing which extends down to a depth of 50 feet or more, the minimum separation between field and well may be reduced to 50 feet.

The length of line required in the disposal field depends on the ability of the soil to absorb sewage. The subsurface conditions are checked first by digging a hole about 5 feet below the final grade to observe the type of soil encountered and whether any ground-

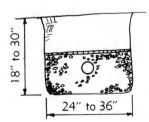

18" to 30"

24" to 36"

Figure 18-71. Disposal trench.

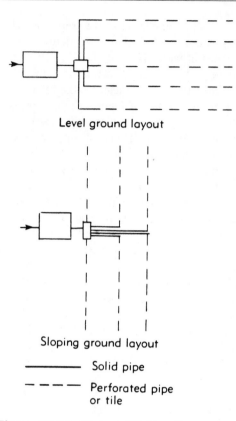

Level ground layout

Sloping ground layout

—————— Solid pipe

— — — — Perforated pipe
            or tile

Figure 18-72. Disposal field patterns.

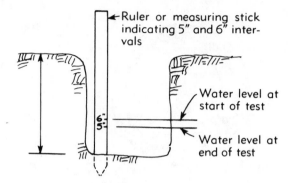

Figure 18-73. Percolation pit.

water is evident. Next, a soil percolation test is made to measure the ability of the soil to absorb sewage. The procedure for making the soil percolation test is:

1. Dig a hole about 12 inches in diameter and 30 inches deep, keeping the sides of the hole vertical (Figure 18-73).

2. Presoak the hole by filling it with water and allowing it to completely seep away. The hole should be presoaked several hours before the test and again at the time of the test.

3. After presoaking, remove any loose soil

that might have fallen in from the sides of the hole.

4. Carefully fill the hole to a depth of 6 inches with clean water with as little splashing as possible.

5. Record the time (in minutes) that it takes for the water level to drop 1 inch (from 6 inches to 5 inches).

6. Now, repeat the test a minimum of three times until the time it takes the water to drop 1 inch for two successive tests is approximately the same. Then take the last test as the stabilized rate of percolation; it is the time recorded for this test which will be used to size the disposal field.

Individual residences are sized from Figure 18-74, which lists the time required for the water to drop 1 inch and the square feet of trench required *per bedroom* for the various times.

Example: Assuming a soil percolation time of 4 minutes, from Figure 18-74, the square feet required per bedroom is 70. For a three-bedroom residence, a total of 70 square feet (per bedroom) x 3 bedrooms = 210 square feet is required. Using a trench width of 3 feet, find the lineal feet of line required by dividing the square feet required by the width of the trench; in this design, 210 square feet ÷ 3 = 70 lineal feet, the minimum length of line.

Seepage pits (also referred to as *leaching pits* and illustrated in Figure 18–69) may be used instead of tile fields. They are preferable where the soil becomes more porous below a depth of 2 or 3 feet and where the property does not have sufficient space for all buildings, driveways, parking areas, and the drain field. They cannot be used in areas with high water tables since the bottom of the pit must be at least 2 feet above the water table.

The seepage pit is usually made of concrete block or is a precast concrete unit (Figure 18–75). Typically, 8-inch thick block are used, and they are laid with the cells (holes) placed horizontally to allow the effluent to seep into

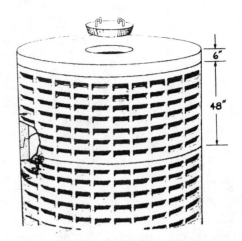

Figure 18–75. Precast seepage pit.

the ground (Figure 18–69). The tapered cells of the block are set with the widest area to the outside to reduce the amount of loose material behind the lining that might fall into the pit. The typical precast concrete seepage pit shown in Figure 18–75 is the same unit used for drywells. The bottom of the pit is lined with coarse gravel a minimum of 1 foot deep before the block or concrete is placed. Between the block or concrete and the soil is a minimum of 6 inches of clean crushed stone or gravel. Straw is placed on top of the gravel to keep sand from filtering down and reducing the effectiveness of the gravel. The top of the pit should have an opening with a watertight cover to provide access to the pit if necessary. The construction of the pit above the inlet pipe should be watertight.

When more than one seepage pit is used, the pipe from the settling tank must be laid out so that the effluent will be spread uniformly to the pits. To provide equal distribution, a distribution box with separate laterals (Figure 18–76)—each lateral feeding no more than two pits—provides the best results. The dis-

ABSORPTION AREAS FOR
INDIVIDUAL RESIDENCES

Time Required for Water to Fall 1 Inch (Minutes)	Effective Absorption Area Required in Bottom of Disposal Trenches (Square Feet per Bedroom)
2 or less	50
3	60
4	70
5	80
10	100
15	130
30	180
60	240
Over 60	(¹)

¹Special design
NOTE: A minimum of 150 sq. ft. should be provided for each dwelling unit.
Extracted from American Standard National Plumbing Code (ASA A 40.8 - 1955) with permission of the publisher, The American Society of Mechanical Engineers.

Figure 18–74. Residential absorption rate.

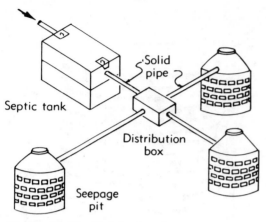

Figure 18-76. Distribution box to seepage pits.

tance between the outside walls of the pits should be a minimum of 3 pit diameters and not less than 10 feet.

The size of the seepage pit is based on the outside area of the walls plus the area of the bottom of the pit. The areas for pits of various diameters are given in Figure 18–77. Many designers exclude the bottom area of the pit from the absorption area required to allow for a safety factor, while others calculate the total area available and then size the system to allow some safety factor. The latter approach is used in this text.

ABSORPTION AREAS						
Diam.	Depth					
	4'	5'	6'	7'	8'	10'
4'	62.8	75.3	87.9	100.5	113	138.1
5'	82.4	97.8	113.5	129.2	144.9	176.3
6'	103.7	122.5	141.4	160.2	179	197.9
8'	150.7	175.8	200.9	226	251.3	276.3

Figure 18-77. Seepage pit absorption areas.

18.27
SEWAGE PROBLEMS

The uncertainty of exactly how much the soil will actually absorb is reflected in the values given in the various tables and charts. Even so, many designers prefer to oversize the system slightly to allow for poor absorption and also to allow for future increased amounts of effluent either because more people are using the facility than anticipated or because of an addition to the individual residence or an addition of various water-using fixtures which may not have been included in the original design.

One of the most wasteful uses of the private system is to connect a washing machine to it. Many times the tying-in of a washing machine to an older system has resulted in more water flow than the ground could handle through the system installed. It is suggested that when a washing machine is installed (especially where there are several children in the family), a drywell should be installed. Problems may also occur if all of the family washing is done on one day. This puts a tremendous additional flow into the system for a brief period, and as a result, the system may back up into the house. One of the simplest solutions to this is to spread the washing out over several days, giving the ground a chance to absorb the water. Other solutions are to connect the washer to a drywell, to increase the size of the disposal field, or to add a seepage pit.

The connection of gutters, storm drains, and roof drains may also cause periodic overloads on the private sewage system. When these are connected, the designer must increase the size of the system to accommodate the periodic additional flow. Most designers prefer to run such connections into drywells if no storm drain system is available.

Chapter 19

Heating and Air-Conditioning Systems

19.1
TYPES OF SYSTEMS

The selection of a heating system and the possible inclusion of air conditioning will depend on local climate conditions, degree of comfort desired, client's budget, and fuel costs. An increasing number of clients want a system which incorporates total year-round air conditioning. There are three basic methods of delivering heat to a space:

1. Forced air.
2. Hot water.
3. Radiant electric.

Of these three, only one, forced air, is used for central air conditioning in residences.

In *forced air systems,* the air is heated or cooled in a central unit and then delivered to the room through supply ducts. Air is returned to the central unit for treatment (to heat, cool, add humidity, or purify) and then recirculated through the rooms.

Hot water heating systems heat the water in a central unit and pass it through pipes to a heating device in the room. As the water goes through the pipes and devices, it cools, and then it goes back through the central unit to be reheated. Cooling with water follows the same principle.

Radiant electric heat delivers the heat by electricity running through a cable, and the resistance produced gives off the heat. The radiant heating devices are actually in the room and may be ceiling, floor or baseboard units.

In addition, *infrared heaters* use a lamp (bulb) which transfers heat by radiation to any people or objects which its heat rays come in contact with. It is effectively used when it is necessary to warm people, yet the surrounding air need not be heated or, perhaps, cannot be effectively heated. They are commonly used in bathrooms where for a short period of time (such as when a person steps out of a hot shower in the winter) extra heat is needed for comfort.

19.2
HEATING SYSTEM COMBINATIONS

It is not unusual to use different types of systems to heat and cool a building.

Residential designs may use combinations of systems such as hot water for heat and forced air for cooling, hot water or forced air heat with electric supplements (particularly in bathrooms or kitchens) and finned tube units with supplemental hot water unit heaters in areas such as a kitchen or basement.

19.3
FUELS

The most commonly used sources of building heat are the sun, electricity, gas, oil, and coal. Use of the sun as an energy source is discussed in Section 19.17. The decision on which of the other three fuels to use is based on availability and cost of operation. In this section, we will review the advantages and disadvantages of each fuel and the method used to determine the cost of operation.

Electricity

Electricity is used as a fuel for a variety of heating systems, including baseboard radiant heat and electric coils in the ceiling and/or walls. In addition, there are electric furnaces for forced hot air systems, and it is used with heat pumps both to operate the system and to provide supplemental electric resistance heat for the system. Electricity has simplicity as its advantage. It requires no chimney to remove toxic gases, and when baseboard strips and ceiling and wall coils are used, the system has individual room controls, providing a high degree of flexibility and comfort. Electric systems cost significantly less to install than other systems (including electric furnaces for forced hot air), and all electric systems (except heat pumps) require much less upkeep and maintenance than those using the other fuels.

The primary disadvantage of electric heat is its yearly cost of operation when compared with those of the other fuels. Electricity rates vary tremendously throughout the country,

and the rates are continuing to climb. It is necessary to determine the rates in the geographical area of construction to do a cost analysis. Electric heat is quite popular, and generally most economical, in the southern regions since winter is shorter and heat bills are generally lower. It is also used extensively in apartments, offices and similar buildings where the developer is primarily interested in building the units as inexpensively as possible and where the cost of heating is usually paid by the person renting the space. Its minimal need for maintenance and repairs is also a factor in such construction. Builders of some homes, especially those which they may want to be able to sell at the lowest price, may use electric baseboard heat.

Gas

Gas, also a popular fuel for heating, is used to heat the water in hot water systems and the air in forced air systems. Gas fuels available include natural gas, which is piped to the residence or building, and propane gas, which is delivered in pressurized cylinders in trucks and tanks and stored in tanks at or near the building. Since natural gas is simply available as needed, with no storage or individual delivery required, it is considered simpler to use. However, it is not available in many areas, and the more suburban the area, the less likely it is that natural gas will be available. So the designer must first determine if natural gas is available. Secondly, at the present time, there is a shortage of natural gas. While this shortage may be alleviated in the future, at the present many areas will not permit any new natural gas customers. In periods of shortages, the residential customer can be reasonably assured that he will have sufficient natural gas for his use, but industrial and commercial customers cannot be so assured. While the reasons for such shortages may be debated as to whether they are real or contrived—caused by government, industry or both—the designer is concerned with one thing: Is it available, and will it continue to be available?

The primary advantage of natural gas has been its relatively low cost and its simple and clean burning, which reduces the maintenance required on the heating unit. As with all of the fuels, as the costs go up in the future, it is difficult to say which will be the most economical.

Propane gas is equally as clean as natural gas, and its availability is not limited to areas where supply pipes have been installed. To date the cost of propane gas has generally been higher than that of natural gas.

Oil

Oil is one of the most popular fuels, and it is used extensively in the Northeast. The primary reason for its use has been its availability and historic low cost. It does require delivery by trucks to storage tanks located in or near the building, and the heating unit will generally require more maintenance than a gas heating unit. The selection of oil as a heating fuel has diminished some since the oil embargo in the early 1970s. In addition, the cost of this fuel has risen dramatically since that time. Costs of other fuels have also risen, but the fear of not having oil if another embargo is imposed is one of the most important concerns.

Oil is available in various weights (Figure 19-1) with various heating values and at different costs. Generally, the lower the number, the more refined it is and the higher the cost. Number 2 oil is commonly used in residences,

Commercial standard number	Weight (lb./gal.)	Btu per gallon	Btu average
1	6.675-7.076	132,900-138,800	136,000
2	6.870-7.481	135,800-144,300	140,000
4	7.529-8.212	145,000-153,000	149,000
5	7.627-8.328	146,200-154,600	150,000
6	7.909-8.448	149,700-156,000	154,000

Figure 19-1. Fuel characteristics.

while numbers 4, 5, and 6 are commonly used in commercial and industrial projects.

Coal

Coal is rarely used for residential heating in new construction, and its use in industrial and commercial construction fluctuates.

Its primary advantage is that it is available and there are ample supplies, so that a shortage seems unlikely at this time. Generally, its cost is competitive with those of other fuels. Its disadvantages lie in the amount of space required for storage and the fact that it does not burn as completely as oil or gas, thus producing more pollution.

Coal's decreased use for heating residences is due to the handling required in delivery and the inconvenience of having a coal bin as part of a basement. Also, originally the coal had to be shoveled into the furnace by the occupant of the house which is sufficient reason to change oil, gas or electricity. Now, coal heating units are fed by automatic stokers which require no hand shoveling.

Since coal is used so little as a residential heating fuel, it will not be considered further.

19.4
COOLING PRINCIPLES

While the principle of providing heat to a source (water or air) which is then used to heat the space is easily understood, the principles of providing cooling should be discussed. The fundamental principles on which the cooling process is based are derived from physics:

1. As a gas is compressed, it will liquefy at a given point, and as its liquefies, it will *release* a large amount of latent *heat* from within the gas/liquid.
2. As the pressure on the liquid is lowered, it vaporizes back to a gas, and as it boils through the vaporizing process, it *absorbs* a large amount of latent heat into the liquid/gas.

The refrigerant medium in the cooling system is cycled through three components (Figure 19–2):

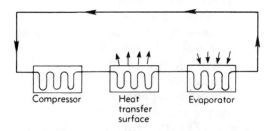

Figure 19-2. Refrigerant cycle.

1. A *compressor* which will compress the refrigerant, causing it to liquefy.
2. A heat transfer surface which will distribute the *heat released* to a surrounding medium such as water or air. This heat transfer surface is called a *condenser*.

3. A second heat transfer surface which will *extract heat* from the surrounding medium, such as water or air, as it is *absorbed* into the refrigerant. This heat transfer surface is called an *evaporator*.

The refrigerant is run continuously through the cycle while the system is in operation:

1. The refrigerant is compressed to a liquid in the compressor, generally located in or near the condenser.
2. The liquid passes through the condenser which allows the latent heat to be released. The condenser is often located on the exterior of the building. For most residences and small commercial projects, it is located on the ground; however, it may be located on the roof, often the case on larger projects. The heat is released through the condenser to a surrounding medium. This surrounding medium is the outside air for most installations, but water, such as a pond, can be used. In the typical installation, the condenser unit has a fan which pushes the air past the refrigerant to take as much heat away from it as possible.
3. The refrigerant then passes out of the condenser to the second heat transfer surface (the evaporator) which will extract heat from the surrounding medium. So, as the liquid vaporizes to a gas, it draws heat out of the surrounding medium as it passes through the evaporator. The surrounding medium may be air or water. In a forced air system, air would be the medium, and this drawing of heat from the air and into the refrigerant causes the air to cool. As the air is forced back through the system, it is cool air. When water is used as the medium, the heat is drawn from the water, making it

cool or chilled; the water is then circulated through the system to cool the space.

These basic principles apply for all types of cooling systems. Some equipment, such as a room air conditioner, may be designed to combine all the components in one unit (Figure 19-3), or they may be separate pieces of equipment. These principles are also the basic principles of heat pump operation (Section 19.19).

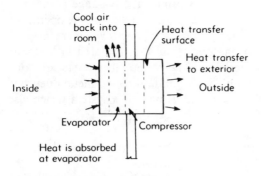

Figure 19-3. Single unit cooling.

19.5

FORCED AIR SYSTEMS

A motor-driven fan is used to circulate filtered, heated, cooled, or air-conditioned air from a central heating unit through supply ducts to each of the rooms. As the air is delivered through the ducts and into the room through the supply outlet, the air from the space (room) is being returned through return grilles, into ducts and back through the central unit to be heated or cooled and sent back to the space. The ducts may be circular or rectangular, and their size depends on the amount of treated air which must flow through them to maintain the desired temperature of the

room. A variety of duct systems, or basic designs, may be used; several of the most common are shown (Figure 19-4).

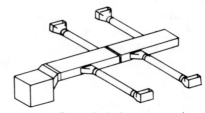

Extended plenum supply

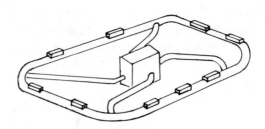

Perimeter-loop system

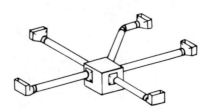

Individual supply system

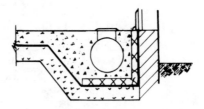

Figure 19-4. Typical heating duct systems.

Forced air heating systems are economical and generally easy to install. Filters are put in the system to reduce the amount of dust in the air. The unit may be located wherever convenient in the building, including the basement, crawl space, attic, utility room, or garage.

Humidifiers are recommended for use in most heating systems. They provide extra comfort at a minimal cost, making them a good investment. The humidifier will fit right into the duct system as it is being installed.

19.6
DUCTS AND FITTINGS

The ductwork is used to take the forced treated air from the furnace to the supply outlet and return registers take air back to the unit.

The most commonly used materials for ducts are galvanized iron and aluminum. Both materials are relatively lightweight and easily shaped to whatever size duct is required, either round or rectangular. Minimum metal thicknesses required of these materials vary according to the size of the duct required and are shown in Figure 19-5. These ducts may have to be insulated as discussed in Section 19.7.

Another very popular ductwork material is glass fiber, molded duct board. This is available in a variety of sizes, both round and rectangular. Its principal advantage is that it is less expensive and installed in one operation, as opposed to metal ducts which are installed in one operation and insulated in a second operation. The round fiber ducts are compatible with standard round metal ducts and may be used as part of a system which also

Round Ducts Diameter, In.	Minimum Thickness		Minimum Weight of Tinplate
	Galv. Iron, U.S. Gage	Aluminum, B&S Gage	
Less than 14	30	26	
14 or more	28	24	IX (135 lb)
Rectangular Ducts Width, In.	Minimum Thickness		Minimum Weight of Tinplate
	Galv. Iron, U.S. Gage	Aluminum, B&S Gage	
Ducts Enclosed in Partions			
14 or less	30	26	
Over 14	28	24	IX (135 lb)
Ducts Not Enclosed in Partions			
Less than 14	28	24	—
14 or more	26	23	—

Note: The table is in accordance with Standard 90B of the National Board of Fire Underwriters.' Industry practice is to use heavier gage metals where maximum duct widths exceed 24 in. (see also NBFU No. 90A).'

Reprinted with permission from ASHRAE, Systems Handbook, 1976

Figure 19-5. Metal duct thicknesses.

Figure 19-6. Fiber duct connections.

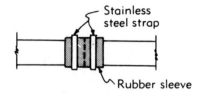

Figure 19-8. Asbestos-cement connection.

uses metal ducts. The round fiber duct makes use of metal fittings to connect, reduce and make elbows, as shown in Figure 19-6. Another advantage of this type of duct is excellent acoustic properties to ensure a quiet system. Fiber duct is also sometimes used for the section of ductwork connecting the main trunk supply and the return lines to the furnace (Figure 19-7).

Where the ducts will be in and under a concrete slab, an asbestos-cement round duct is most commonly used. Ducts placed below a slab must be made from a material which is not subject to moisture transmission or corrosion by concrete, will not float as the concrete is poured and is noncombustible. The asbestos-cement duct is available with inside diameters from 4 to 36 inches. The ducts are joined with an impermeable rubber sleeve and two stainless steel straps (Figure 19-8). The tees, wyes, elbows, reducers, and end caps are also made of asbestos-cement. This type of duct material is much more expensive than the fiber glass or metal and is rarely used except, for example, for installations in or under a concrete slab.

Often the ducts are made an integral part of the construction of the building, especially the returns. The spaces between joists may also be used for air returns, with the bottom usually formed from sheet metal nailed to the bottom of the joists (Figure 19-9). If the joist runs through cold spaces, it may be desirable to insulate the underside of the return. Also, the designer must check to be certain that the joists run in the direction which the return air must run.

A wide variety of fittings may be used to make all of the reductions, branch take-offs, turns and bends required in many duct systems. Ideally, the best system layout has the fewest and simplest duct fittings since fittings restrict the flow of the forced air and increase the friction in the system.

Supply ducts should be equipped with an adjustable locking-type damper (Figure

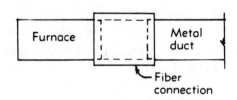

Figure 19-7. Fiber duct as connection.

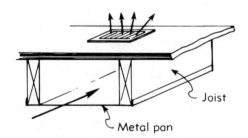

Figure 19-9. Joists as ducts.

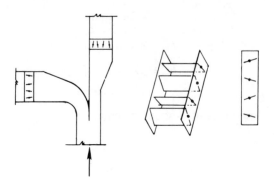

Figure 19-10. Adjustable dampers.

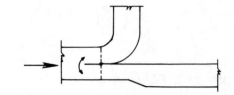

Figure 19-11. Splitter damper.

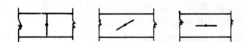

Figure 19-12. Squeeze dampers.

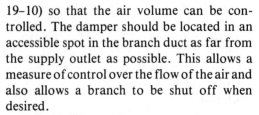

Figure 19-13. Turning vanes.

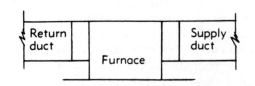

Figure 19-14. Boiler-duct connection.

19-10) so that the air volume can be controlled. The damper should be located in an accessible spot in the branch duct as far from the supply outlet as possible. This allows a measure of control over the flow of the air and also allows a branch to be shut off when desired.

Splitter dampers (Figure 19-11) are used to direct part of the air into the branch where it is taken off the trunk. They do not give precise volume control.

Squeeze dampers (Figure 19-12) are placed in a duct to provide a means to control the air volume in the duct.

Turning vanes may be used to direct the flow of air smoothly around a corner or a bend, as shown in Figure 19-13.

The entire system should be checked to be certain that proper attention has been given to the elimination of as much noise as possible from the system. The following suggestions will help keep noise to a minimum.

1. The furnace and metal ducts should be connected with a flexible fire-resistant fabric. In this manner, any noises or vibrations will not be transmitted directly through the system (Figure 19-14).

2. All electrical conduits and pipes should have flexible connections to the furnace.

3. Do not locate the return air immediately adjacent to the furnace.

4. Do not install a fan directly below the return air grille.

19.7
DUCT INSULATION

Ducts which are located in heated spaces do not need to be insulated. But ducts which run through enclosed, unheated spaces or in spaces which are exposed to outdoor temperatures should be insulated. While glass fiber ducts are made of an insulating material (the glass fiber), sheet metal ducts must be wrapped in an insulation. Thicknesses of insulation recommended are:

Supply Ducts

1. When located in enclosed, unheated spaces—1 inch of insulation.
2. When located in a space in which it is exposed to outdoor temperatures—2 inches of insulation.

Return Ducts

When not located in a heated space, use 1 inch of insulation.

The most commonly used insulation on residential and small commercial buildings is fiberglass, with a facing of reinforced aluminum foil vapor-barrier which goes to the outside.

19.8
SUPPLY AND RETURN LOCATIONS

This information given on the location of supply and return ducts is for installations where only heating will be supplied or in areas where heat is required much of the time while any cooling requirements are small. (For example, in an upstate New York residence, heat will probably be required regularly for about six to seven months, October 15–May 15, and air conditioning intermittently for two months.)

The supply registers should be located in the floor, 4 inches out from the baseboard, or be very low in an exterior wall (Figure 19–15) and near or under windows, in exterior walls and near exterior doors. In effect, put the heat supplying registers as close as possible to the

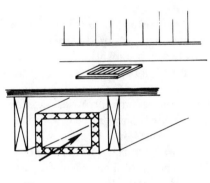

Floor register

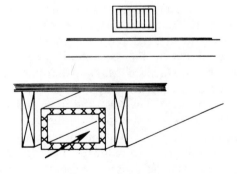

Low wall register

Figure 19–15. Register location.

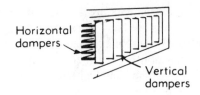

Figure 19-16. Register dampers.

spots where the most heat is lost. These registers should have both vertical and horizontal dampers (Figure 19–16) so that the air will be directed downward at the floor by the horizontal dampers and diffused to the sides by the vertical dampers.

Returns are often located on interior walls, in hallways and in exposed corners.

A low baseboard location is required for a return on the floor, and a centrally located return, perhaps one for a small residence and two for a large one, provides satisfactory results. If there is only one central return, it is important to put it in a location where it will be able to draw return air from as much of the building as possible. This is why they are frequently located in hallways. More expensive, individual room exhausts may be used if the designer feels that the layout of rooms may cause an uneven return of air or that the air flow from the rooms to the return (under doorways or through adjoining rooms) may cause a problem. For example, if there is a central return in the hallway and the door to a bedroom is closed all night, it will be very difficult for air to flow from the room to the return, unless the bottom of the door is trimmed (undercut) up about an inch or so. Unless air is drawn from the room as the supply keeps bringing air into the room, a slight pressure is built up in the room, reducing the amount of warm air coming through the supply outlets, and the room will tend to be cool. The same situation will occur in any other

room closed off from the return. If the designer is aware that such a problem may exist, the solution would be to put a separate return in the room.

The returns are covered with grilles which are put over the opening primarily to "cover it up" so there won't be a big hole in the wall. There are no movable dampers. The type of grille used will determine the required grille size (in conjunction with the cfm required for the system). This is because air can only flow through the openings in the grille (called the *face openings*) and not through the material. Therefore, the percentage of face opening for the grille used must be determined from the manufacturer's specifications. The register must be proportionally larger than the return duct, based on the percentage of face opening of the grille required. For example, if a grille with 50 percent face opening is used on a 20-inch × 14-inch duct carrying 1,400 cfm, it will be necessary to have a grille twice as large as the duct branch (or trunk).

19.9
FURNACE LOCATION

Furnaces for forced air systems are available in various designs so that they may be located in the basement, crawl space, attic, or first floor of the building. In addition, each is usually classified in terms of the direction in which the air is delivered. The three basic types are *upflow, counter flow* and *horizontal,* and there are several variations depending on the actual installation (Figure 19–17).

As shown in Figure 19–17, a lowboy-style (upflow) furnace is installed in the basement of a building, and the air is delivered to the building through ducts from the top of the boiler.

The highboy (upflow) furnace is used primarily in single-level homes when the ducts

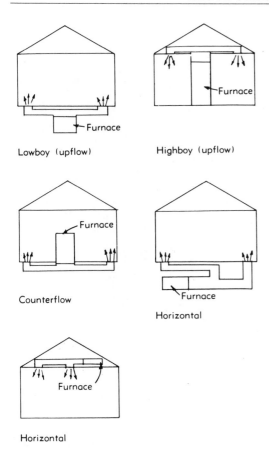

Figure 19-17. Furnace locations.

are placed in the attic. This may be because of the slab on grade construction or because there is limited crawl space. The furnace may be located in a closet or recessed area in the wall. The air enters the furnace through a low side entry or through the bottom, and it leaves through the top.

The counterflow (downflow) furnace is used primarily where it is preferable to have the furnace in the building, on the first floor, and not in a crawl space. The air enters the furnace at the top, and it leaves through the bottom. It is usually used with crawl spaces and slabs on grade. While it may be used in a

building with a basement, such buildings usually have the furnace located in the basement.

The horizontal furnace is used primarily in homes with crawl spaces or concrete slab floors (i.e. no basement). The furnace may be set off the ground in the crawl space, placed in the attic or suspended from posts in an attic or utility room. The air enters the furnace at one end, and it leaves through the opposite end.

Combinations of systems are often used in larger homes. In a two-story home it may be desirable to heat the first floor with an upflow furnace in the basement and the second floor with a horizontal furnace in the attic.

_____ 19.10 _____

AIR CONDITIONING EQUIPMENT

The forced air heating and cooling system may obtain its heat from a furnace, just as is used in forced air systems which supply only heat, with an added-on package of cooling coils placed next to the furnace bonnet, and a condenser located outside the building (Figure 19–18). The other unit which is becoming increasingly popular is the heat pump (Section 19.19) which produces both warm and cool air.

In this section, the furnace with a separate cooling unit (commonly called an air condi-

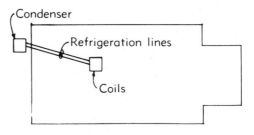

Figure 19-18. Split system.

tioner) will be discussed. The size of the cooling unit is rated according to its cooling capacity in Btuh, often referred to as "tons." One ton is equal to 12,000 Btuh (so, a 3-ton unit would have a capacity of 36,000 Btuh). The rating of the unit selected should be adequate to provide cooling Btuh equal to, or slightly more than, what the heat gain calculations call for. The wide range of sizes commonly available allows for the selection of a unit with the rated cooling capacity close to the required cooling calculated. Selection of a much larger unit will result in less efficient operation, and thus higher costs (due to the inefficiency of the on-off cycles, the time it takes to begin to cool and the warming of the system as it is off, only to be cooled again as it is turned back on). But if a unit is too small, it may not be able to provide sufficient cool air. This becomes especially critical if any of the design assumptions (such as the amount of moisture, insulation or the size or type glass used) varies, and the designer is not aware of the change. Also, the "tightness" of the construction, how well it is built, is assumed by the designer to be average. If it is not, there may be more heat gain than was calculated.

The unit selected for use should also be checked for its energy efficiency. Most manufacturers have more than one type of unit available. It is important to get the most efficient model available. The tables in Figure 19-19 give the manufacturer's specifications for "standard" and "high-efficiency" models.

The efficiency of the models is checked by comparing the EER (energy efficiency ratio) listed for each model. This EER rating is obtained by dividing the total Btuh of the unit by its watts; the higher the number, the more efficient the unit. Note that the standard 24,000-Btuh model in Figure 19-19 has an EER of 6.3, while the highest-efficiency 24,000-Btuh model has an EER of 8.9. This indicates that the highest-efficiency model is

Cooling capacity (Btuh)	Watts	EER (Btuh per watt)
24,000	2750	8.9
29,000	3100	9.4
36,000	4050	9.0
42,000	5000	8.4

Highest efficiency

Cooling capacity (Btuh)	Watts	EER (Btuh per watt)
24,000	3800	6.3
30,000	4600	6.3
36,000	5800	6.2
42,000	7100	5.9

Standard

Figure 19-19. Energy efficiency ratios.

slightly more than 40 percent more efficient that the standard model. This means that the fuel bill for cooling will be about 40 percent less when using the highest-efficiency model, compared to the standard. High-efficiency units typically cost 50 percent more than the standard, but in terms of dollars, it may only be $200 to $300. The EER ratings shown are typical, but each manufacturer must be checked since they will vary.

A typical duct layout for a residence is shown in Figure 19-20.

_____ 19.11 _____

HOT WATER HEATING SYSTEMS

For heat, water is heated in a boiler to the preset temperature (180° to 210°F). Then a circulating pump is automatically cut on, and the hot water is circulated through the system of pipes, passing through any of a variety of

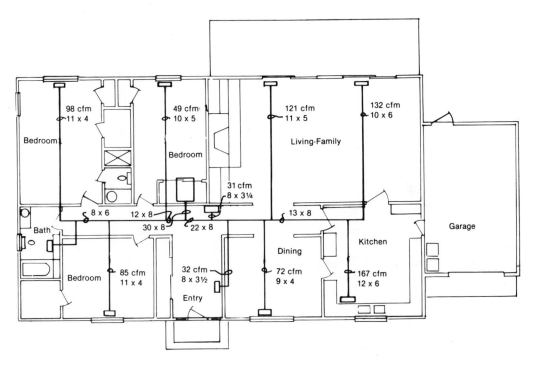

Figure 19-20. Typical duct layout.

convector types. As it circulates, it gives off the heat, primarily through the convectors; then the pipes return the water to the boiler to be reheated and circulated again. Most of the convectors used can be regulated slightly by opening or closing adjustable dampers. A compression tank is included in the system to adjust for the varying pressure in the system since water expands as it is heated.

Circulating hot water systems are also referred to as *hydronic* heating systems, and the four different hot water piping systems commonly used are series loop, one-pipe and two-pipe systems and radiant panels.

Series Loop Systems

Most commonly used for residences and small buildings, the convectors in the series loop system are fed by a single pipe which goes

through the convector and makes a loop around the building, or one portion of the building (Figures 19–21 and 19–22). The pipe acts as supply and return with the water getting cooler as it progresses through the system. To the designer this means that larger convectors are required to obtain the same amount of heat at the end of the loop, as com-

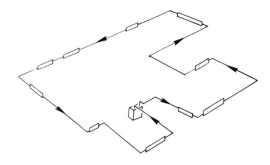

Figure 19-21. Series loop isometric.

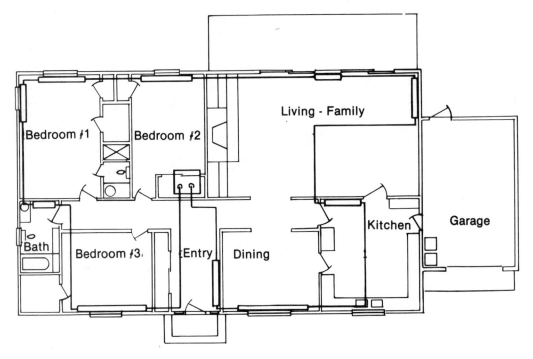

Figure 19-22. Series loop plan.

pared to convectors at the beginning, because the water is cooler. Also, the longer the run of piping and the more convectors it serves, the cooler the water will become. For more even heat, the building may be broken into zones (Figure 19-23), each with its own series loop

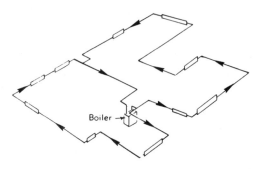

Figure 19-23. Series loop (zoned) isometric.

system activated by its individual thermostat. It is economical to install, but any control of individual convectors is minimal. Only heating is supplied with this system.

One-Pipe Systems

As shown in Figures 19-24 and 19-25, the one-pipe system has a supply pipe going around the building, or a zone of the building. A portion of the hot water is diverted through a special tee at each convector so that a portion of the hot water is diverted through the convector where it gives off heat to the room, thereby cooling the water. Then the water is returned to the supply pipe. Upon entering the supply pipe, it will slightly reduce the temperature of the water in that pipe. The primary

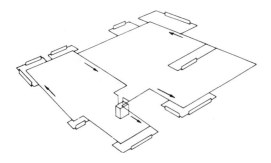

Figure 19-24. One-pipe (zoned) isometric.

advantage of this system over the series loop is that each individual convector may be controlled by a valve to turn it on, off or in between. The convectors may be placed above the pipe (upfeed) or below it (downfeed). The upfeed is more effective since the water tends

to be diverted more easily in that direction. This system is more expensive than the series loop since it requires additional piping, fittings and valves. It may also be zoned for larger buildings where the temperature drop over a long pipe run would be excessive.

Two-Pipe Systems

For large installations the supply of hot water needs to be kept separate from that water which has been cooled by passing through a convector. To accomplish this, one pipe is used to supply the hot water, which then empties into another pipe used for return. This type of system provides the water at as high a temperature as possible. The return may be classified as *reverse return* (Figure 19-26) or

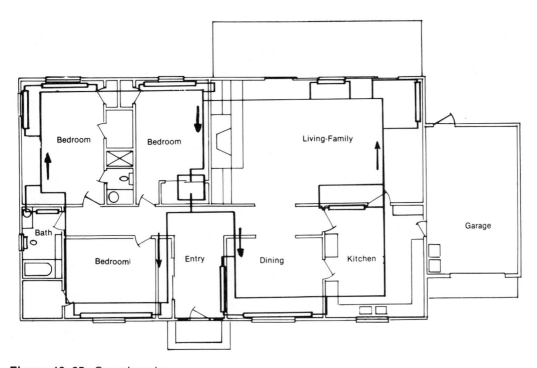

Figure 19-25. One-pipe plan.

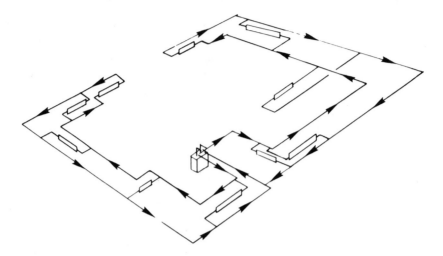

Figure 19-26. Two-pipe (zoned) reverse return.

direct return (Figure 19-27). The reverse return results in a more even flow of water because the supply and return are of equal length, resulting in equal friction losses. The direct return requires slightly less piping. Individual convector control is available by the installation of valves. This is the best system available and also the most expensive.

Radiant Panels

The hot water may be circulated through pipes located, usually, in the floor or ceiling of a building. The pipes are laid in a coil or grid arrangement (Figure 19-28); for a floor system they are embedded in concrete, whereas for a ceiling they are attached to the fram-

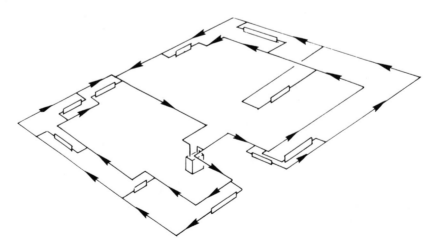

Figure 19-27. Two-pipe (zoned) direct return.

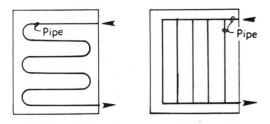

Figure 19-28. Radiant panels.

ing and plastered or drywalled over. They may also be used on walls, but this is not common. The basic disadvantage with this type of system is that while it provides uniform heat over an entire room, heat is not uniformly lost; most of it is lost at exterior walls, usually at the window. This situation tends to cause drafts and a feeling of being cold near the windows. The floor panels have a tendency to make the occupant's feet hot and uncomfortable, while the heat coming down from the ceiling tends to stay high (since heat rises), and a person's legs under a table will probably feel cool. (This is also true of electric ceiling and wall radiant panels.)

19.12
PIPING AND FITTINGS

The piping used for hot water heating is usually copper, but steel pipe is sometimes used. Copper is preferred because it is lightweight and easy to work with—characteristics described in the section on plumbing. The types of copper piping used for hot water heating systems are Type L pipe and tubing and Type M pipe.

The most commonly used fittings for the system are the same as those used in plumbing systems.

19.13
BOILER AND SYSTEM CONTROLS

Boiler

The boiler (furnace, Figure 19-29) heats the water for circulation through the system. It may be rectangular or square (occasionally even round) and made of steel or cast iron. A boiler is rated by the amount of heat it can produce in an hour. The maximum amount of heat the system can put out is limited first by the size of the boiler selected. Boiler efficiency increases when the boiler runs for long periods of time, so the unit selected should not be oversized or it will run intermittently and thus be less efficient. Hot water boilers may use oil, gas, propane gas, coal, or electricity as fuel. In residences oil and gas boilers are most commonly used.

For a hot water heating system, the temperature that the water is heated to in the boiler is of prime importance since it has a direct relationship to the amount of heat which the radiation or convector units will give off. Typically, the thermostat should be

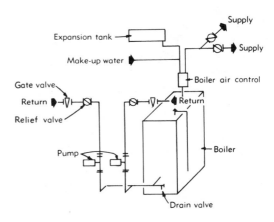

Figure 19-29. Typical boiler installation.

Figure 19-30. Boiler thermostat control.

Figure 19-31. Thermostat.

set at about 180°F but temperatures as high as 220°F are possible. The thermostat which controls the water temperature (Figure 19-30) is located somewhere on the boiler (usually in plain sight, sometimes behind a small sheet-metal cover plate) and should be checked, after installation, by the designer.

Thermostat

The thermostat which controls the air temperature in the space (Figure 19-31) is placed in the building. With hot water heat, there will be one thermostat for each zone. The temperature desired is set on the thermostat, and when the temperature in the room falls below the desired temperature, the thermostat turns on the boiler and the circulating pump. When the desired temperature is reached, the thermostat will turn the boiler and the pump off. The thermostat has a temperature differential within it which will call for heat and then cut it off. Typically, this temperature differential is about 1°F, which means that the thermostat will call for heat at 69°F and shut it off at

70°F. This differential may be adjusted on a small calibration setting inside the thermostat housing (Figure 19-32). The less the differential, the more comfortable the space will feel. But it also means that the boiler will be turned on and off more frequently (to maintain a

Figure 19-32. Thermostat calibration.

close tolerance of temperature) which will lower the efficiency of the boiler operation.

Thermostats used for hot water systems are usually either the single-setting thermostat (Figure 19–31) or the day-night thermostat (Figure 19–33).

The single-setting thermostat is the one most commonly used. The desired temperature is set and the thermostat will operate the cycles of the heating system. The temperature may be changed at any time by simply adjusting the temperature setting on the thermostat.

A day-night thermostat permits the setting of one temperature for the daytime and another (usually lower) for nighttime. With this thermostat a clock is set for a given time (say, 8 A.M.) and at that time the temperature will go to the daytime setting (perhaps 72°F); then at the other time set (say, 5 P.M.) the temperature will go to the nighttime setting (perhaps 60°F). The times and temperatures are set on the thermostat and may be changed as desired.

The day–night thermostat is used in residences, stores, offices, apartments, institutions, and commercial and industrial projects. In a residence, the temperature differential may be 3°F or 4°F, set to begin to cool just before bedtime and to heat as the occupants arise. In apartments where the heat is furnished and paid for by the apartment owner and not the renter, this type of thermostat is commonly used to control heat costs for the owner. In this case, the thermostat used would have two parts, one located in the apartments to sense the temperature and the second located in a place the renters cannot enter which controls the temperature desired and the times of operation.

Thermostats are usually placed about 5 feet up on a wall. The location should be carefully checked to be certain that the thermostat will provide a true representation of the temperature in the spaces it serves. Guidelines used in thermostat location include the following:

1. Always mount it on an inside wall (on an outside wall, the cold air outside will affect the readings).

2. Keep it away from the cold air drafts (such as near a door or window), away from any possible warm air drafts (near a radiation unit, fireplace or stove) and away from any direct sunlight.

Expansion Tank

The expansion tank (also called a *compression tank*) allows for the expansion of the water in the system as it is heated. It is located above the boiler.

Automatic Filler Valve

When the pressure in the system drops below 12 psi, this valve opens, allowing more water into the system; then the check valve automat-

Figure 19-33. Day-night thermostat.

ically closes as the pressure increases. This maintains a minimum water level in the system.

Circulating Pump

The hot water is circulated from the boiler through the pipes and back to the boiler by a pump. The pump provides fast distribution of hot water through the system, thus delivering heat as quickly as possible. The circulating pump size is based on the delivery of water required, in gpm, and the amount of friction head allowed. Circulating pumps are not used on gravity systems.

Flow Control Valve

The flow control valve closes to stop the flow of hot water when the pump stops, so that the hot water will not flow through the system by gravity. Any gravity flow of the water would cause the temperature in the room to continue to rise.

Boiler Relief Valve

When the pressure in the system exceeds 30 psi, a spring-loaded valve opens, allowing some of the water to bleed out of the system and allowing the pressure to drop. The valve should be located where the discharge will not cause any damage.

Boiler Rating

Boiler manufacturers list both gross and net Btuh ratings of the boiler. The gross Btuh is the heat input to the boiler by the fuel, and the net Btuh is the usable heat output which is available for use in the hot water system.

_____ 19.14 _____

HOT WATER HEATING DEVICES

Some type of heating device must be used to distribute the heat efficiently from the hot water to the room being heated. These devices are classified as *radiant* and *convector,* according to the way in which they transfer the heat.

Radiant heating units have the heat transfer surface exposed so that the heat is transferred by radiation to the surrounding objects and by natural convection to the surrounding air. A typical radiant unit is shown in Figure 19–34.

Convector heating units have the heat transfer surface enclosed in a cabinet or other enclosure (Figures 19–35 and 19–36). The transfer of heat occurs primarily through convection as air flows through the enclosure, and past the heat transfer surface, by gravity.

Unit heaters use a fan to distribute the air through the space. The unit usually consists of a heating coil and a fan enclosed in a cabinet (Figure 19–37). The heat may be supplied by hot water or even steam when available. Often this type of unit is suspended from the ceiling in areas such as a warehouse, or storage area or any large room situation, especially one with high ceilings. Such a unit is sized according to heat output, and the layout of these units to provide adequate coverage of the space with warm air must be carefully checked. This type of unit is also effective when the room is deep with a relatively small outside wall area for a convector or radiation

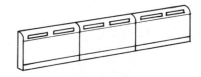

Cast iron baseboard

Figure 19-34. Radiant heating device.

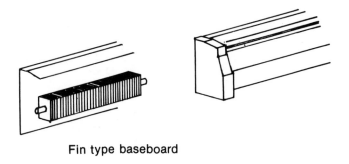

Fin type baseboard

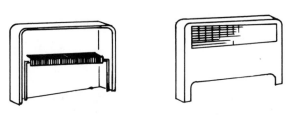

Fin type radiator

Figure 19-35. Convector heating devices.

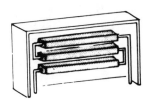

Figure 19-36. High output convector.

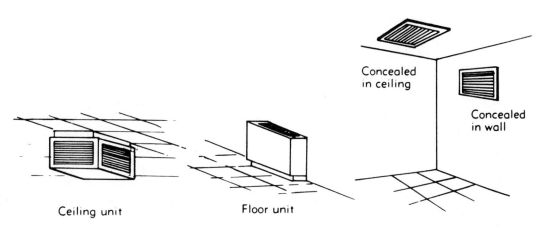

Concealed
in ceiling

Concealed
in wall

Ceiling unit

Floor unit

Figure 19-37. Unit heaters.

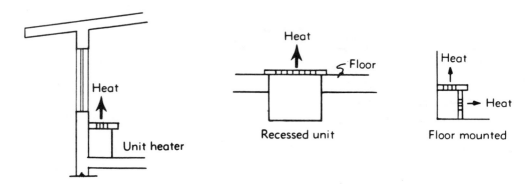

Figure 19-38. Floor unit heaters.

unit. A typical situation would be that of a room (Figure 19-38) where the unit heater is placed under the glass area in the exterior wall and the air is fan-blown through the room. Unit heaters may be recessed, surface mounted, or suspended from the ceiling or wall, or recessed or surface mounted on the floor (Figure 19-38).

Heating Device Ratings

All radiant, convector and unit heaters are rated in terms of the amount of heat they will give off in an hour; the capacity ratings are given in Btuh for small devices and in MBH (thousands of Btuh) for larger devices. The heating capacities vary, depending on the type of pipe, the size of the finned tube, the number of fins per foot of radiation and the temperature of the water. Each manufacturer's specifications (or technical data report) should be checked for heating capacity of a particular unit.

Baseboard radiation units may be cast-iron radiation units or finned-tubed convectors. A typical manufacturer's specification rating is shown in Figure 19-39. The various types of elements available from this manufacturer

are shown in the left column and the heat ratings, for the various water temperatures, on the right. Notice that the heating capacity of the unit increases as the water temperature increases. Comparing the copper-aluminum elements, the capacity for the 2¾-inch x 5-inch x 0.020 x 40/feet 1¼-inch tube element (the second entry on left) at 180°F is 850 Btuh, while at 200°F the capacity is 1030 Btuh. The capacity is increased by 180 Btuh simply by increasing the temperature of water that will flow through the element. This 180-Btuh increase represents a 21 percent gain in the heating capacity at no increase in the cost of boiler or heating device, only an increase in the temperature of the water. The effects of the fins and tube on the heating capacity of an element can be seen by comparing the first and third listings under the copper-aluminum elements. The smaller top element (2¾-inch x 3¾-inch x 0.011 x 50/feet 1-inch tube) has a capacity of 840 Btuh at 180°F, while the third listing (2¾-inch x 5-inch x 0.020 x 50/feet 1¼-inch tube) has a capacity of 910 Btuh. This is an increase of 70 Btuh or about 8.33 percent. The copper-aluminum elements may be compared with the steel elements by checking the second copper-aluminum listing against the second steel element listing. The second

Copper-Aluminum Elements	Hot Water Output* 1 gal. flow rate (for 5 gal. flow rate use factor 1.067)				
	220°	210°	200°	190°	180°
2¾" x 3¾" x .011 x 50/ft. 1" tube	1240	1120	1020	930	840
2¾" x 5" x .020 x 40/ft. 1¼" tube	1250	1120	1030	940	850
2¾" x 5" x .020 x 50/ft. 1¼" tube	1340	1200	1100	1000	910
Steel Elements					
2¾" x 5 x 24 ga. x 40/ft. 1" tube (IPS)	1020	920	840	770	690
2¾" x 5 x 24 ga. x 40/ft. 1¼" tube (IPS)	1040	940	860	780	710

*Based on 65° entering air.

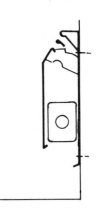

Figure 19-39. Finned tube ratings.

copper-aluminum listing (2¾-inch x 5-inch x 0.020 x 40/feet 1¼-inch tube) has a capacity of 850 Btuh at 180°F, and the second steel element listing has a capacity of 710 Btuh at 180°F. The copper-aluminum element capacity is 140 Btuh or 19.7 percent higher than that of the steel element.

As initial selections of heating devices are made, the various ratings must be checked. It may be necessary to use different sizes of heating devices in different rooms (usually depending on the amount of wall space available for mounting the devices), but once a basic decision is made as to type of material (steel or copper-aluminum), this will usually be used throughout.

To make an economical selection, the designer should also consider the relative

costs of the devices per foot and then compare these costs to the heating capacities of the units. Prices vary considerably, but unless the steel elements, installed, cost about 20 percent less than the copper-aluminum elements, they will not be as economical to install. It is part of the designer's responsibility to provide the best system at the lowest cost.

19.15
BOILERS

In residences the boiler (furnace) which provides the heat (either hot water or forced air) is commonly located in the basement or crawl space or in a first-floor utility room, but it may also be in the attic. Boilers are most commonly fired by natural gas or oil, and occasionally by electricity, coal, or bottled gas.

The boiler selected must be large enough to supply all of the heat loss in the spaces as calculated. The boiler must also have enough extra capacity to compensate for heat loss which will occur in the pipes which circulate the hot water through the system (called pipe loss or pipe tax). In addition, an extra reserve capacity is required so that the boiler can provide heat quickly to warm up a "cooled off" space. This "pickup" allowance also provides some extra heating capacity when it is necessary to increase the temperature of a space several degrees, not merely to maintain a temperature. For example, a space not used after 5 P.M. may set its thermostat back to 60°F before leaving. Then in the morning as the thermostat is raised to 68° or 70°F, it is necessary to have some extra capacity in the boiler to provide this extra heat as quickly as possible. Many buildings have automatic, timed thermostats which are used to regulate the temperatures throughout the day.

Typically, the designer allows about 33⅓ percent for the heat loss in the pipes and an additional 15 to 20 percent as a pick-up allowance, in addition to the calculated heat loss of the building. Spaces which are heated intermittently, such as studio areas or workshops, will have to have a much larger pick-up allowance, perhaps as much as 50 percent. This is because the temperature will often be allowed to drop quite low (perhaps 50°F) when the space is not in use, and then when the heat is raised to 70°F, it must heat up quickly.

While most designers simply allow a certain percentage for piping loss, it can be calculated—but only once the piping layout is finalized and the installation decided upon and specified. From the piping layout it would be necessary to determine the length of pipe and the various temperatures of the air through which the pipe travels. Where the pipe travels through an unheated basement or attic, the heat loss would be much greater than the heat loss of the same pipe traveling through a heated space.

Once the piping loss and the pickup allowance have been taken into account, the boiler size may be selected. The boiler selected should be as close to the total Btuh as possible to provide the most efficient operation. Keep in mind that the outside design temperature used will not actually occur very often, that the boiler will be operating at less than full capacity during most of the time and that the pickup allowance also gives a little extra capacity. Also, oversizing the boiler will probably be of little value during any unusually long cold spells with temperatures well below the outside design temperature unless the radiation or convector units selected are able to put out additional heat.

_____ 19.16 _____

ELECTRIC HEATING SYSTEMS

The advantages of an electric heating system include low installation cost, individual room control, quiet operation, and cleanliness, and when cable or panels are used, there are no exposed heating units. The primary disadvantage of electric heat is its high cost of operation in almost all areas. A specialist is required to give honest figures for comparing electric heat costs with those of other types of fuels. Electrically operated boilers (furnaces) are not discussed in this chapter since they are not different systems but rather a forced air furnace with electricity as a fuel.

Most electric ratings are given in watts and Btuh. There are 3,413 Btuh for every 1,000 watts (1 kilowatt = 1,000 watts), and often the ratings will be given or noted as MBH (thousand Btuh's) and kW (kilowatts), such as 6.8 MBH and 2 kW.

Overall Systems

Baseboard. Electric baseboard units have a heating element enclosed in a metal case. This system offers individual room control, but furniture arrangement and draperies must not interfere with the operation of the units by blocking the natural flow of air. This system is economical for a builder to install and is especially popular in low-cost housing and housing built for speculation (to try to sell), although the cost of operation is generally higher than most other fuels.

Resistance Cable. With this system, electric heating cable is stapled to the drywall in a grid pattern and covered with plaster or gypsum board. Individual thermostats control the heat in each room. However, since the cable is usually installed on the ceiling, the disadvantages of heat rising and cold feet must be considered.

Drawings are not usually done for this type of heating system; instead, the amount of heat required is noted for each space. The cable system allows complete freedom in furniture and drapery placement.

Panels. The prefabricated ceiling and wall panels used in this system have the heating wire sandwiched in rubber with an asbestos board backing. They are only ¼ inch thick and may be plastered, painted, or wallpapered. The panels come in a standard size (usually 2 feet x 4 feet) and cannot be cut. They are often used in hung suspension ceilings. A panel system has the same basic advantages and disadvantages as a resistance cable system, including flexibility of furniture and drapery arrangement.

Unit Heaters

There are a variety of electric unit heaters available, which may be used to supplement other heat sources in a space or to completely heat a room. In residences, such a unit heater is often installed in the bathroom to supplement other heat sources, since a temperature higher than 72°F is necessary to feel comfortable when washing and after bathing.

Unit heaters are also commonly installed in spaces where the heat is only used periodically, such as a basement, workshop, or garage.

So that the warm air will be quickly spread throughout the area to be heated, unit heaters are equipped with a fan. It is preferable if the fan switch is the type which will not start until after the unit comes on and the air has warmed to a preset temperature; in this way, cold air is not pushed around the room.

Units are generally available for use in walls (recessed into the wall), and ceiling recessed units are also available.

System Combinations

Quite often, in order to provide the best results, the design may incorporate more than one type of system. For example, a building may be predominantly heated with baseboard units, but in spaces where there is little free wall space, such as a kitchen, it may be desirable to use a unit heater in the ceiling or the wall under the cabinets. Another possibility for the kitchen might be to use resistance cables or panels in the ceiling. Bathrooms may be heated with unit heaters to provide added comfort.

Codes and Installation

While a hot water heating system is installed by the heating contractor, an electrical heating system is installed by an electrical contractor. ASHRAE publishes guides which discuss the use of the units for proper heating conditions and the installation, but the *National Electric Code* (NEC) also governs the installation of the resistance cable. The designer must have both the NEC code and the ASHRAE available for reference.

Baseboard System Design

A baseboard electric system transfers heat to the space primarily by convection. It consists of baseboard units which may be mounted on the wall or recessed into the wall (Figure 19–40). In selecting a baseboard unit, the designer must consider the direction in which the air will be discharged from the baseboard unit (Figure 19–41).

Baseboard units are rated in watts and Btuh. These ratings, which may vary for different manufacturers, are given *for the length specified* and not per lineal foot. For example, in Figure 19–42, the very first listing specifies

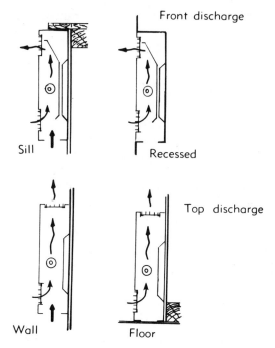

Figure 19–41. Baseboard air discharge.

a 36-inch (3-foot) length and a rating of 500 watts and 1,707 Btuh for the 3-foot length. The manufacturer often has several models, sizes, and ratings available, and the ratings will probably vary with other manufacturers. In Figures 19–43 and 19–44, two separate series are listed; both have the same outside dimensions and the *B* series has a rating about 100 percent higher for each length than the *A*

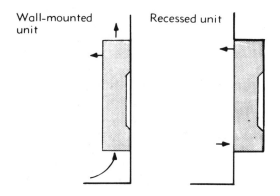

Figure 19–40. Baseboard mounting.

Length	Watts	Btuh
36" (3'-0")	500	1,707
52" (4'-4")	750	2,560
68" (5'-8")	1,000	3,413
100" (8'-4")	1,500	5,120

Figure 19–42. Baseboard ratings.

Length	Watts	Btuh
36" (3'-0")	750	2,560
48" (4'-0")	1,000	3,413
60" (6'-0")	1,250	4,269
72" (6'-0")	1,500	5,120
96" (8'-0")	2,000	6,830

A Series

Figure 19-43. High output baseboard ratings.

series. In addition, many manufacturers make larger units, with higher ratings, which are normally used in commercial, industrial, and institutional buildings. A 3-foot length of this style may range from 750 to 2,250 watts.

In designing an electric baseboard system, it may be necessary to use the larger units in some areas in order to provide the Btuh required.

Resistance Cable System

The amount of heat given off by the cables will vary with the amount of heating cable used in the installation.

The electric cable used for ceiling installations comes in rolls; it is stapled to the ceiling

and then covered with gypsum or plasterboard in accordance with the manufacturer's specifications and the code requirements. Basically, the cable must not be installed within 6 inches of any wall, within 8 inches of the edge of any junction box or outlet, or within 2 inches of any recessed lighting fixtures.

Cable assemblies are usually rated at 2.75 watts per lineal foot, with generally available ratings from 400- to 5,000-watt lengths in 200-watt increments; the manufacturer's specifications should be checked to determine what is available. A typical list of available lengths, watts, and Btuh from one manufacturer is shown in Figure 19-45. The cables have insulated coverings which are resistant to medium temperatures, water absorption, and the effects of aging and chemical reactions (concrete, plaster, etc.); a polyvinyl chloride covering with a nylon jacket is most commonly used. Each separate cable has an individual thermostat, providing flexible control throughout the building.

Length	Watts	Btuh
36" (3'-0")	1,500	5,120
48" (4'-0")	2,000	6,830
60" (5'-0")	2,500	8,538
72" (6'-0")	3,000	10,245
96" (8'-0")	4,000	13,660

B Series

Figure 19-44. High output baseboard ratings.

Btuh	Watts	Length Ft.
1365	400	145
2047	600	218
2730	800	292
3413	1000	362
4095	1200	436
5461	1600	582
6143	1800	654
6826	2000	728
7509	2200	800
8533	2500	910
10,239	3000	1090
11,287	3600	1310
15,700	4600	1672

Figure 19-45. Typical cable ratings.

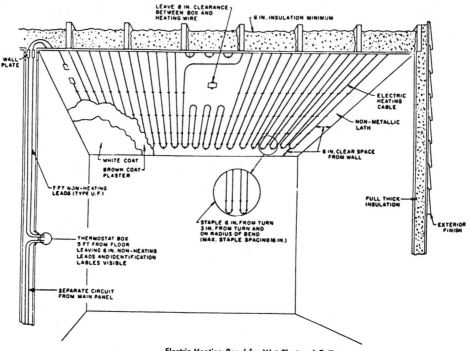

Electric Heating Panel for Wet Plastered Ceiling

Reprinted with permission from ASHRAE, Systems Handbook, 1976

Figure 19-46. Cable heat installation.

A typical cable installation in a plastered ceiling is shown in Figure 19–46. The space between the rows of heating cable is generally limited to a minimum of 1.5 inches, and some manufacturers recommend a 2-inch minimum spacing when drywall construction is used. Another limitation on the spacing of the cable is a 2.5-inch clearance required between cables under each joist (Figure 19–47), and a review of the layout in Figure 19–46 shows that the cable is installed parallel to (in the same direction as) the joists.

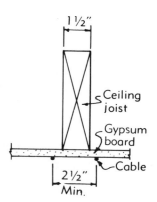

Figure 19-47. Cable detail.

Radiant Panel System Design

The heating rates for radiant ceiling panels are generally given in watts per panel, Btuh per panel or both. As shown in Figure 19–48, the ratings for a 2-foot x 4-foot panel may vary from about 500 to 750 watts per panel, depending on the panel selected and the manu-

Watts	Btuh	Size
500	1707	2'-0"x4'-0"
750	2560	2'-0"x4'-0"
560	1911	2'-0"x4'-0"
700	3019	2'-0"x5'-0"
500	1707	2'-0"x3'-0"
750	2560	2'-0"x3'-0"
1000	4313	2'-0"x3'-0"

Figure 19-48. Radiant ceiling panel ratings.

facturer. Since 1 kW = 3,413 Btuh, 500 watts = 1,707 Btuh and 750 watts = 2,560 Btuh, it is important that the designer get accurate ratings by checking the engineering specifications for the type of panel which will actually be used on the project.

19.17
SOLAR ENERGY AND HEAT PUMPS

The idea of harnessing the sun's energy for use in homes and factories is nothing new. Work began on solar furnaces more than two centuries ago—but with little success. In the United States solar hot water heaters were used in the early 1900s in Florida, Arizona, and California. The introduction of mass-produced hot water heaters which were low in cost and which used inexpensive oil, natural gas, and electricity all but stopped the further development of the solar hot water heaters. The higher initial cost of the solar unit made it uneconomical, except where the price of oil or electricity was high.

Since the 1930s limited experimentation continued in the application of solar energy, and interest increased markedly after World War II. But by 1960, the basic obstacle which had to be overcome was that the solar units were not "economically feasible" when compared with the low cost of other fuels. Not being "economically feasible" means that the amount of money saved on fuel costs is not sufficient to pay the increased cost of a solar system over a set period (say, 10 years).

In the 1970s, several major factors occurred to change the public's feelings toward solar systems.

1. The "oil embargo" of the early 1970s made the public aware that the flow of oil into the country was largely controlled by "others" and that the flow could be cut off at any time.

2. The price increase of oil after the embargo caused the prices of other fuels dependent on oil (such as electricity) to skyrocket. Also, this rise in the cost of oil resulted in similar price rises for all other fuels (natural gas, propane gas, and coal).

3. The public realized, and the government admitted, that the supply of fuels used (oils, gas, and coal) is limited and that if the present use rate continues, one day we will "run out of gas." As a result the government has begun to put together a national energy policy while assuring the public that the energy crisis is real.

4. The energy crisis which occurred in the mid-1970s when natural gas was in short supply put thousands of people out of work due to a lack of natural gas to heat buildings and run machinery. While part of this crisis was brought on by the reluctance of firms to explore and produce natural gas, it made the government and the public face the fact that there is a limit to the amount of natural gas.

As a result of all this, solar energy systems were deemed to be economically feasible for use in hot water heaters and, in many areas, for heating and cooling. Also, many people are now willing to pay a little extra if it means that they can reduce the constantly increasing amount being paid for other fuels. The price of fuels will continue to go up, and cheap fuel is now gone forever.

Uses

Solar energy is being used to:

1. Heat hot water for use in the building.
2. Provide heating for the building.
3. Provide cooling for the building.
4. Provide heating and cooling for the building.
5. Provide heating and hot water for the building.
6. Provide heating, cooling and hot water for the building (Figure 19–49).

As outlined above, quite often the solar energy package will provide more than one service, usually at very little extra cost.

In this chapter, we will review some solar systems available, how they work and what they have to offer.

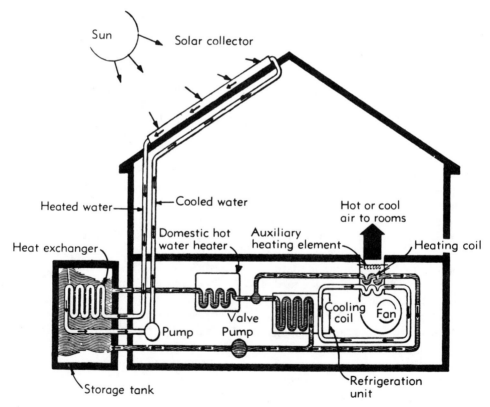

Figure 19-49. Solar heating and cooling system.

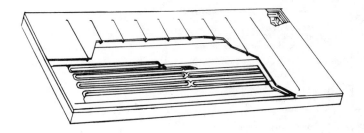

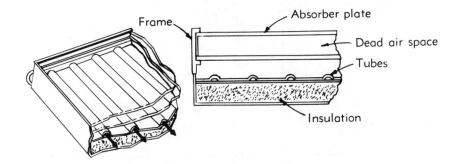

Liquid collector

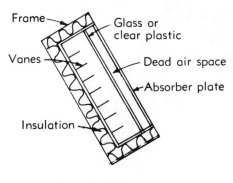

Air collector

Figure 19-50. Flat plate collectors.

Collectors and Storage

All systems are made up of two basic components—collectors and storage areas.

Collectors. First, all systems must have some type of collector. The flat-plate collector (shown in Figure 19-50) is most commonly used to gather heat from the sun. Basically, all

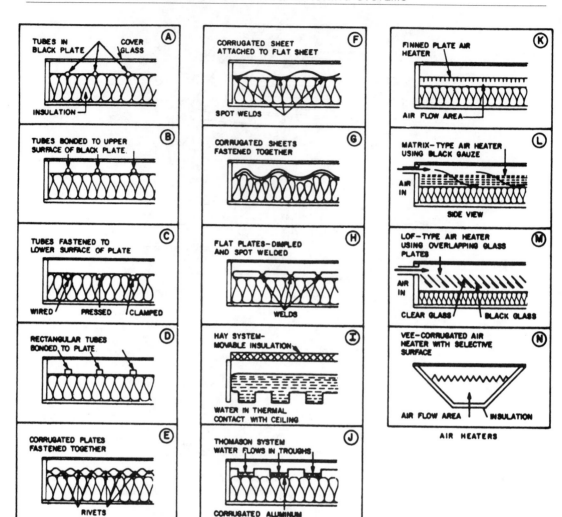

WATER HEATERS

AIR HEATERS

Variations of Solar Water and Air Heaters

Figure 19-51. Collector designs. (Reprinted with permission from ASHRAE, *Applications Handbook,* 1974.)

collectors are similar in that they have a casing (or frame), insulation, plate, heat transfer medium (liquid or gas), and glazing. There is no one design that is used, however, and several different collector designs are shown in Figure 19–51. But it is important to note there are two basic types of heat transfer mediums—*liquid* and *air*—and this makes for

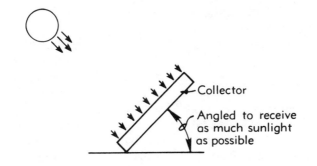

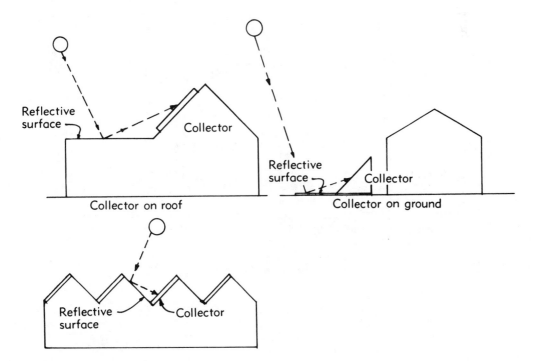

Figure 19-52. Reflection to collector.

two basic types of collectors—liquid and air.

The collector is located outside and is angled to receive the maximum amount of sunshine possible (Figure 19–52). To increase the amount of the sun's rays which hit the collector, some systems even use a reflective surface in front of the collector (Figure 19–52), with the reflective panel placed either on an adjoining flat roof or in front of the collector. The collector is commonly located on the roof, and many units are also designed to be placed on the ground.

Storage Mediums.   Once the heat exchange medium (liquid or air) is warmed in the collector, the heat it absorbs is transferred to a storage medium for future use. The storage mediums most commonly used are coarse aggregates (clean, washed rock), water, and a combination of the coarse aggregates and water (Figure 19–53).

Heating, Cooling and Hot Water

The system shown in Figure 19–49 utilizes a liquid heat transfer medium (water or water and anti-freeze), a collector on the roof, and water as the heat storage medium. Such a system is the most versatile, providing the most benefits to the owner. The system shown has auxiliary units for heating, cooling and hot water for the time when the solar system cannot provide sufficient heat (or cooling).

Heating.   The heat transfer medium is then heated in the collector and is then circulated

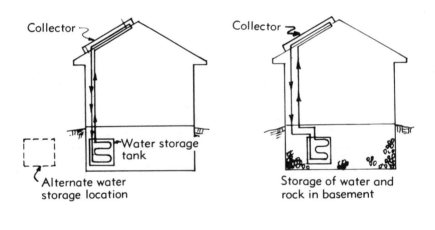

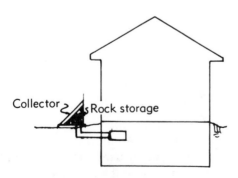

Figure 19-53. Storage medium.

through the storage tank, transferring its heat to the storage medium—in this case, water. This heated water then circulates through the hot water tank to heat all or part of the hot water required for the building. The heated water then passes a fan which forces air past the hot water coils. The air passing the coils is warmed and circulates through the building.

There are many variations on the design; in Figure 19-49, the water from the storage tank passes first through the fan and forced air and then through the hot water tank. A hot water heating system is shown in Figure 19-54.

Cooling.   To provide cooling for the building, the system is, in effect, reversed. While the collectors used to gather the heat ideally face south, the radiators used to give off heat to the exterior face north. (Although some systems utilize the collectors already on the south, it is not as effective.)

In order to use the system for cooling, the storage medium must be made as low in temperature (cool) as possible; then it circulates past the refrigerant in the refrigeration unit, which then circulates past the fan to blow cool air into the building. One method used is to have the heat pump transfer heat from the house to the heat storage medium and into the storage tank, where the temperature of the entire storage medium slowly increases. Then at night, the radiation system turns on and the heat transfer medium cools the storage medium by absorbing the heat as it passes through the tank. Circulating it up through the night radiator collectors facing north, the collectors radiate the heat to the surrounding atmosphere and then the medium goes back to the storage tank to constantly repeat the process of lowering the water temperature in the storage tank so it can be used for cooling.

Another method used with a system of this type is to connect to a well; then if the water in the storage tank reaches a certain temperature (about 65°F), the system will automatically switch to the well as its source of cool water. Since many of the same controls (valves and pumps) are utilized in both the heating and the cooling operations, it is often this dual feature that makes the system economically feasible.

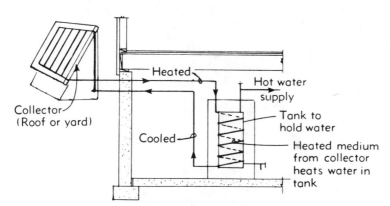

Tank—This tank could be used to just hold water heated by the collector medium or it could be a direct hot water heater used to supplement the collector during times when the collector cannot provide all of the hot water required.

Figure 19-54. Hot water system.

Another method used converts the sun's energy into the energy source used to power the cooling equipment. In this type of installation, storage batteries are charged during the daytime hours to operate a heat pump.

The use of solar energy for cooling is not nearly as advanced as its use for heating. A great deal more research is required before economical, dependable solar cooling is available.

Hot Water Heating

Hot water solar systems are similar to the system described except that the storage medium (water) is used to heat the hot water which is then circulated through the system (Figure 19-54). An auxiliary heat source is still required to supplement the solar system. This system could also be used for cooling (similar to Figure 19-49) if the devices in the building are selected to effectively handle chilled water. Realistically, this is rarely the case in residential work.

Hot Air System (Heating)

A hot air heating system uses air as the heat transfer medium and rocks as the heat storage medium. As the diagram in Figure 19-52 shows, the collector may be located on the ground. The air in the collector is heated as it is pulled across the vanes in the collector, and it is then pushed over the rocks in the storage medium, transferring much of the heat collected to the rocks. This continuous flow of air is constantly warmed in the collector and then cooled as it transfers its heat to the rocks (Figure 19-55).

When the thermostat in the building calls for heat, the distribution fans come on and circulate the cooler air from the building over the rocks in the storage area, and this warmed air is then circulated back through the building (Figure 19-56).

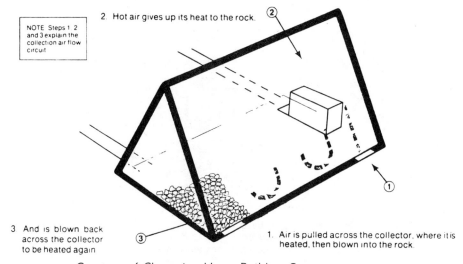

NOTE Steps 1 2 and 3 explain the collection air flow circuit

2. Hot air gives up its heat to the rock.

3 And is blown back across the collector to be heated again

1. Air is pulled across the collector, where it is heated, then blown into the rock.

Courtesy of Champion Home Builders Company

Figure 19-55. Hot air collection.

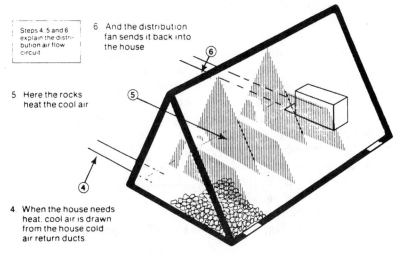

Steps 4, 5 and 6 explain the distribution air flow circuit

6 And the distribution fan sends it back into the house

5 Here the rocks heat the cool air

4. When the house needs heat, cool air is drawn from the house cold air return ducts.

Courtesy of Champion Home Builders Company

Figure 19-56. Hot air storage and transfer.

When both the collection and the distribution are on, the air flow from the building passes through the collector to provide heat directly from the collector. This commonly occurs when the building thermostat calls for heat during the daytime while the collector is also activated to gather the heat from the sun. With a system of this type, it is usually recommended that there be continuous air circulation in the building. A solar system of this type is designed to continuously circulate air in the building; this circulated air is relatively cooler than that from the forced air system using a furnace. In the solar system, the air will usually feel just slightly warm, or perhaps even a little cool to the touch since the air temperature will be lower than the body temperature. Studies of systems using continuous air circulation show that less heat is required when such a system is used.

Also important in this system is the fact that it is easily adaptable to existing forced air systems. In such an installation, the collector could be set on the ground.

19.18
SUPPLEMENTARY HEAT

Almost all solar systems are more expensive to install than conventional boiler or furnace systems. The conventional system is sized to deliver all of the heat required down to a particular design temperature. It is quite inexpensive to size the boiler or furnace to provide all of the heat required to meet those demands, even with an extra 10 percent pickup load added on. But keep in mind two things:

1. During most of the heating season, the temperatures are well above the design temperature used. (Many times it might reach the low during the night, but it warms up 15° to 25°F during the day.)

2. The solar system costs much more for the materials used.

This means that providing all of the heat required to keep a house warm in a northern city on *the coldest day ever recorded there* would require a solar unit about four times as large as one that would provide 90 percent of the heat required during the entire heating season.

For these reasons many of the solar systems have auxiliary or supplemental heating systems. In this situation the solar system would provide a certain percentage of the heat required (perhaps 40, 50, 60 percent or more), and the supplemental heating boiler or furnace would provide the rest. In this manner the most economically feasible installation can be obtained, and the problem of allocating huge areas to contain the storage medium is overcome.

A solar system with supplemental heat can be designed to save a significant amount of the total heat bill, but it is important to realize that when the *total annual fuel bill* is being discussed, the savings will be highest in the milder months and the heat bills will increase in the colder months. So it is the "average" being discussed, as illustrated in Figure 19–57.

While this type of system may not satisfy the "purist" in terms of total solar systems, it provides a very "cost effective" system which will save large amounts of fuel. In addition, it makes solar systems easily adaptable to existing installations where the existing system may be used as the supplemental system. It is most important not only that fuels be conserved in new construction but also that systems be developed that are adaptable to existing heating systems (hot water and air).

19.19
HEAT PUMPS

Use of refrigeration to provide cooling is discussed in Section 19.4, as are the basic principles which apply to such use. The use of refrigeration to provide both heating and cooling is accomplished with a device called a *heat pump*. This increasingly popular form of heating and cooling requires an explanation of the principles involved and its advantages and disadvantages.

First, the discussion of the cooling principles of refrigerants in Section 19.4 must be carefully read and understood. In effect, it states that the refrigerant can take heat from

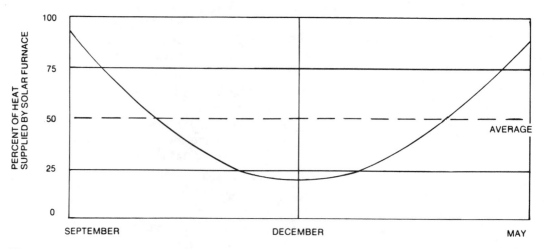

Figure 19–57. Yearly heat supplied.

one place (air or water) and move it to another by use of an evaporator, compressor, and condenser.

Now, using those basic cooling principles and a valve control, it is possible to reverse the cycle, causing the refrigerant to absorb heat from an outside surrounding medium (usually air, but water is sometimes used) and to release this heat inside the building (in the air or water being used to heat the building).

The reversing valve control allows the refrigerant to provide cooling or heating, depending on the direction of flow after it leaves the compressor. If the refrigerant flows to the condenser, it provides cooling; if it flows toward the evaporator section, it provides heating.

The most commonly used surrounding mediums are the outside air and the forced air in the ductwork.

Outside air is the surrounding medium most likely to be used to draw heat from in the winter for heating and to release heat to in the summer for cooling. The temperature of this outside air affects the efficiency of the unit in providing heating and cooling Btuh. Typically, in the winter as the outside air temperature goes down, the amount of heat which the pump will produce goes down also. In effect, the more heat needed in the space, the less Btuh the unit will provide.

It is important to realize that a heat pump is not electric heat. Actually, it is a type of solar heating which draws heat from the surrounding air. It requires electrical power to run the equipment, but it is more efficient than electric heat. The efficiency of the heat pump can be determined from the manufacturer's engineering data. While there are many sizes and capacities of units available, we will review the typical data for a 3-ton (36,000-Btuh) cooling capacity unit. Since each manufacturer's units vary, so will their data; and as more efficient units are developed, the result will be higher values from the heat pump.

Note in Figure 19–58 that at a 60°F outside temperature, the unit will provide 41,000 Btuh for heating. As the temperature goes down to 40°F, the unit will supply 32,000 Btuh. When it gets cold—say, 20°F—outside, the unit will provide 21,000 Btuh. It is obvious that as the temperature goes down, so does the ability of the heat pump to produce heat.

It is when the heat pump alone is producing the heat that the unit is economical to operate. The coefficient of performance [(Btuh/watts) x 3.14] is not generally given by the manufacturer, but it has been calculated for the 3-ton unit in Figure 19–58. This coefficient of performance offers a comparison of the heat pump and electric resistance heat. The table shows a COP of 2.46 at 60°F, dropping to 2.17 at 40°F and to 1.71 at 20°F. Due to this fluctuation, it is difficult to give an actual comparison with other fuels and systems. It

Typical 3-ton (36,000 Btuh) heat pump Cooling capacity, 62° outside wet bulb, 85° air temperature entering evaporator.

Outside air temperature db°F	Cooling Btuh	Watts	EER
85	36,000	5,250	6.85
95	34,000	5,400	6.30
105	32,000	5,750	5.56

Heating capacity

Outside air temperature	Heating Btuh	Watts	COP
60	41,000	5,300	2.46
50	38,000	5,100	2.37
45	35,000	4.900	2.37
40	32,000	4,700	2.17
30	26,000	4,300	1.92
20	21,000	3,900	1.71
10	15,000	3,500	1.36
0	10,000	3,100	1.02

Figure 19-58. Typical three-ton heat pump.

depends very much on the geographic area and how many hours it operates at different temperatures.

For comparison, electric heat has a COP of 1.0. For the heat pump to be competitive, it needs to operate at a COP of about 2.9, and at 40°F it does this. But at lower temperatures, it becomes less and less efficient.

Since the heat pump puts out 21,000 Btuh at 20°F, the rest of the heat is supplied by electric heat resistance units, which mount inside the evaporator blower discharge area; or else electric duct heaters, which are placed right in the ductwork. These supplemental units run on electricity and provide the standard 3,413 Btuh per 1,000 W (watts) (1 kW). These electric resistance units are generally available in increments of 3, 5, 10, 15 and 20 kW. Many of the smaller heat pump units have space to install only the smaller sizes, perhaps to a limit of 10 kW, and the manufacturer's information must be checked. The large units, such as the 3-ton unit being discussed here, will take the 15-kW unit.

The 15-kW unit would be made up of three 5-kW elements, which would operate at three different stages set to provide the required heat to the space. The first stage is controlled by the room thermostat, and it operates on a 2°F differential from the setting on the room thermostat. This means that if the room thermostat is set, say, at 70°F, as the thermostat calls for heat, the heat pump goes on and begins to send heat to the space. If the temperature of the room falls below 68°F, then the first 5-kW unit would come on and begin to provide additional heat to supplement the heat pump. The second and third stages of electrical resistance heat are activated by outdoor thermostat settings, which are adjustable (from 50° to 0°F).

This differential of 2°F between the thermostat setting and the room temperature,

which activates the first stage, many mean that in cold weather the room temperature may stabilize at 68°F. This is because 68°F is the temperature at which the first stage shuts off; to get a 70°F reading on those days, it may be necessary to raise the thermostat reading to about 72°F.

The cooling efficiency of the heat pump should also be checked. Many times the less expensive units have much lower efficiency ratings than the individual cooling units. It is important that the most efficient model be selected. For example, the 3-ton unit being discussed in this section has its engineering data shown in Figure 19–58. As the temperature goes up, the cooling Btuh provided goes down; at 85°F it produces 36,000 Btuh for cooling using 5,250 W, while at 95°F it produces 34,000 Btuh using 5,450 W. Its EER ranges from 6.85 to 5.56 as listed in the table. The EER most commonly used for comparison of cooling units is based on its performance at a 95°F temperature; based on that, this unit's EER would be 6.24.

In reviewing the operation and efficiency of the heat pump:

1. The heat pump is more efficient in warmer climates than in cooler areas. In Florida and southern California, for example, heat pumps are reasonably efficient. As you approach "mid-South" states, from North Carolina and Virginia on the east to northern California on the west, the efficiency begins to fall off. In northern states these units would depend on the electric resistance heat for so much of the time that efficiency is minimal.

2. Technology will continue to improve the efficiency of the heat pump. It is part of the designer's responsibility to keep abreast of developments that will provide increased efficiency.

3. A comparison of the heat pump with the other fuels and systems must be made for each geographical area, and its complexity suggests that a computer be used. This type of analysis is available in many areas. However, the designer should be aware that it is being made available by the same people who are selling heat pumps and perhaps by those who sell the electricity to run the unit. This is not to infer that it would not be "technically accurate"; only that the utmost caution should always be used when using calculations, analyses, claims, and the like from any interested party (such as the company selling the unit). The best idea is to check actual installations and compare the costs with those of comparable installations. But be certain that they are comparable installations, that the thermostats are set at about the same temperatures, and that the actual electrical and fuel bills are available (don't rely on word of mouth).

4. While heat pumps are more efficient than electric resistance heat, in northern and mid-South states actual use has shown that it is not less expensive to heat with a heat pump than with oil and natural gas fuels.

5. While the currently available heat pump is a solar unit, it will probably soon be available with a solar collector panel which can be attached to it to provide greater efficiency. This may be one of the first practical, small and mid-sized solar system projects. Currently, at least one manufacturer is working on its development.

6. The shortage of fuel oil due to the embargo in the early 1970's, the shortage of natural gas available to the user, and the general uncertainty over future supplies and prices of these fuels have caused many owners and designers to consider using the heat pump in many areas where it might not ordinarily be considered. In effect, most people feel that, one way or another, at least with the heat pump, the fuel (the electricity) will be available to run the unit and provide the heating and cooling required.

7. It is more efficient and generally less expensive, especially during the heating season, to use one large unit instead of two smaller units, unless one section of the house (or building) will be closed off. For example, based on the data in Figure 19-59, if a building requires a 4-ton cooling unit at 20°F, the heat pump will provide 28,000 Btuh for heating while using 4,800 W. Using two 2-ton units, at 20°F each unit will provide 14,000 Btuh using 2,700 W, for a total of 28,000 Btuh using 5,400 W. Again, this is the type of analysis

4-ton heat pump, heating capacity

Outside air temperature db°F	Heating Btuh	Watts
60	52,000	5,900
40	41,000	5,300
20	28,000	4,800
0	20,000	4.300

2-ton heat pump, heating capacity

Outside air temperature db°F	Heating Btuh	Watts
60	31,000	3,850
40	22,000	3,300
20	14,000	2,700
0	8,200	2,200

Figure 19-59. Four-ton and two-ton heat pumps.

that the designer must consider. Now, if part of the building were to be closed off—say, a large bedroom wing—with the temperature set quite low (about 50°F), it might be more economical to have two units—in this case, one unit to serve the bedrooms and one for the rest of the building.

Chapter 20

Electrical Systems

20.1
CODES

All buildings require electrical systems to provide power for the lights and to run various appliances and equipment. The safety of the system is of prime importance, and minimum requirements are included in building codes. Most applicable codes have separate electrical sections, or else completely separate electrical codes are prepared. In addition, many local codes make reference to the *National Electric Code* (NEC) or are based on the national code. The designer must first determine what code is applicable to the locale in which the building will be built, and then be certain that the electrical design is in accordance with the code. The *National Electric Code* is used as a basis for this portion of the text. Generally, these codes place limitations on the type and size of the wiring to be used, the circuit size, the outlet spacings, the conduit requirements, and the like. The tables used in this text are from the 1981 NEC. Be certain that you always have a copy of the latest edition available for your use.

20.2
UNDERWRITERS LABORATORIES (UL)

UL is an independent organization which tests various electrical fixtures and devices to determine if they meet minimum specifications as set up by UL. The device to be tested is furnished by the manufacturer, and if the test shows that it meets the minimum specifications, it will be put on the UL official published list, referred to as "listed by Underwriters Laboratories, Inc." The approved device may then have a UL label on it. A typical UL seal is shown in Figure 20–1, and many consumers will not buy any electrical device which does not have a UL label.

20.3
LICENSES

Most municipalities have laws which require that any person who wishes to engage in the business of installing electrical systems must be licensed (usually by the state or the province). This generally means that the person must have a minimum number of years of experience working with a licensed electrician and must pass a written test which deals primarily with the electrical code being used and with methods of installation.

Figure 20–1. Typical Underwriters Laboratories seal.

By requiring a license it is assured that the electrician knows, at a minimum, the code requirements and the installation procedures. Some areas have no laws requiring that only licensed electricians may install electrical systems, and, in effect, there is no protection for the consumer against an unskilled electrician. Always insist on licensed electricians for all installations.

_____ 20.4 _____
PERMITS

Many municipalities require a permit before any electrical installation may be made on the project. Depending on the municipality, a complete electrical drawing may be required and may even be reviewed by the municipality before installation may begin, while others may require no drawings at all. In general, most municipalities that require electrical permits also require licensed electricians.

In addition, these municipalities will probably have electrical inspectors, trained personnel who check the project during regularly scheduled visits. Typically, they will want to inspect the installation after the rough wiring is in and before it is concealed in the construction, and again when all of the fixtures and devices are installed and wired back to the panel and the service and meter installed.

On large projects it may be necessary for many electrical inspections, since the work may be done in stages. For example, conduit which will be encased in concrete may have to be checked before the concrete is poured, and conduit to be built into the masonry walls will have to be checked before the walls are begun. These types of covering will occur throughout the project. Be certain that the installer and the designer are aware of scheduled inspections and of what will be inspected. Also, it is

important that close coordination and cooperation be maintained with the inspector since he could slow down the progress of the work if he does not make his inspections promptly. Whenever possible, he will need to know as early as possible when inspections will be necessary.

_____ 20.5 _____
TERMINOLOGY

Circuit: Two or more wires which carry electricity from the source to an electrical device and back.

Circuit Breaker: A switch which automatically stops the flow of electricity in a particular circuit when the circuit is overloaded.

Conductor: The wire used to carry electricity.

Conduit: A channel or tube which is designed to carry the conductors in locations where the conductors need protection.

Convenience Outlet: An outlet which receives the plugs of electrical devices such as lamps, radios, clocks, etc. Also referred to as a *receptacle.*

Fixtures: The lighting fixtures used. They may be wall or ceiling mounted, recessed or surface mounted. Also included are table and floor lighting fixtures.

Ground: To minimize injuries from shock and possible damage from lightning, the electrical system should be equipped with a wire (called a ground) that connects to the earth.

Service Entrance: The wires, fittings and equipment which bring the electricity into the building.

Service Panel: The main panel which receives the electricity at the service entrance, breaks it down and distributes it through the various circuits.

Switch: The control used to turn the flow of electricity on or off to the electrical device to which it is connected.

20.6
AMPS, OHMS, VOLTS

The design of electrical systems in a building requires that the designer have a "working familiarity" with amperes, ohms and volts. These three electrical terms are often used by the designer to determine the total electrical load requirements of the building, and they are all related.

Ampere (Amp, Amperage—A): A unit (or measure) of the flow of electrons passing through a circuit (current).

Volt (Voltage—V): The unit of electrical pressure required to push the amperage through the circuit.

Ohm (Ω): The unit of electrical resistance which resists the flow of electrons through the circuit. Ohm's law states that the current in an electrical circuit is equal to the pressure divided by the resistance.

$$\text{Amps} = \frac{\text{Volts}}{\text{Ohms}}$$

Watt (W): The unit of electrical energy or electrical power. It indicates how much power has been used.

$$\text{Watts} = \text{Amps} \times \text{Volts}$$

Kilowatt (kW): 1,000 watts; for example, 9,500 W equals 9.5 kW.

20.7
SERVICE ENTRANCE

The service entrance furnishes electricity to a house; a three wire service bringing in 120/240 volt, single-phase power is generally standard. These three wires attach to the house at a mast (Figure 20–2), are installed underground (Figure 20–3) and run through a metal conduit through the meter and into the service panel. The meter and service panel should be placed as close to the mast as possible. This will place the main breaker as close to the meter as possible.

The location of the service entrance is generally controlled by economics. Often the service is located near the point of greatest power usage (in a house, this is usually the kitchen) since larger wires are needed where high usage is indicated, and the closer the entrance to the high-usage area, the less of the larger, more expensive wiring is required (Figure 20–4).

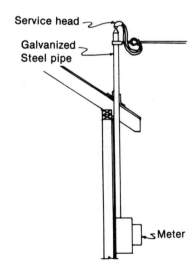

Service head

Galvanized Steel pipe

Meter

Figure 20-2. Overhead service.

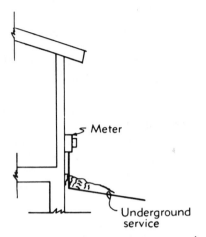

Figure 20-3. Underground service.

Electric meters are weatherproof and should be located on the outside of the house so readings can be taken by the power company without disturbing anyone or when no one is at home.

In many locales service is brought to the buildings in electrical wires below ground (Figure 20-3). In this manner, unsightly wires, masts and fittings are concealed. When the entire community is served by underground service, the power poles and lines are never missed. However, this is more expensive than running services above ground.

The sizing of the service entrance, based on the amount of power it must supply, should be guided by the following information:

100 amperes: Adequate power to provide general-purpose circuits, water heater, electric laundry, and cooking.

150 amperes: Adequate power to provide general-purpose circuits, water heater, electric laundry and cooking and, for a small house, the air conditioning and heating for the house.

200 amperes: Adequate power to provide general-purpose circuits, water heater, elec-

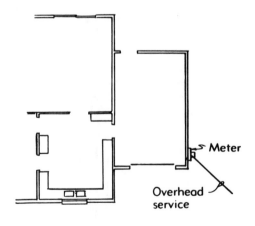

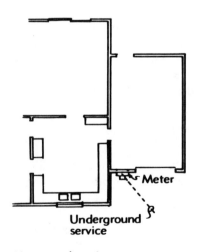

Alternate location

Figure 20-4. Service location.

tric laundry and cooking, air conditioning, and heating.

Once all the circuits, fixtures, appliances, and type of heating are determined, the size of the service entrance may be calculated by experienced personnel. It is best to estimate a little high when selecting the size of the service entrance since it is much more expensive to increase the size later.

20.8
SERVICE PANEL

This distribution box is the main panel (Figure 20-5) that receives the service electricity, breaks it down, and distributes it through branch circuits to the places where the electricity is needed. Inside the panel is a main disconnect switch which cuts off power to the entire building and the circuit breakers (or fuses) which control the power to the individual circuits serving the house.

The circuit breakers (or fuses) are protective devices which will automatically cut off power to any circuit which is overloaded or short-circuited. The circuit breaker may be turned back on by flipping its surface-mounted switch; if the circuit is still overloaded, the switch will immediately flip off again. Fuses burn out when the circuit is overloaded and must be replaced to activate the circuit again. Circuit breakers and fuses are sized as to the amperage they will carry, and when replaced, the replacement should have the same rating.

The service panel is sized to match the service coming in, commonly 100, 150 or 200 amps. The number of circuits required in the building will determine the size of the panel. Panels are rated in amperage and poles, the poles being how many circuits it will handle. All 120-volt circuits, as required for most lighting, convenience outlets, and appliances,

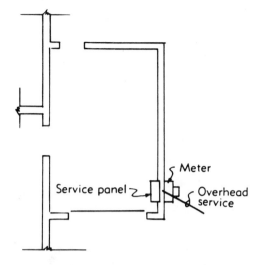

Figure 20-5. Service panel.

require one pole each. All appliances such as ranges, clothes dryers, hot water heaters, large air conditioners, and many motors require 240 volts, and two poles are required. The panel selected should have extra poles so that additional circuits may be run if and when they may be desired.

The panel may be surface mounted or recessed and should be conveniently located for easy servicing. It should not be located anywhere there is a possibility of water being on the panel or on the floor around it. Typically, it is located in a garage, corridor, basement, or utility room of the building.

20.9
FEEDER CIRCUITS

On large buildings where the wiring for circuits would have long runs, a feeder circuit may be run from the service panel to a sub-distribution panel (Figures 20-6 and 20-7). Locating the subdistribution panel conveniently to service the larger feeder conductors

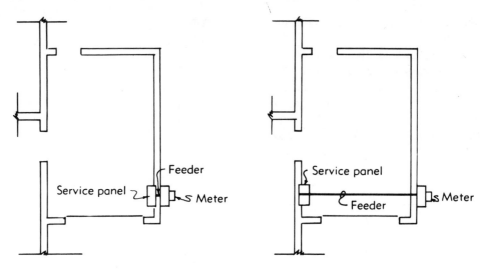

Figure 20-6. Feeder circuit.

(wires) will allow a minimal voltage drop when compared with the excessive voltage drop that occurs when branch circuits are in excess of 75 to 100 feet long.

20.10
BRANCH CIRCUITS

The branch circuit connects the service panel to the electrical device it supplies (Figure 20-8). It may supply power to a single device such as a water heater, range, or air condi-tioner, or it may service a group or series of devices such as convenience outlets and lights. The branch circuit may have a variety of capacities, such as 15, 20, 30, 40 or 50 amps, depending on the requirements of the elec-trical devices serviced. The general-purpose circuits for lights and convenience outlets are usually sized at 15 or 20 amps. One general-purpose branch circuit will provide a max-imum of 20 amps × 120 volts = 2,400 watts, of which the Code allows 80 percent. There is a tendency for people to put higher wattage bulbs (luminaires) in lighting fixtures as they

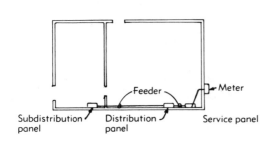

Figure 20-7. Feeder circuit.

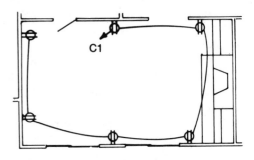

Figure 20-8. Typical branch circuit.

need replacing; a group of light fixtures totalling 1,200 to 1,600 watts may be placed on a circuit and still allow for additional future usage. The same is true with convenience outlets; generally no more than six should be on a circuit.

The wiring layout for most buildings shows the branch circuit arrangements. If not shown on the layout, it will be left to the electrical contractor to decide on groupings (Figure 20–9).

20.11
RECEPTACLES

Receptacles, also referred to as convenience outlets (Figure 20–10), are used to plug in lights and small appliances around the house.

Each room should be laid out with no less than one outlet per wall and with outlets no more than 10 feet apart. The amount, location, and type will vary with the room, depending on both the design of the room and the furniture layout. Duplex outlets (two receptacles) are most commonly used, but single and triple receptacles are also used. Also used are strips which allow movement of the receptacle to any desired location; these strips are available in 3-foot and 6-foot lengths and may even be used around the entire room. When specified, one of the receptacles may be controlled by a wall switch. This is particularly desirable in rooms where portable fixtures are used. Typical room layouts are shown in Figure 20–11.

Some room designs make it difficult to locate the outlets on the walls. Plans such as

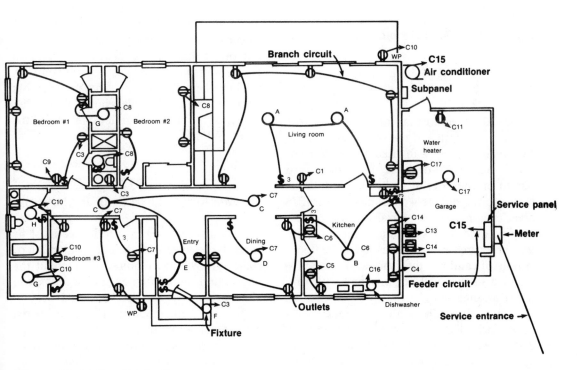

Figure 20-9. Circuit labelling.

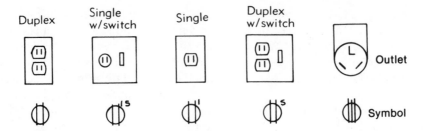

Figure 20-10. Typical receptacles.

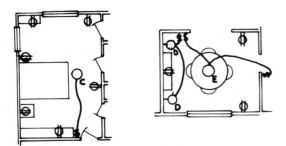

Figure 20-11. Typical room layouts.

Figure 20–12 require that furniture be located out from the wall and the space between the furniture and the wall used for pedestrian traffic out to the deck. If the outlets were placed on the exterior wall, the cords from lamps to the outlets would cross the traffic area, creating a safety hazard. Floor outlets may be used in this situation; these outlets may be located anywhere desired in the floor. A note of caution—they should be carefully planned so they will not interfere with furniture arrangement.

Ranges, dryers, large air conditioners, and other such electrical devices which require 240-volt service require special outlets. These special three-prong outlets (Figure 20–10) are designed so that conventional 120-volt de-

vices cannot be plugged into them. The symbol used will vary only slightly from that used for 120-volt convenience outlets in that there are three straight lines through the circle instead of two. Typically each one of these special outlets is on a circuit itself.

Two commonly used specialty outlets are the split-wired outlet and the weatherproof outlet. A split-wired outlet (Figure 20–13) is any outlet which has the top outlet on a different circuit than the bottom or which has one outlet that may be switched off and on from a wall switch. The weatherproof outlet is used in all exterior locations because it resists damage from weather. It is noted on the plan by use of the standard outlet symbol and the letters WP next to it.

_____ 20.12 _____

LUMINAIRES (LIGHTING FIXTURES)

The luminaires used in a residence must be carefully selected with the client. Lighting of some type will be required throughout the house, and quite often exterior lighting will also be required. The size, type, and location of fixtures throughout must be coordinated with the style of the house and the client's preferences. The lighting may come from built-in and surface-mounted ceiling and wall luminaires; floor and table lamps may be used

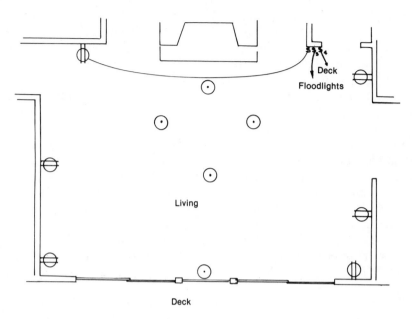

Figure 20-12. Floor receptacles.

as a supplement, or they may be preferred throughout.

In the living areas of the house the client might prefer various types of accent and indirect lighting.

Fluorescent lighting (tube) is more efficient than incandescent lighting (bulb), and it is effective as indirect lighting in valances and coves around rooms and also where high lighting levels are desired, such as in bathrooms, work areas in the kitchen, and workshop areas. Objections to fluorescent lighting include its higher initial cost, slow starting, and a tendency to flicker.

Many clients prefer a permanent ceiling fixture that can be turned on from the wall switch, while others perfer that the wall switch activate a convenience outlet, and they use a table or floor lamp in that outlet to light the room. Small rooms frequently have one cen-

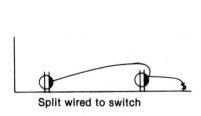

Split wired to switch

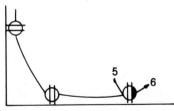

Split wired to different Circuits

Figure 20-13. Split-wired outlets.

tral ceiling fixture activated by a wall switch. This provides the general illumination required and may be supplemented by wall, floor, and table fixtures. Shallow closets generally don't require fixtures if there is average room illumination. However, many clients want fixtures in the closets, and, if so, a ceiling fixture may be used. The closet fixture may be operated by a pull chain, but, again, many clients prefer either a wall switch just outside the door or a door-operated switch which turns the light on when the door is opened and shuts it off when the door is closed. The lighting symbols used are shown in Figure 20-14. For those fixtures activated by a switch, a line must be shown connecting the switch and the fixtures it controls.

_____ 20.13 _____

SWITCHES

Wall switches are used to control lighting fixtures in the various rooms of the house, and they may also be used to control convenience outlets. The switch most commonly used is the *toggle switch* which has a small arm which is pushed up and down. Also used is the *mercury switch,* a completely noiseless switch which has mercury in a sealed tube; when the switch is turned on, the mercury completes the circuit.

Wall switches are located 4 feet above the floor and a few inches in from the door frame on the latch side (doorknob side) of the entry door. Generally, they are located just inside the room.

Rooms with two entrances often have two switches controlling the fixture, referred to as three-way switches. When the fixture is to be controlled from three locations, it will require the use of two 3-way switches and one 4-way switch (Figure 20-15).

A dimmer control is used to vary the intensity of the light from very low to bright. A delayed action switch is used when it is desirable to have the lights go out about a minute after the switch is turned off.

Low-Voltage Switching

The flexibility desired in home lighting systems has led to the development of low-voltage wiring which allows flexibility in the control of fixtures through the use of centrally located remote-control switches which control any or all of the lights in the house.

Low-voltage wiring (about the size of wiring used to wire a door chime) connects the wall switch to a relay center. When the switch is turned on, it sends a low voltage (usually 24 V) to the relay center; this triggers a relay switch which activates the line voltage (120 V)

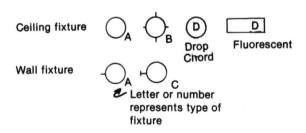

Figure 20-14. Fixture symbols.

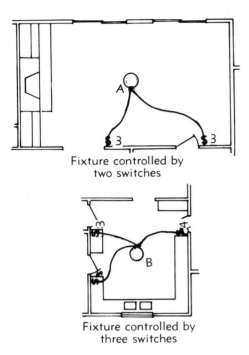

Fixture controlled by
two switches

Fixture controlled by
three switches

Figure 20-15. Switches.

at the light. The savings in wiring costs, the convenience of a centrally located remote-control selector switch, and the elimination of any possible electric shocks at the switch make this increasingly popular in homes.

20.14
CONDUCTORS

The wires used to supply electricity throughout the system are called *conductors.* Copper has traditionally been used as the conductor material, and the wiring practices developed over the years have been based on its use. Aluminum conductors are also available and are allowed in many, but not all, codes. Further engineering study is now going on to be certain that all aluminum conductor

installations will be as safe as possible. One of the biggest reasons for using aluminum conductors is that their costs range from one-third to one-half those of copper conductors. Aluminum conductors must be larger than copper conductors that carry the same amperage. This means that for installations requiring that the conductors be placed in conduit, the larger aluminum conductors may require larger conduits. (Conduits are discussed in Section 20.15.)

The various types of insulation which are placed on the conductor wires have been standardized and are listed in Figure 20-16. The types of conductors are referred to by the type letter assigned to the insulation used. For example, RHW conductor has a moisture- and heat-resistant rubber insulation, with a maximum operating temperature of 75°C (167°F), and it may be used in both dry and wet locations. The maximum operating temperature has an effect on the allowable ampacities which the NEC will allow for the conductor (Figure 20-29); the higher the maximum operating temperature, the higher the allowable ampacity. All individual conductors must be protected by raceways. Individual conductors in raceways are used extensively in commercial and industrial installations where changes in the electrical requirements are likely to occur.

When two or more conductors, each insulated separately, are grouped together in one common covering, they are referred to as *cables* (Figure 20-17). These cables are used extensively in electrical wiring, particularly in residences. They are designated according to type of insulation used and where they are used; generally, they do not have to be protected by raceways.

All conductor sizes are given by an AWG or MCM number. Standard available conductor sizes are given in Figure 20-28. Those conductor sizes based on the American Wire

Trade Name	Type Letter	Max. Operating Temp.	Application Provisions	Insulation
Heat-Resistant Rubber	RH	75°C 167°F	Dry locations.	Heat-Resistant Rubber
Heat-Resistant Rubber	RHH	90°C 194°F	Dry locations.	
Moisture and Heat-Resistant Rubber	RHW	75°C 167°F	Dry and wet locations. For over 2000 volts insulation shall be ozone-resistant.	Moisture and Heat-Resistant Rubber
Heat-Resistant Latex Rubber	RUH	75°C 167°F	Dry locations.	90% Unmilled, Grainless Rubber
Moisture-Resistant Latex Rubber	RUW	60°C 140°F	Dry and wet locations.	90% Unmilled, Grainless Rubber
Thermoplastic	T	60°C 140°F	Dry locations.	Flame-Retardant, Thermo-plastic Compound
Moisture-Resistant Thermoplastic	TW	60°C 140°F	Dry and wet locations.	Flame-Retardant, Moisture-Resistant Thermo-plastic
Heat-Resistant Thermoplastic	THHN	90°C 194°F	Dry locations.	Flame-Retardant, Heat-Resistant Thermo-plastic
Moisture- and Heat-Resistant Thermoplastic	THW	75°C 167°F 90°C 194°F	Dry and wet locations. Special applications within electric discharge lighting equipment. Limited to 1000 open-circuit volts or less. (Size 14-8 only as permitted in Section 410-31.)	Flame-Retardant, Moisture- and Heat-Resistant Thermo-plastic

For insulated aluminum and copper-clad aluminum conductors, the minimum size shall be No. 12.

Figure 20–16. Conductor insulation.

Trade Name	Type Letter	Max. Operating Temp.	Application Provisions	Insulation
Moisture- and Heat-Resistant Thermoplastic	THWN	75°C 167°F	Dry and wet locations.	Flame-Retardant, Moisture- and Heat-Resistant Thermo-plastic
Moisture- and Heat-Resistant Cross-Linked Synthetic Polymer	XHHW	90°C 194°F 75°C 167°F	Dry locations. Wet locations.	Flame-Retardant, Cross-Linked Synthetic Polymer
Moisture-, Heat- and Oil-Resistant Thermoplastic	MTW	60°C 140°F 90°C 194°F	Machine tool wiring in wet locations as permitted in NFPA Standard No. 79. (See Article 670.) Machine tool wiring in wet locations as permitted in NFPA Standard No. 79. (See Article 670.)	Flame-Retardant, Moisture-, Heat- and Oil-Resistant Thermo-plastic
Silicone-Asbestos	SA	90°C 194°F 125°C 257°F	Dry locations. For special application.	Silicone Rubber
Fluorinated Ethylene Propylene	FEP or FEPB	90°C 194°F 200°C 392°F	Dry locations. Dry locations — special applications.	Fluorinated Ethylene Propylene
Modified Fluorinated Ethylene Propylene	FEPW	75°C 90°C	Wet locations. Dry locations.	Modified Fluorinated Ethylene Propylene
Modified Ethylene Tetrafluoro-ethylene	Z	90°C 194°F 150°C 302°F	Dry locations. Dry locations — special applications	Modified Ethylene Tetrafluoro-ethylene
Modified Ethylene Tetrafluoro-ethylene	ZW	75°C 167°F 90°C 194°F 150°C 302°F	Wet locations. Dry locations. Dry locations — special applications	Modified Ethylene Tetrafluoro-ethylene

Figure 20-16. (Continued).

Trade Name	Type Letter	Max. Operating Temp.	Application Provisions	Insulation
Varnished Cambric	V	85°C 185°F	Dry locations only. Smaller than No. 6 by special permission.	Varnished Cambric
Asbestos and Varnished Cambric	AVA	110°C 230°F	Dry locations only.	Impregnated Asbestos and Varnished Cambric
Asbestos and Varnished Cambric	AVL	110°C 230°F	Dry and wet locations.	
Asbestos and Varnished Cambric	AVB	90°C 194°F	Dry locations only.	Impregnated Asbestos and Varnished Cambric
Asbestos	A	200°C 392°F	Dry locations only. Only for leads within apparatus or within raceways connected to apparatus. Limited to 300 volts.	Asbestos
Asbestos	AA	200°C 392°F	Dry locations only. Only for leads within apparatus or within raceways connected to apparatus or as open wiring. Limited to 300 volts.	Asbestos
Asbestos	AI	125°C 257°F	Dry locations only. Only for leads within apparatus or within raceways connected to apparatus. Limited to 300 volts.	Impregnated Asbestos
Asbestos	AIA	125°C 257°F	Dry locations only. Only for leads within apparatus or within raceways connected to apparatus or as open wiring.	Impregnated Asbestos
Paper		85°C 185°F	For underground service conductors, or by special permission.	Paper

For insulated aluminum and copper-clad aluminum conductors, the minimum size shall be No. 12.

Figure 20-16. (Continued).

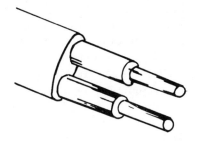

Figure 20-17. Cable.

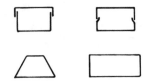

Figure 20-18. Surface raceway shapes.

Gauge (from No. 16 to No. 4/0) are listed as AWG—for example, No. 12 AWG. The larger the AWG number, the smaller the conductor. The cross-sectional area of any conductor is listed in *circular* mils, with a circular mil defined as the area of a circle one mil in diameter; then the area is the square of the diameter (in mils). All conductors larger than 4/0 AWG are sized in direct relation to the circular mil and are labeled MCM or one thousand (M) circular mils. A wire with an area of 500,000 circular mils would be called 500 MCM.

20.15
CONDUIT

A channel which is designed exclusively, and used solely, for holding wires, cables, or bus bars is called a *conduit* or *raceway*. The conduit may be made of metal or insulating material. Metal conduits include rigid and flexible electric metallic tubing (EMT) and cellular metal raceways, and all may be concealed in the construction or exposed in the space.

Electrical metallic tubing (EMT) is a thin-walled metallic conduit which weighs about one-third less than rigid metal conduits. It may be used anywhere except where it will be subject to severe physical damage during or after installation, in cinder concrete or fill underground, or in any hazardous locations. This type of conduit connects to its fittings with set screws, saving the time often required to put screw threads on rigid metal conduits.

Surface metal raceways (Figure 20-18) must be installed in dry locations and must not be concealed in the construction. This type of raceway is used extensively with metal partitions (Figure 20-19) where the raceway has a backplate which attaches to the metal stud and a snap-on cover. The NEC limits conductor sizes used in this type of raceway to No. 6 AWG and smaller. The number of conductors allowed must be taken from the raceway manufacturer's data sheet and is based on the amount which Underwriters Laboratories will allow. This type of installation is used extensively when remodeling, adding to existing systems and building new additions where room arrangements will be subject to changes.

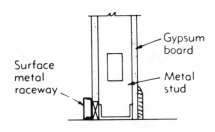

Figure 20-19. Surface metal raceway.

20.16
PRELIMINARY INFORMATION ON ELECTRICAL SYSTEM DESIGN

Before actually beginning the design layout of the project, the designer will need to accumulate certain information:

1. Determine whether electrical service is available. If it is not, arrangements must be made with the power company to extend service to the building site. Large housing projects may require more voltage or more wattage than existing service can supply. Each of these situations requires coordination with the power company as early in the design stage as possible. Costs which may have to be paid by the owner should be thoroughly discussed, written and given to the owner.

2. Obtain a list from the owner of all the types of equipment, appliances, etc. which will require electricity. While the electrical designer will know the electrical requirements of much of the equipment, it may be necessary to find the manufacturer's specifications for certain items, such as motor sizes and power required.

3. Working with the architectural designer, locate all of the equipment and appliances on the floor plan. There are times when the type of equipment used and its location must be approved by governmental agencies.

4. Review with the architect where the basic mechanical equipment, such as the service entrance, the power and lighting panels, and the conduit or cable, will be located.

5. Discuss with the owners their future plans for adding to the building, remodeling, constructing other buildings, increasing future equipment requirements, and anything else that could affect the size and location of the electrical service. Many times the service entrance must be sized to anticipate future expansion as well as present building plans. Once the basic information has been gathered, the designer can begin to design the system.

System Layout

The system layout is made on the floor plan of the building.

1. Locate on the floor plan all connections to be made for appliances and equipment. This calls for close coordination with the architectural designer since each appliance and piece of equipment must be known in order to locate the receptacles properly. Many times the plans do not show everything which must be connected. For example, a garbage disposal must be connected to power and may even have a switch on the wall to turn it on and off; yet it is seldom shown on the floor plan. Specifically, the electrical designer should ask the architectural designer to list all appliances and equipment in a letter or on the drawing for complete coordination. A checklist of typical appliances that require electrical connections is listed in Figure 20–20. In addition, the electrical designer will need to know the voltage and amperage required. Typical requirements for various appliances are shown in Figure 20–20, but these may vary among manufacturers and should be checked.

Also, check the number of appliances or equipment. For example, quite often two water heaters are used in residences and both will need connections. Also, in some residences and many apartment buildings, more than one heating and/or cooling unit may be used.

Device	Watts
Air conditioner, central	3,000 to 5,000
Air conditioner, room	800 to 1,500
Clothes dryer	4,000 to 8,000
Garbage disposal	300 to 500
Heat pump	3,000 to 6,000
Humidifier	80 to 200
Iron, hand	600 to 1,200
Lamp, incandescent	10 to 250
Lamp, fluorescent	15 to 60
Radio	40 to 150
Range	8,000 to 14,000
Range, oven	4,000 to 6,000
Range, top	4,000 to 8,000
Television	200 to 400
Water heaters	2,000 to 5,000

Figure 20-20. Residential appliance check-list.

2. Locate the lighting fixtures.

Make a list of what types of fixtures will be used throughout and how many watts each will use. This list will be used later when grouping circuits, and it must also be included in the specifications or on the drawings as a "Fixture Schedule" (Figure 20-21). The list is also used by the electrical estimator when he is determining the cost to be charged for the work, then by the electrical purchasing agent when he orders the material for the project and then by the electrician who actually installs the work. Many times the architectural designer decides what fixtures are to be used, often in coordination with the electrical designer.

When a minimum amount of illumination is required, the electrical designer may have to calculate the number and types of lighting fixtures which should be used.

3. The circuit layout may be an *individual branch circuit* which feeds only one receptacle, light, appliance, or piece of equipment or a *branch circuit* which supplies power to two or more receptacles, lights, appliances or equipment. An example of an individual branch circuit would be the circuit to a clothes dryer (240 V) and back to the electrical panel. The NEC states that an individual branch circuit should supply any load required to service that single item.

15- and 20-ampere branch circuits are used for receptacles, light fixtures, and small appliances. They are limited according to what will be connected to them.

a. When the circuit serves fixed appliances, light fixtures, or portable appliances, the total of the fixed appliances shall be no more than 50 percent of the branch circuit rating. Assuming a 15-amp, 120-V branch circuit, it would have a maximum rating of 15 amps x 120 V = 1,800 W (refer to Section 20.6 for formula explanation). In this case, the fixed appliances would be limited to 900 W, leaving the other 900 W available to supply the light fix-

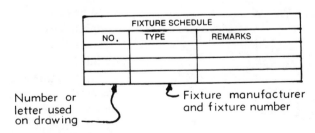

Number or letter used on drawing ——— Fixture manufacturer and fixture number

Figure 20-21. Fixture schedule.

tures or portable appliances also served by the branch circuit. A 20-amp, 120-V branch circuit would have a maximum of 2,400 W.

b. When portable appliances will be used on a circuit, the limit for any one portable appliance is 80 percent of the branch circuit rating.

c. Receptacles are computed at a load of 1½ amps each, and limited to 80 percent of the rating. This limits a branch circuit serving only receptacles to its rating divided by 1½ amps. For example, a 15-amp circuit is limited to a maximum of 8 outlets and a 20-amp circuit to 10 outlets.

d. A minimum of two 20-amp circuits is required for small appliances in the kitchen, laundry, dining room, family room, and breakfast or dinette area. These are in addition to the other receptacles required, and no lights or fixed appliances should be connected to these circuits. A typical layout of the two circuits is shown in Figure 20–22.

e. A minimum of one 20-amp circuit is required as an individual branch to the laundry room receptacle. Any other

special requirements, such as an electric clothes dryer requiring 240-V service, must also be added.

The designer must be certain that he doesn't exceed the code requirements. In actual practice, the designer tends to be a little more conservative, generally limiting a 15-amp branch to 1,000 to 1,200 W and a 20-amp branch to 1,300 to 1,600 W. Receptacles are generally limited to about six on a circuit. This allows the circuit to take additional loads, such as when higher wattage bulbs are used to replace those originally installed and calculated. In addition, more and more small appliances and equipment are being purchased and connected to receptacles (for example, air purifiers, humidifiers, stereos, and the like). The designer must also try to anticipate any future requirements. Such a layout allows for the extension of a circuit if it is necessary to add a light or a receptacle instead of adding a whole new circuit from the panel. If it is possible that the occupant will want to install individual air conditioners, an individual branch circuit to each receptacle required may be desired.

30-, 40-, 50-, and 60-ampere branch circuits will be used for fixed appliances, equipment, and heavy-duty lampholders (in other than residential occupancies). Generally, the electrical requirements of the connected load must be determined, and the total load connected to the branch circuit should be limited to 80 percent of the branch circuit rating.

For example, if the motor to operate the air conditioner requires 22 amps, it will require a minimum branch circuit of 22 amps x 1.25 percent = 27.5 amperes, and a 30-ampere branch cir-

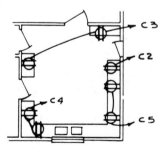

Figure 20–22. Typical kitchen circuit.

cuit will be used. This means that the electrical designer will need the manufacturers' data for all equipment and appliances selected to be certain the proper branch circuit sizes are used. Many times the fuse sizes required for individual branch circuits are listed in the manufacturers' data.

4. Lay out the switches required to control the lights, appliances, equipment and any desired receptacles. The discussion of switches in Section 20.13 generally outlines where they are most commonly used and the symbols used.

5. Calculate the electrical load, the total of all general lighting, appliance and equipment loads in the building.

   a. *The general lighting load* is calculated for all types of occupancies (Figure 20–31) based on the unit load given in the table (in watts) times the square footage of the building.

      The square footage shall be determined using the outside dimensions of the building involved, and the number of stories. For dwellings do not include any open porches, garages, or carports. Any unfinished or unused spaces do not have to be included in the square footage *unless* they are adaptable for future use.

   b. *The appliance and laundry circuit load* is calculated next.

      Since the code requires two 20-amp branch circuits, the load would be based on 1,500 W (from the code) for each branch circuit. In addition, one 20-amp circuit is required for laundry room appliances. This results in a total of three 20-amp branch circuits for appliances.

   c. *Subtotal* the general lighting, appliance and laundry branch circuit loads.

   d. *The demand load* allowed by the code takes into account the fact that all of the electrical connections will not be in use at one time. While there are limits to this reduction for certain types of occupancies, in a dwelling the first 3,000 W are taken as 100 percent, and from 3,000 to 120,000 W, only 35 percent of the load is calculated (from Figure 20–32).

      The loads of all other appliances and equipment (motors) must be added to this demand load in order to determine the total service load on the system.

   e. To determine the *appliance and equipment load,* all appliances and equipment which will not be on the lines discussed above must be listed along with their electrical requirements. While typical ratings are given in Figure 20–20, it is most important that the manufacturer's data be used in the design.

      The demand load for an *electric range,* consisting of an oven and a countertop cooking unit, is taken from Figure 20–33.

      The demand load for a *clothes dryer* is the total amount of power required according to the manufacturer's data.

      The demand for *fixed appliances* (other than the range, clothes dryer and air conditioner and space heating equipment) is taken as 100 percent of the total amount they require; *except* that when there are four or more of these fixed appliances (other than those omitted), the demand load can be taken as 75 percent of the fixed appliance load.

      Motors, such as those used in *central air conditioners,* have their demand loads calculated as 125 percent of the motor rating.

The demand load for all of the lighting and appliances has now been calculated, and it should be tabulated.

6. Size the minimum service entrance based on the demand load from step 5. Service entrances and the typical sizes used for residential work are discussed in Section 20.7. The service entrance size is found by dividing the demand load for the building by the voltage serving the building. Most commonly 240-V service is used.

7. Sizing the feeder (the circuit conductors between the service equipment and the branch overcurrent device—circuit breaker or fuse) is the next step. The feeder size is based on the total demand load calculated. The size is then selected from the table in Figure 20–29. The table lists the size of the conductor in the left column and the different insulation temperature ratings across the top. Generally, conductors with insulation ratings of 60°, 75°, and 95°C (140°, 167° and 185°F) are most commonly used.

For example a No. 6 copper conductor, 60°C insulation, will carry 55 amps without subjecting the insulation to damaging heat. The same No. 6 conductor with 75°C insulation can safely carry 65 amps. These values must be adjusted using a correction factor from the lower portion of the table if the room temperatures are within certain ranges. For example, if the No. 6 conductor with 60°C insulation will carry 55 amps and if the room temperature is 45°C (113°F), the load-carrying capacity would be corrected by a factor of 0.58; 55 amps x 0.58 = 31 amps, the maximum safe allowable amperage.

The size of the *neutral feeder conductor* may be determined as 70 percent of the demand load calculated for the range (step 5) plus all other demand loads on the system.

Now size the feeder conduit. The minimum conduit size is determined for new work by using the table in Figure 20–30. The conductor sizes are listed in the left column, and conduit (or tubing) sizes across the top. The numbers within the body of the table indicate the maximum number of conductors of any given size which can be put in the conduit in accordance with the NEC. First, find the conductor size being used in the left column. Then move to the right until the number shown is the same as, or larger than, the number of conductors to be placed in the conduit. Then move vertically upward and read the conduit size.

8. Determine the minimum number of lighting circuits by dividing the general lighting load by the voltage, finding the amperage required and dividing the amperage into circuits.

9. Lay out all branch circuits on the drawing. The branch circuits for receptacles and switching are discussed in Sections 20.10 and 20.16. The symbols used for the circuits are shown in Figure 20–9. Remember that all of these general-use receptacles and all lighting will use 120-V service.

A typical circuit layout is shown in Figure 20–9. Note that each circuit is numbered, beginning with 1. It will be necessary to know the total number of 120- and 240-V circuits required.

10. Select the lighting panel based on the number of circuits and the required amperage. Be certain that all the pole space is not taken up so there is room for expansion. Remember that each 120-V circuit takes up one pole, while each 240-V circuit takes up two poles.

11. Lay out the panel circuits, either on the drawing or in tabulated form as shown in Figure 20–23. In large designs, with more than one panel, this provides the electrician with a schedule of what circuits will

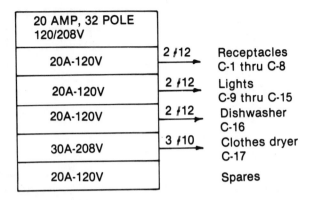

Figure 20-23. Circuits and panels.

be served from what box. It is also used to note the size of the conductors used for circuits (Figure 20–24).

While a lighting panel layout is not often done for a residence, it is helpful to both the electrician and the designer if one is included.

12. Size the conductors for all of the branch circuits, and note them on the panel drawing or in the tabulation. The conductors are sized just as the feeder conductor was sized in step 7.

Branch circuit conductors to general-purpose receptacles and light fixtures must be a minimum of No. 14 AWG when used with a 15-amp overcurrent device (circuit breaker or fuse). Because

No. 14 AWG conductors are limited to a maximum load of 1,725 W and a maximum circuit length of 30 feet, most designers use 20-amp breakers and No. 12 AWG conductors.

Next, the conductor for each individual branch circuit is sized, based on the amperage required and the type of conductor being used. The Code requires limiting of such loads to 80 percent of the ratings. To allow for this, the calculated amperage is multiplied by 1.25.

Example: Air conditioner—11,250 W:

$$11,250 \text{ W} \div 240 \text{ V} = 46.9 \text{ amps}$$
$$46.9 \times 1.25 = 58.7 \text{ amps}$$

150 AMP, 24 POLE SERVICE PANEL	
Recep. & Lights	Recep. & Lights
Recep. & Lights	Recep.
Recep.	Recep.
Recep. & Lights	Recep. & Lights
Recep.	Recep. & Lights

C1 — 2 # 12
C3 — 2 # 12
C5 — 2 # 12
C7 — 2 # 12
C9 — 2 # 12

2 /12 — C2
2 /12 — C4
2 # 12 — C6
2 # 12 — C8
2 # 12 — C10

Figure 20-24. Conductor notation.

From Figure 20-29, use No. 6 AWG, RHW, copper.

Water heater—3,800 W:

$$3,800 \text{ W} \div 240 \text{ V} = 15.8 \text{ amps}$$
$$15.8 \times 1.25 = 19.75 \text{ amps}$$

From Figure 20-29, use No. 12 AWG, RHW, copper.

Note the conductor sizes for each branch circuit on the panel drawing as shown in Figure 20-25.

Whenever there will be more than three conductors in a raceway, the allowable ampacity of each conductor must be reduced. The amount of reduction required is shown in Figure 20-29, note 8.

For example, if six No. 10 AWG copper conductors are placed in a 1-inch conduit, according to the reduction values in Figure 20-29, the allowable ampacity will be reduced to 80 percent of the allowable values given in Figure 20-29. The allowable ampacity for No. 10 AWG copper conductors (Figure 20-29) is 30. Since this must be reduced by 80 percent:

$$30 \times 0.80 = 24 \text{ amps (allowed ampacity)}$$

_____ 20.17 _____

OPTIONAL CALCULATION FOR A ONE-FAMILY RESIDENCE

There is an optional method of calculating the demand load for a residence, based on percentages of the loads listed in Figure 20-26. Under the optional method of calculating the demand load for a one-family residence, a much smaller service is required (117.6 amps compared with 143 amps), and a 150-amp service would still be used. Based on amps, the feeder required (from Figure 20-26) is a No. 1 AWG, RHW copper conductor.

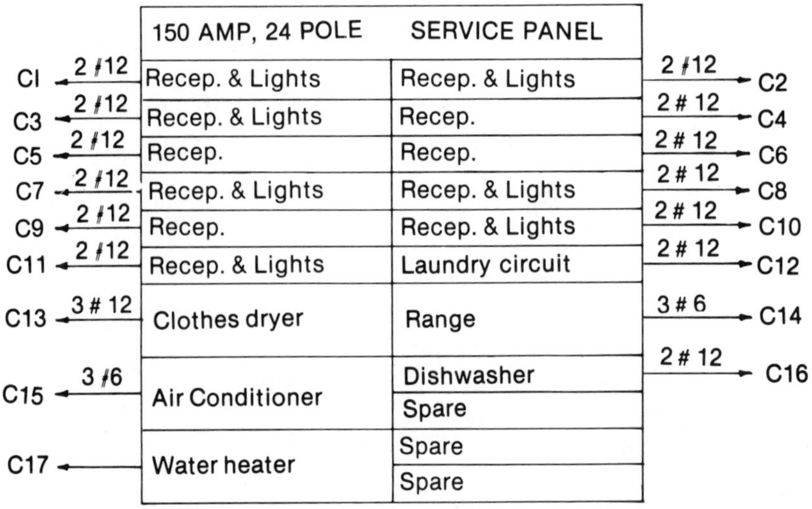

	150 AMP, 24 POLE	SERVICE PANEL	
C1 — 2 /12	Recep. & Lights	Recep. & Lights	2 /12 — C2
C3 — 2 /12	Recep. & Lights	Recep.	2 # 12 — C4
C5 — 2 /12	Recep.	Recep.	2 # 12 — C6
C7 — 2 /12	Recep. & Lights	Recep. & Lights	2 # 12 — C8
C9 — 2 /12	Recep.	Recep. & Lights	2 # 12 — C10
C11 — 2 /12	Recep. & Lights	Laundry circuit	2 # 12 — C12
C13 — 3 # 12	Clothes dryer	Range	3 # 6 — C14
C15 — 3 /6	Air Conditioner	Dishwasher	2 # 12 — C16
		Spare	
C17 —	Water heater	Spare	
		Spare	

Figure 20-25. Conductor notation.

Optional Calculation for Dwelling Unit

Load (in kW or kVA)	Demand Factor Percent
Air conditioning and cooling, including heat pump compressors	100
Central electric space heating	65
Less than four separately controlled electric space heating units	65
First 10 kW of all other load	100
Remainder of other load	40

Reproduced by permission from the National Electrical Code, NFPA 70-81, 1981 edition, Copyright National Fire Protection Association, 470 Atlantic Avenue, Boston, Mass. 02210, 1981.

Figure 20-26. Optional residential demand load.

Neutral feeder calculations are the same as for the feeder *except* for the reduced range load.

20.18
CONDUIT SIZING

Conduits and raceways are discussed in Section 20.15, and in this section, the sizing of rigid metal conduit and electrical metallic tubing is covered. The size of the conduit is directly related to the number and size of the conductors which will be placed in it.

All of the tables to be used are based on the NEC's maximum percentages of the conduit that can be filled with conductors. Based on these percentages, the table in Figure 20-30 gives the maximum number of conductors which can be placed in the trade (standard) sizes of conduit or tubing listed.

Example:

1. What size conduit is required to run three No. 3 AWG conductors in?

   From Figure 20-30, 1¼-inch conduit or tubing is required for three No. 3 AWG conductors.

2. How many No. 1 AWG conductors may be placed in 2-inch conduit or tubing?

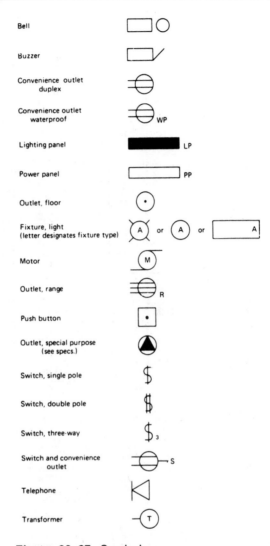

Bell	
Buzzer	
Convenience outlet duplex	
Convenience outlet waterproof	WP
Lighting panel	LP
Power panel	PP
Outlet, floor	
Fixture, light (letter designates fixture type)	A or A or A
Motor	M
Outlet, range	R
Push button	
Outlet, special purpose (see specs.)	
Switch, single pole	
Switch, double pole	
Switch, three-way	
Switch and convenience outlet	S
Telephone	
Transformer	T

Figure 20-27. Symbols.

A maximum of five No. 1 AWG conductors may be placed in 2-inch conduit or tubing.

Remember that when more than three conductors are placed in any raceway (including conduit or tubing), the allowable ampacity of each conductor is derated (reduced) in accordance with note 8 for Figure 20-29.

Properties of Conductors

Size AWG, MCM	Area Cir. Mils	Concentric Lay Stranded Conductors			Bare Conductors		DC Resistance Ohms/M Ft. At 25°C, 77°F.		
		No. Wires	Diam. Each Wire Inches		Diam. Inches	*Area Sq. Inches	Copper		Aluminum
							Bare Cond.	Tin'd. Cond.	
18	1620	Solid	.0403		.0403	.0013	6.51	6.79	10.7
16	2580	Solid	.0508		.0508	.0020	4.10	4.26	6.72
14	4110	Solid	.0641		.0641	.0032	2.57	2.68	4.22
12	6530	Solid	.0808		.0808	.0051	1.62	1.68	2.66
10	10380	Solid	.1019		.1019	.0081	1.018	1.06	1.67
8	16510	Solid	.1285		.1285	.0130	.6404	.659	1.05
8	16510	7	.0486		.1458	.0167	.653	.679	1.07
6	26240	7	.0612		.184	.027	.410	.427	.674
4	41740	7	.0772		.232	.042	.259	.269	.424
3	52620	7	.0867		.260	.053	.205	.213	.336
2	66360	7	.0974		.292	.067	.162	.169	.266
1	83690	19	.0664		.332	.087	.129	.134	.211
0	105600	19	.0745		.372	.109	.102	.106	.168
00	133100	19	.0837		.418	.137	.0811	.0843	.133
000	167800	19	.0940		.470	.173	.0642	.0668	.105
0000	211600	19	.1055		.528	.219	.0509	.0525	.0836
250	250000	37	.0822		.575	.260	.0431	.0449	.0708
300	300000	37	.0900		.630	.312	.0360	.0374	.0590
350	350000	37	.0973		.681	.364	.0308	.0320	.0505
400	400000	37	.1040		.728	.416	.0270	.0278	.0442
500	500000	37	.1162		.813	.519	.0216	.0222	.0354
600	600000	61	.0992		.893	.626	.0180	.0187	.0295
700	700000	61	.1071		.964	.730	.0154	.0159	.0253
750	750000	61	.1109		.998	.782	.0144	.0148	.0236
800	800000	61	.1145		1.030	.833	.0135	.0139	.0221
900	900000	61	.1215		1.090	.933	.0120	.0123	.0197
1000	1000000	61	.1280		1.150	1.039	.0108	.0111	.0177
1250	1250000	91	.1172		1.289	1.305	.00863	.00888	.0142
1500	1500000	91	.1284		1.410	1.561	.00719	.00740	.0118
1750	1750000	127	.1174		1.526	1.829	.00616	.00634	.0101
2000	2000000	127	.1255		1.630	2.087	.00539	.00555	.00885

* Area given is that of a circle having a diameter equal to the overall diameter of a stranded conductor.

The values given in the table are those given in Handbook 100 of the National Bureau of Standards except that those shown in the 8th column are those given in Specification B33 of the American Society for Testing and Materials, and those shown in the 9th column are those given in Standard No. S-19-81 of the Insulated Power Cable Engineers Association and Standard No. WC3-1969 of the National Electrical Manufacturers Association.

The resistance values given in the last three columns are applicable only to direct current. When conductors larger than No. 4/0 are used with alternating current, the multiplying factors in Table 9 compensate for skin effect.

Figure 20-28. Properties of conductors.

Allowable Ampacities of Insulated Conductors
Rated 0-2000 Volts, 60° to 90°C

Not More Than Three Conductors in Raceway or Cable or Earth
(Directly Buried), Based on Ambient Temperature of 30°C (86°F)

Size	Temperature Rating of Conductor, See Table 310-13								Size
	60°C (140°F)	75°C (167°F)	85°C (185°F)	90°C (194°F)	60°C (140°F)	75°C (167°F)	85°C (185°F)	90°C (194°F)	
AWG MCM	TYPES †RUW, †T, †TW, †UF	TYPES †FEPW, †RH, †RHW, †RUH, †THW, †THWN, †XHHW, †USE, †ZW	TYPES V, MI	TYPES TA, TBS, SA, AVB, SIS, †FEP, †FEPB, †RHH †THHN, †XHHW*	TYPES †RUW, †T, †TW, †UF	TYPES †RH, †RHW, †RUH, †THW †THWN, †XHHW, †USE	TYPES V, MI	TYPES TA, TBS, SA, AVB, SIS, †RHH, †THHN, †XHHW*	AWG MCM
	COPPER				ALUMINUM OR COPPER-CLAD ALUMINUM				
18	....	....	....	14	....	....	....	....	....
16	....	....	18	18	....	....	....	....	
14	20†	20†	25	25†	....	....	....	....	
12	25†	25†,	30	30†	20†	20†	25	25†	12
10	30†	35†	40	40†	25†	30†	30	35†	10
8	40	50	55	55	30	40	40	45	8
6	55	65	70	75	40	50	55	60	6
4	70	85	95	95	55	65	75	75	4
3	85	100	110	110	65	75	85	85	3
2	95	115	125	130	75	90	100	100	2
1	110	130	145	150	85	100	110	115	1
0	125	150	165	170	100	120	130	135	0
00	145	175	190	195	115	135	145	150	00
000	165	200	215	225	130	155	170	175	000
0000	195	230	250	260	150	180	195	205	0000
250	215	255	275	290	170	205	220	230	250
300	240	285	310	320	190	230	250	255	300
350	260	310	340	350	210	250	270	280	350
400	280	335	365	380	225	270	295	305	400
500	320	380	415	430	260	310	335	350	500
600	355	420	460	475	285	340	370	385	600
700	385	460	500	520	310	375	405	420	700
750	400	475	515	535	320	385	420	435	750
800	410	490	535	555	330	395	430	450	800
900	435	520	565	585	355	425	465	480	900
1000	455	545	590	615	375	445	485	500	1000
1250	495	590	640	665	405	485	525	545	1250
1500	520	625	680	705	435	520	565	585	1500
1750	545	650	705	735	455	545	595	615	1750
2000	560	665	725	750	470	560	610	630	2000
CORRECTION FACTORS									
Ambient Temp. °C	For ambient temperatures over 30°C, multiply the ampacities shown above by the appropriate correction factor to determine the maximum allowable load current.								Ambient Temp. °F
31-40	.82	.88	.90	.91	.82	.88	.90	.91	86-104
41-45	.71	.82	.85	.87	.71	.82	.85	.87	105-113
46-50	.58	.75	.80	.82	.58	.75	.80	.82	114-122
51-60	....	.58	.67	.71	....	.58	.67	.71	123-141
61-70	....	.35	.52	.58	....	.35	.52	.58	142-158
71-80	....	....	.30	.41	....	....	.30	.41	159-176

† The load current rating and the overcurrent protection for conductor types marked with an obelisk (†) shall not exceed 15 amperes for 14 AWG, 20 amperes for 12 AWG, and 30 amperes for 10 AWG copper; or 15 amperes for 12 AWG and 25 amperes for 10 AWG aluminum and copper-clad aluminum.

* For dry locations only. See 75°C column for wet locations.

Figure 20-29. Allowable ampacities of insulated conductors.

Allowable Ampacities of Insulated Conductors Rated 0-2000 Volts, 60° to 90°C

Single conductors in free air, based on ambient temperature of 30°C (86°F).

Size AWG MCM	Temperature Rating of Conductor, See Table 310-13								Size AWG MCM
	60°C (140°F) TYPES †RUW, †T, †TW	75°C (167°F) TYPES †FEPW, †RH, †RHW, †RUH, †THW, †THWN, †XHHW, †ZW	85°C (185°F) TYPES V, MI	90°C (194°F) TYPES TA, TBS, SA, AVB, SIS, †FEP, †FEPB, †RHH, †THHN, †XHHW*	60°C (140°F) TYPES †RUW, †T, †TW	75°C (167°F) TYPES †RH, †RHW, †RUH, †THW, †THWN, †XHHW	85°C (185°F) TYPES V, MI	90°C (194°F) TYPES TA, TBS, SA, AVB, SIS, †RHH, †THHN, †XHHW*	
	COPPER				ALUMINUM OR COPPER-CLAD ALUMINUM				
18	. . . .	. . . .	. . . .	18	. . . .	. . . .	. . . .	. . . .	. . . .
16	. . . .	. . . .	23	24	. . . .	. . . .	. . . .	. . . .	. . . .
14	25†	30†	30	35†	. . . .	. . . .	. . . .	. . . .	. . . .
12	30†	35†	40	40†	25†	30†	30	35†	12
10	40†	50†	55	55†	35†	40†	40	40†	10
8	60	70	75	80	45	55	60	60	8
6	80	95	100	105	60	75	80	80	6
4	105	125	135	140	80	100	105	110	4
3	120	145	160	165	95	115	125	130	3
2	140	170	185	190	110	135	145	150	2
1	165	195	215	220	130	155	165	175	1
0	195	230	250	260	150	180	195	205	0
00	225	265	290	300	175	210	225	235	00
000	260	310	335	350	200	240	265	275	000
0000	300	360	390	405	235	280	305	315	0000
250	340	405	440	455	265	315	345	355	250
300	375	445	485	505	290	350	380	395	300
350	420	505	550	570	330	395	430	445	350
400	455	545	595	615	355	425	465	480	400
500	515	620	675	700	405	485	525	545	500
600	575	690	750	780	455	540	595	615	600
700	630	755	825	855	500	595	650	675	700
750	655	785	855	885	515	620	675	700	750
800	680	815	885	920	535	645	700	725	800
900	730	870	950	985	580	700	760	785	900
1000	780	935	1020	1055	625	750	815	845	1000
1250	890	1065	1160	1200	710	855	930	960	1250
1500	980	1175	1275	1325	795	950	1035	1075	1500
1750	1070	1280	1395	1445	875	1050	1145	1185	1750
2000	1155	1385	1505	1560	960	1150	1250	1335	2000

CORRECTION FACTORS

Ambient Temp. °C	For ambient temperatures over 30°C, multiply the ampacities shown above by the appropriate correction factor to determine the maximum allowable load current.								Ambient Temp. °F
31-40	.82	.88	.90	.91	.82	.88	.90	.91	86-104
41-45	.71	.82	.85	.87	.71	.82	.85	.87	105-113
46-50	.58	.75	.80	.82	.58	.75	.80	.82	114-122
51-60	. . . .	.58	.67	.71	. . . .	.58	.67	.71	123-141
61-70	. . . .	.35	.52	.58	. . . .	.35	.52	.58	142-158
71-80	. . . .	. . . .	.30	.41	. . . .	. . . .	.30	.41	159-176

† The load current rating and the overcurrent protection for conductor types marked with an obelisk (†) shall not exceed 20 amperes for 14 AWG, 25 amperes for 12 AWG, and 40 amperes for 10 AWG copper, or 20 amperes for 12 AWG and 30 amperes for 10 AWG aluminum and copper-clad aluminum.
* For dry locations only. See 75°C column for wet locations.

Reproduced by permission from the National Electrical Code, NFPA 70-81, 1981 edition, Copyright National Fire Protection Association, 470 Atlantic Avenue, Boston, Mass. 02210, 1981

Figure 20-29. (Continued).

Notes to Tables

1. **Explanation of Tables.** For explanation of Type Letters, and for recognized size of conductors for the various conductor insulations, see Section 310-13. For installation requirements, see Sections 310-1 through 310-10, and the various articles of this Code. For flexible cords, see Tables 400-4 and 400-5.

2. **Application of Tables.** For open wiring on insulators and for concealed knob-and-tube wiring, the allowable ampacities of Tables 310-17 and 310-19 shall be used. For all other recognized wiring methods, the allowable ampacities in Tables 310-16 and 310-18 shall be used, unless otherwise provided in this Code.

3. **Three-Wire, Single-Phase Dwelling Services.** In dwelling units, conductors, as listed below, shall be permitted to be utilized as three-wire, single-phase, service-entrance conductors and the three-wire, single-phase feeder that carries the total current supplied by that service.

Conductor Types and Sizes
RH-RHH-RHW-THW-THWN-THHN-XHHW

Copper	Aluminum and Copper-Clad AL	Service Rating in Amps
AWG	AWG	
4	2	100
3	1	110
2	1/0	125
1	2/0	150
1/0	3/0	175
2/0	4/0	200

4. **Type MC Cable.** The ampacities of Type MC cables are determined by the temperature limitation of the insulated conductors incorporated within the cable. Hence the ampacities of Type MC cable may be determined from the columns in Tables 310-16 and 310-18 applicable to the type of insulated conductors employed within the cable.

5. **Bare Conductors.** Where bare conductors are used with insulated conductors, their allowable ampacities shall be limited to that permitted for the insulated conductors of the same size.

6. **Mineral-Insulated, Metal-Sheathed Cable.** The temperature limitation on which the ampacities of mineral-insulated, metal-sheathed cable are based is determined by the insulating materials used in the end seal. Termination fittings incorporating unimpregnated, organic, insulating materials are limited to 85°C operation.

7. **Type MTW Machine Tool Wire.** The ampacities of Type MTW wire are specified in Table 200-B of the Standard for Electrical Metalworking Machine Tools and Plastics Processing Machinery (NFPA 79-1980).

8. **More than Three Conductors in a Raceway or Cable.** Where the number of conductors in a raceway or cable exceeds three, the ampacity shall be as given in Tables 310-16 and 310-18, but the maximum allowable load current of each conductor shall be reduced as shown in the following table:

Number of Conductors	Percent of Values in Tables 310-16 and 310-18
4 thru 6	80
7 thru 24	70
25 thru 42	60
43 and above	50

Where single conductors or multiconductor cables are stacked or bundled without maintaining spacing and are not installed in raceways, the maximum allowable load current of each conductor shall be reduced as shown in the above table.

Exception No. 1: When conductors of different systems, as provided in Section 300-3, are installed in a common raceway the derating factors shown above shall apply to the number of power and lighting (Articles 210, 215, 220, and 230) conductors only.

Exception No. 2: The derating factors of Sections 210-22(c), 220-2(a) and 220-10(b) shall not apply when the above derating factors are also required.

Exception No. 3. For conductors installed in cable trays, the provisions of Section 318-10 shall apply.

9. **Overcurrent Protection.** Where the standard ratings and settings of overcurrent devices do not correspond with the ratings and settings allowed for conductors, the next higher standard rating and setting shall be permitted.

Exception: As limited in Section 240-3.

10. **Neutral Conductor.**

(a) A neutral conductor which carries only the unbalanced current from other conductors, as in the case of normally balanced circuits of three or more conductors, shall not be counted when applying the provisions of Note 8.

(b) In a 3-wire circuit consisting of 2-phase wires and the neutral of a 4-wire, 3-phase wye-connected system, a common conductor carries approximately the same current as the other conductors and shall be counted when applying the provisions of Note 8.

(c) On a 4-wire, 3-phase wye circuit where the major portion of the load consists of electric-discharge lighting, data processing, or similar equipment, there are harmonic currents present in the neutral conductor and the neutral shall be considered to be a current-carrying conductor.

11. **Grounding Conductor.** A grounding conductor shall not be counted when applying the provisions of Note 8.

12. **Voltage Drop.** The allowable ampacities in Tables 310-16 through 310-19 are based on temperature alone and do not take voltage drop into consideration.

Figure 20-29. (Continued).

Maximum Number of Conductors in Trade Sizes of Conduit or Tubing

(Based on Table 1, Chapter 9)

Type Letters	Conductor Size AWG, MCM	½	¾	1	1¼	1½	2	2½	3	3½	4	4½	5	6
RHW,	14	3	6	10	18	25	41	58	90	121	155			
	12	3	5	9	15	21	35	50	77	103	132			
	10	2	4	7	13	18	29	41	64	86	110	138		
	8	1	2	4	7	9	16	22	35	47	60	75	94	137
RHH (with outer covering)	6	1	1	2	5	6	11	15	24	32	41	51	64	93
	4	1	1	1	3	5	8	12	18	24	31	39	50	72
	3	1	1	1	3	4	7	10	16	22	28	35	44	63
	2		1	1	3	4	6	9	14	19	24	31	38	56
	1		1	1	1	3	5	7	11	14	18	23	29	42
	0		1	1	1	2	4	6	9	12	16	20	25	37
	00			1	1	1	3	5	8	11	14	18	22	32
	000			1	1	1	3	4	7	9	12	15	19	28
	0000				1	1	2	4	6	8	10	13	16	24
	250				1	1	1	3	5	6	8	11	13	19
	300				1	1	1	3	4	5	7	9	11	17
	350				1	1	1	2	4	5	6	8	10	15
	400					1	1	1	3	4	6	7	9	14
	500				1	1	1	1	3	4	5	6	8	11
	600					1	1	1	2	3	4	5	6	9
	700					1	1	1	1	3	3	4	6	8
	750					1	1	1	1	3	3	4	5	8

Reproduced by permission from the National Electrical Code, NFPA 70-81, 1981 edition, Copyright National Fire Protection Association, 470 Atlantic Avenue, Boston, Mass. 02210, 1981.

Figure 20-30. Conductors in conduit.

General Lighting Loads by Occupancies

Type of Occupancy	Unit Load per Sq. Ft. (Watts)
Armories and Auditoriums	1
Banks	3½**
Barber Shops and Beauty Parlors	3
Churches	1
Clubs	2
Court Rooms	2
*Dwelling Units	3
Garages — Commercial (storage)	½
Hospitals	2
*Hotels and Motels, including apartment houses without provisions for cooking by tenants	2
Industrial Commercial (Loft) Buildings	2
Lodge Rooms	1½
Office Buildings	3½**
Restaurants	2
Schools	3
Stores	3
Warehouses (storage)	¼
In any of the above occupancies except one-family dwellings and individual dwelling units of multifamily dwellings:	
Assembly Halls and Auditoriums	1
Halls, Corridors, Closets	½
Storage Spaces	¼

For SI units: one square foot = 0.093 square meter.

* All receptacle outlets of 20-ampere or less rating in one-family and multifamily dwellings and in guest rooms of hotels and motels [except those connected to the receptacle circuits specified in Section 220-3(b)] shall be considered as outlets for general illumination, and no additional load calculations shall be required for such outlets.

** In addition a unit load of 1 watt per square foot shall be included for general purpose receptacle outlets when the actual number of general purpose receptacle outlets is unknown.

Reproduced by permission from the National Electrical Code, NFPA 70-81, 1981 edition, Copyright National Fire Protection Association, 470 Atlantic Avenue, Boston, Mass. 02210, 1981

Figure 20-31. General lighting load.

Lighting Load Feeder Demand Factors

Type of Occupancy	Portion of Lighting Load to Which Demand Factor Applies (wattage)	Demand Factor Percent
Dwelling Units	First 3000 or less at	100
	Next 3001 to 120,000 at	35
	Remainder over 120,000 at	25
*Hospitals	First 50,000 or less at	40
	Remainder over 50,000 at	20
*Hotels and Motels — Including Apartment Houses without Provision for Cooking by Tenants	First 20,000 or less at	50
	Next 20,001 to 100,000 at	40
	Remainder over 100,000 at	30
Warehouses (Storage)	First 12,500 or less at	100
	Remainder over 12,500 at	50
All Others	Total Wattage	100

* The demand factors of this table shall not apply to the computed load of feeders to areas in hospitals, hotels, and motels where the entire lighting is likely to be used at one time; as in operating rooms, ballrooms, or dining rooms.

Reproduced by permission from the National Electrical Code, NFPA 70-81, 1981 edition, Copyright National Fire Protection Association, 470 Atlantic Avenue, Boston, Mass. 02210, 1981

Figure 20-32. Demand factors.

Note 1. Over 12 kW through 27 kW ranges all of same rating. For ranges individually rated more than 12 kW but not more than 27 kW, the maximum demand in Column A shall be increased 5 percent for each additional kW of rating or major fraction thereof by which the rating of individual ranges exceeds 12 kW.

Note 2. Over 12 kW through 27 kW ranges of *unequal ratings*. For ranges individually rated more than 12 kW and of different ratings but none exceeding 27 kW an average value of rating shall be computed by adding together the ratings of all ranges to obtain the total connected load (using 12 kW for any range rated less than 12 kW) and dividing by the total number of ranges; and then the maximum demand in Column A shall be increased 5 percent for each kW or major fraction thereof by which this average value exceeds 12 kW.

Note 3. Over 1¾ kW through 8¾ kW. In lieu of the method provided in Column A, it shall be permissible to add the nameplate ratings of all ranges rated more than 1¾ kW but not more than 8¾ kW and multiply the sum by the demand factors specified in Column B or C for the given number of appliances.

Note 4. Branch-Circuit Load. It shall be permissible to compute the branch-circuit load for one range in accordance with Table 220-19. The branch-circuit load for one wall-mounted oven or one counter-mounted cooking unit shall be the nameplate rating of the appliance. The branch-circuit load for a counter-mounted cooking unit and not more than two wall-mounted ovens, all supplied from a single branch circuit and located in the same room, shall be computed by adding the nameplate rating of the individual appliances and treating this total as equivalent to one range.

Note 5. This table also applies to household cooking appliances rated over 1¾ kW and used in instructional programs.

See Table 220-20 for commercial cooking equipment.

Reproduced by permission from the National Electrical Code, NFPA 70-81, 1981 edition, Copyright National Fire Protection Association, 470 Atlantic Avenue, Boston, Mass 02210, 1981

Demand Loads for Household Electric Ranges, Wall-Mounted Ovens, Counter-Mounted Cooking Units, and Other Household Cooking Appliances over 1¾ kW Rating. Column A to be used in all cases except as otherwise permitted in Note 3 below.

NUMBER OF APPLIANCES	Maximum Demand (See Notes) COLUMN A (Not over 12 kW Rating)	Demand Factors Percent (See Note 3) COLUMN B (Less than 3½ kW Rating)	COLUMN C (3½ kW to 8¾ kW Rating)
1	8 kW	80%	80%
2	11 kW	75%	65%
3	14 kW	70%	55%
4	17 kW	66%	50%
5	20 kW	62%	45%
6	21 kW	59%	43%
7	22 kW	56%	40%
8	23 kW	53%	36%
9	24 kW	51%	35%
10	25 kW	49%	34%
11	26 kW	47%	32%
12	27 kW	45%	32%
13	28 kW	43%	32%
14	29 kW	41%	32%
15	30 kW	40%	32%
16	31 kW	39%	28%
17	32 kW	38%	28%
18	33 kW	37%	28%
19	34 kW	36%	28%
20	35 kW	35%	28%
21	36 kW	34%	26%
22	37 kW	33%	26%
23	38 kW	32%	26%
24	39 kW	31%	26%
25	40 kW	30%	26%
26-30	15 kW plus 1 kW for each range	30%	24%
31-40	15 kW plus 1 kW for each range	30%	22%
41-50	25 kW plus ¾ kW for each range	30%	20%
51-60	25 kW plus ¾ kW for each range	30%	18%
61 & over	25 kW plus ¾ kW for each range	30%	16%

Figure 20-33. Demand load for electric ranges.

_____ 20.19 _____
VOLTAGE DROP

The NEC limits the amount of voltage drop (the loss of voltage due to resistance in the conductors) for power, heating, and lighting, or any combination of these loads, to 5 percent. In addition, the maximum total voltage drop for feeders and feeder circuits should be no more than 3 percent, leaving 2 percent for branch circuit loss.

In most residences and small buildings, the voltage drop in the feeder is small because the length of the conductor is short.

The voltage drop can be calculated by using the formula:

$$\text{Voltage drop} = \frac{I \times L \times R}{1{,}000}$$

where

  I = current carried in the conductor (amps)
  L = length of current-carrying conductor (feet)
  R = resistance of conductor (ohms per 1,000 feet)
  1,000 = constant (converting R per 1,000 feet to R per foot)

Example: Given a residence, check the voltage drop in the branch circuit to the farthest circuit from the panel 90 feet away.

The circuit has four receptacles on it, and its load is about 1,440 (4 duplex outlets at 180 W each = 4 x 2 x 180 = 1,440 W). This 1,440 W gives an amperage of 1,440 W ÷ 120 V = 12.0 amps. The approximate length of the branch circuit is 90 feet, and the resistance of a No. 12 AWG conductor is 1.68 ohms per 1,000 feet.

$$\frac{12.0 \text{ amps} \times 90 \text{ ft} \times 1.68 \text{ ohms}}{1{,}000} = 1.81\text{V}$$

The percentage of voltage drop, in this case, would be 1.81 V (voltage drop) divided by 120 (voltage) or about 1.5, well within the 2 percent drop that is considered good design practice.

Part IV

Estimating

Chapter 21

Introduction to Estimating

21.1
GENERAL INTRODUCTION

Building construction estimating is the determination of probable construction costs of any given project. There are many items that influence and contribute to the cost of a project; each item must be compiled and analyzed. Since the estimate is prepared before the actual construction of the project, a great deal of study must be put into the construction documents; this makes estimating one of the most important phases of any contractor's business.

In most instances, it is necessary to submit an estimate on the cost of the project, often in competitive bidding against other firms. The competition in construction bidding is keen, sometimes with ten or more firms bidding on a single project. In order to stay in business, a contractor must be low bidder on a certain number of projects; yet his prices cannot be so low that it is impossible to make a profit on them.

Since the estimate is prepared from the working drawings and specifications of a building, the ability of the estimator to visualize all of the different phases of con-

struction becomes a prime ingredient in his success.

The working drawings usually contain information relative to design, location, dimensions, and construction of the project, while the specifications are a written supplement to the drawings and include information pertaining to materials and workmanship. The working drawings and specifications *must* be considered together when an estimate is being prepared. The two complement each other; often there is an overlap in the information they convey. The bid submitted must be based on the drawings *and* specifications. You, as the estimator, are responsible for everything contained in the specifications as well as for that which is covered on the drawings. Read everything thoroughly; recheck all items whenever it is necessary. The plans must be measured and checked carefully, and all listings of quantities and materials must be done as clearly and accurately as possible. Each item must be listed with as much information about it as is reasonably possible.

Estimating the cost of a project is a process subject to many variables that may affect the actual construction of the project. These include weather, transportation, soil conditions, labor strikes, material availability, and subcontractors available. Regardless of the variables involved, the estimator must strive to prepare as accurate an estimate as possible. Estimating is not a guessing game. Carefully organized work, based on the estimator's best judgment and records of past projects completed, will result in accurate bids.

21.2
TYPES OF BIDS

Basically, there are two bidding procedures by which the contractor gets to build a project for an owner:

1. Competitive bidding.
2. Negotiated bidding.

In competitive bidding each contractor submits lump-sum bids (prices) in competition with other contractors. In most cases the lowest lump-sum bidder is awarded the contract, as long as the bid form and proper procedures have been followed and it is a reputable firm.

Negotiated bidding involves the contractor working with the owner (or through the owner's architect-engineer) to arrive at a mutually acceptable price for the construction of the project. It often involves negotiations back and forth on materials used, sizes, finishes, and similar items which affect the price of the project. The owner may negotiate with as many contractors as he wishes. This type of bidding is often used when the owner knows what contractor he would like to have build the project, in which case a competitive bidding would waste time. The biggest disadvantage of this arrangement is that the contractor may not feel the need to work quite as hard to get the lowest possible prices as he

might when a competitive bidding process is used.

Cost-plus agreements are also used in many projects, and they have the owner pay the contractor all of the costs of construction plus some kind of percentage or fee. This type arrangement is thoroughly discussed in Section 22.4. Both competitive and negotiated bids require extensive material, labor, and equipment estimates. The cost-plus arrangement may not require so detailed an estimate if it is for a small project, such as a residence. However, on large projects the contractor may be required to submit detailed estimates for the work as the project progresses.

21.3
COST ANALYSIS METHODS

In competitive bidding, when you are submitting a proposal on a project you want to be low bidder and still make a profit, you can't afford to take chances with shortcuts. In order to obtain accurate estimates, you must take the extra time to figure the work in as much detail as is required to provide accurate material and labor costs. Shortcuts may be acceptable when preparing preliminary estimates, but you should never use them when you are involved in competitive bidding.

1. *Detailed Method.* The detailed method includes determination of the quantities and costs of everything required to complete the work including materials, labor, equipment, insurance, bonds, and overhead, as well as an estimate of profit. This is the only method that should be used for competitive bidding. Each item of the project should be broken down into its parts and estimated. Each piece of work has a distinct labor requirement that should be estimated. Then, during and

after the job, it is possible to compare estimated and actual costs of doing the work.

In this manner the operation in question can be analyzed with an eye to cost control on future projects. This type of estimate also requires that costs be specifically allowed for the various types of equipment that may be required on the job.

The detailed estimate must establish the estimated quantities and costs of material, the time required for costs of labor, the equipment required and its cost, the items required for overhead and the cost of each item, and the percent of profit desired considering the investment, the time to complete, and the complexity of the project.

2. *Preliminary Methods: Area and Volume Methods.* The *volume method* involves computing the number of cubic feet contained in the building and multiplying that volume by an assumed cost per cubic foot. Using the *area method,* you compute the square footage of the building and multiply that area by an assumed cost per square foot. This latter method requires skill and experience in adjusting the unit cost to the varying conditions of each project. The chance for error is high. As a note of warning, it should be added that most of the unit prices published in manuals do not include the site improvements, landscaping, utilities outside the building, unusual foundations, or cost of the land or furnishings.

In using published unit costs, select a project most nearly like the one you are estimating with regard to size, proposed use, site and soil conditions, mechanical conditions, and climate. Always add a contingency allowance of 5 to 10 percent.

These methods are preliminary (approximate) methods, and are used by architects and engineers for first consultations with clients and by contractors to guide their thinking in the first approach to a probable cost of a building. The contractor will also compare the approximate methods with his final detailed bid and, if a large discrepancy exists, he must recheck his bid.

In using the area and volume methods, you *must* keep comparative data completely up-to-date in your office. At the end of each job you should update all records.

21.4
THE CONTRACT DOCUMENTS

The bids submitted for any construction project are based on the contract documents. If the estimator is to prepare a complete and accurate estimate, he must become familiar with all of the documents. The documents are listed and briefly described in this article. Further explanations of the portions and how to bid them are contained in later chapters.

The Contract Documents comprise the *Owner-Contractor Agreement,* the *General Conditions of the Contract,* the *Supplementary General Conditions,* and the *Drawings* and *Specifications,* including all *Addenda* incorporated in the documents before their execution. All of these taken together form the *contract.*

Working Drawings: Those actual plans (drawings, illustrations) from which the project is to be built. They contain the dimensions and locations of building elements and materials required and delineate how they fit together.

General Conditions: These define the rights, responsibilities, and relations of all parties to the construction contract.

Supplementary General Conditions: (*Special Conditions*) Since conditions vary by locality and project, these are used to amend or supplement portions of the General Conditions.

Specifications: Specifications are written instructions concerning project requirements that describe the quality of materials to be used and the results to be provided by the application of construction methods.

Agreement: The document that formalizes the construction contract, it is the basic contract and incorporates by reference all of the other documents and makes them part of the contract; it states the contract sum.

Addenda: The addenda statements or drawings that modify the basic contract documents after they have been issued to the bidder, but prior to the taking of bids. They may provide clarification, correction, or changes in the other documents.

21.5
SOURCES OF ESTIMATING INFORMATION

For matters relevant to estimating and costs, the best source of information is past experience. This is why a careful, accurate accounting system, combined with accuracy in field reports, is so important. All of the information relating to the job can be monitored and thus will be available for future reference. Many firms feed this information into computers, which the company may own, lease, or share time on. In this manner the information is readily available for future reference.

There are several "guides to construction cost" manuals available. However, a word of extreme caution is offered regarding the use of the manuals. They are only *guides:* the figures should *never* be used to prepare an actual estimate. The manuals may be used as a guide in checking current prices and should enable the estimator to follow a more uniform system and save valuable time. The actual pricing in the manuals is most appropriately used in helping the architect check approximate current prices and facilitate his preliminary (which in itself is only approximate) estimate.

In most areas, the local construction firms have grouped together to form associations to promote the industry. Many of these associations hold seminars that deal with the problems of construction today (including use of computers, how to obtain bonding, cold weather concrete, or other pertinent local, regional or national projects). Some have advancement programs whereby employees of member firms are given the opportunity to attend classes (often free) in conjunction with local colleges. If such an organization is located in or near your area, become a part of it.

Technical information can often be obtained from technical or trade organizations. A list of many of these organizations and the addresses of their home offices should be kept on file. Many have regional offices throughout the world, and a note to the home office will give you the appropriate address.

21.6
COMPUTERS

Computers can assist the estimator in several ways. The computer may be used to analyze labor and material costs from past projects, project labor and material costs to future projects, and facilitate information retrieval systems and building product data retrieval.

Computer estimating still involves the estimator taking off the dimensions and compiling material data. This information can then be fed to the computer to obtain the labor, equipment, and material costs that may be used to compile the final estimate. To perform this type of function, the computer must be programmed to proceed through the required steps, and appropriate unit costs must be stored in it. The computer will generate only the material that it is programmed to deliver; no less, no more. A mistake in the program or an incorrect symbol will return incorrect information. The computer is one of the finest tools available to the modern construction office, but it can generate no more than it is programmed to do. It cannot exercise judgment or experience on any given item. It should be used in conjunction with the experienced judgment of the estimator; it cannot replace him.

21.7
ORGANIZATION

The organization that should be maintained by the estimator cannot be overemphasized. The organization required includes the maintaining of complete and up-to-date files; it must include a complete breakdown of costs for each of the projects, both of work done by company forces and of work done by subcontractors. The information should include quantities, materials prices, labor conditions, costs, weather conditions, job conditions, delays, plant costs, overhead costs, and salaries of foremen and superintendents. All data must be filed in an orderly manner. Some large firms microfilm their reports and file them for future reference; other large firms put the information into a computer for quick recall.

The estimate of the project being bid must be systematically done, neat, clear, and easy to follow. The estimator's work must be kept organized to the extent that in an unforeseen circumstance (such as illness or automobile accident), someone else may step in, complete the estimate, and submit a proposal on the project. If the estimator has no system and the work cannot be read and understood, then there is no possible way that anyone can pick up where the original estimator left off. The easiest way for you to judge the organization of a particular estimate is to ask yourself if someone else could pick it up, review it and the contract documents, and be able to complete the estimate. Ask yourself: Are the numbers labeled? Are calculations labeled? Where did the numbers come from? What materials are you estimating?

21.8
DAILY LOG

One of the most important responsibilities of anyone involved in the construction process (or any other business) is the orderly planning of what must be done and the accumulation of information of what was done. One of the most effective, informal approaches to the accumulation of this information is the daily log (Figure 21-1).

The daily log helps plan each day, it lists (in a few words) work to be done, phone calls and personal visits that are made and received, as well as providing a convenient place to keep track of important future dates. Questions that arise concerning the past and the future, business trips, project visits, and meetings with people are only a few examples of the information the log would provide you with. If you had made a call on Wednesday and were

	J U L Y					
S	M	T	W	T	F	S
						1
2	3	4	5	6	7	8
9	10	11	12	13	14	15
16	17	18	19	20	21	22
23	24	25	26	27	28	29
30	31					

	A U G U S T					
S	M	T	W	T	F	S
		1	2	3	4	5
6	7	8	9	10	11	12
13	14	15	16	17	18	19
20	21	22	23	24	25	26
27	28	29	30	31		

Thursday, July 11

9   MTG. W/ HOWARD KOERNER
    @ 8:30 A.M. - RE: DESIGN UNLIMITED - ELECTRICAL ✓ O.K.
10

11   Call Ted Marotta— 371-7384 - re: Blacktop
    at Clifton Knolls park - bids due 9/2 at 2:00pm.
12   NOT IN - CALL ON FRI ✓

1   Milco windows - rec'd bid from Ed c.
    for Design Unlimited ✓

2   Pick up plans and Specification for
3   Dagostino warehouse ✓ Deposit 35⁰⁰ ✓

4   TO DO: BID
5   DAGOSTINO BLOCK WAREHOUSE
     DESIGNS UNLIMITED

Friday, July 12

9   CALL: T. MAROTTA - 371-7384 re: bid date —
10   T. Dennis - 283-1100 re: Convenient
      Market bid date
11

12   LEFT at noon for meeting in Hartford
    with MR. Dennis re: Conv. Market —
1   Back at 5:15

2

3

4   TO DO: BIDS
5

Figure 21-1. Typical daily log.

told the person was away until the following Tuesday, you would note in the log for the following Tuesday to call him again.

The next question is what to include in the log: the answer is anything that concerns you in your work should be included. The log would contain phone calls made and received (include number and with whom you spoke), trips, work scheduled to be completed each day, bid time and date, meetings (include where and with whom), addenda received, drawings and specifications received, and similar items.

The size of the log used depends on where it will be used. The most common type is about 5 x 8 inches so that it may be conveniently carried with you when you are out of the office. In this manner dates of future meetings can be immediately checked, and the miscellaneous items to be completed during the day are not forgotten. The log should be ruled and have the day and date at the top of each page. A typical day as recorded in a daily log is shown in Figure 21-1.

No one really wants to bother keeping a daily log. It's been said to be a bother and to serve no real purpose. True, at first the daily log seems a chore, but only until a person is accustomed to keeping it up to date. If you work in an office where no one else keeps a daily log, you can expect a few chuckles at first, but it never takes long before the jokers end up asking for information you have in your log: When did we bid a particular job? Have you noted on what date I went to a particular meeting? and so on. And be sure to keep the old daily log available for reference.

21.9
SITE INVESTIGATION

The importance of the site visit and the items to be checked vary, depending on the type of project and its location. As a contractor ex-

pands to relatively new and unfamiliar areas, the importance of the preliminary site investigation increases, as does the list of items that must be checked. Examples of the type of information that should be collected are:

1. Site access.
2. Availability of utilities (electric, water, telephone).
3. Site drainage.
4. Transportation facilities.
5. Any required protection or underpinning of adjacent property.
6. Subsurface soil conditions (bring a post hole digger to check this).
7. Local ordinances and regulations, and note any special requirements (permits, licenses, barricades, fences).
8. The local labor situation and local union rules.
9. Availability of construction equipment rentals, the type and conditions of what is available as well as the cost.
10. Prices and delivery information from local material suppliers (request they send you proposals for the project).
11. The availability of local subcontractors (note their names, addresses, and what type of work they usually handle).

21.10
SUBCONTRACTORS

A *subcontractor* is a separate contractor, who is hired by the prime contractor to perform certain portions of the work. Specialty work, such as pile driving, elevator installation, plumbing, heating, and electrical work, is the most common type of work subcontracted out. Often, less specialized work, such as drywall construction, masonry work, and flooring, is also subcontracted out.

The amount of work that the prime contractor will subcontract out varies from project to project. Some federal and state regulations limit the proportion of a project that may be subcontracted out, but this is rarely the case in private work.

If subcontractors will be used, the contractor must be certain to notify them early in the bidding period so they have time to prepare a complete, accurate proposal. If rushed, there is a tendency for the subcontractor to bid high just for protection against what might have been missed.

The use of subcontractors can be economical, but it does not mean that an estimate need not be done for a particular portion of work. Even if the estimator intends to subcontract the work out, he must prepare an estimate of the work. It is possible that the estimator will not receive proposals for a project before the bid date and he will have to use his own estimated cost of the work in totaling his proposal. All subcontractors' proposals are compared with the estimator's price (Figure 21-2); it is important that a subcontractor's price is neither too high nor too low. If either situation exists, the estimator should call the subcontractor and discuss the proposal with him.

The subcontractor's proposal is often phoned in to the general contractor's office at the last minute. This is done because of the subcontractor's fear that the contractor will tell other subcontractors the proposal price and encourage lower bids. This practice is commonly referred to as "bid peddling," and

---

McBill HVAC Systems
1215 Briarhill Rd.
Charleston, South Carolina
October 1, 19--

Ace Construction
501 Hightower St.
Charleston, South Carolina

Re: Dill Residence

Gentlemen:

We propose to furnish and install an electric furnace (34,000 Btuh) and all required duct-work for the above mentioned project for the lump sum of: $2,185.00

Alternate No. 2

We propose to furnish and install a two-ton heat pump and all required ductwork for the above mentioned project for the lump sum of: $3,215.00

Sincerely,

Charles McBill

Figure 21-2. Subcontractor's proposal.

it is discouraged. Even when proposals are sent in writing, many subcontractors will call at the last minute and revise their price. This often leads to confusion and makes it difficult for the estimator to analyze all bids carefully.

In checking subcontractor proposals, note especially what is included and what is left out. Each subsequent proposal may add or delete items. Often the proposals set up certain conditions, such as use of water, heat, or hoisting facilities. The estimator must compare each proposal and select the one that is the most economical.

All costs must be included somewhere. If the subcontractor does not include an item in his proposal, it must be considered elsewhere.

21.11
MATERIALS

For each project being bid the contractor will request quotations from materials suppliers and manufacturers' representatives for all materials required. While on occasion a manufacturer's price list may be used, it is more desirable to obtain written quotations that spell out the exact terms of the freight, taxes, time required for delivery, the materials included in the price, and the terms of payment (Figure 21-3). The written proposals should be checked against the specifications to make certain that the specified material was bid.

```
                                            United Block Company
                                            712 Charles Blvd.
                                            Hartford, Connecticut
                                            October 1, 19--

Ace Construction                            RE: Designs Unlimited
501 Hightower St.
Hartford, Connecticut

Dear Sir:

We are pleased to quote you on the materials required for the above referred
project.

All materials quoted meet the requirements of the drawings and specifications.

        8 x 8 x 16      Concrete Block              .72
        8 x 4 x 16      Concrete Block              .42
        8 x 6 x 16      Concrete Block              .56
        8 x 12 x 16     Concrete Block              .76
        Mortar, per bag                            2.95
        Brick, Tru-red range              120.00 per M

All prices delivered to jobsite, plus sales tax.

Delivery: 24 hour notice.

Sincerely,
```

Figure 21-3. Manufacturer's quotation.

All material costs entered on the workup and summary sheets (Section 21.13) must be based on delivery to the job site, not including tax. This means the total will take into account all necessary freight, storage, transportation insurance, and inspection costs. The taxes should be added on the summary sheet. Remember that the sales tax that must be paid on a project is the tax in force in the area in which the project is being built; sometimes cities have different rates that the county they are in. Take time to check.

21.12
WORKUP SHEETS

The estimator uses two basic sheets in his estimate, the *workup sheet* and the *summary sheet*. The workup sheet is used to make calculations and sketches, and to generally "work up" the cost of each item. Material and labor costs should always be estimated separately. Labor costs vary more than material costs, and the labor costs will vary in different stages of the project. For example, a concrete block will cost less for its first three feet than for the balance of its height, and if the block must be laid from the outside of the building, the labor cost goes up as the scaffold goes up—yet material costs remain the same.

When beginning the estimate on workup sheets, the estimator must be certain to list the project name and location, the date that the sheet was worked on, and the estimator's name. All sheets must be numbered consecutively, and when completed, the total number of sheets noted on each sheet (e.g., if the total number were 56, sheets would be marked "1 of 56" through "56 of 56"). The estimator must be certain he has every sheet,

since if one is lost, chances are that the costs on that sheet will never be included in the bid price. Few people can write so legibly that others may easily understand what they have written; it is therefore suggested that the work should be printed. Never alter or destroy calculations; if you wish to change them, simply draw a line through them and rewrite. Numbers that are written down must be clear beyond a shadow of a doubt. Too often a "4" can be confused with a "9," a "2" with a "7," and so on.

All work done in compiling the estimate must be totally clear and self-explanatory. It should be clear enough to allow another person to come in and follow all work completed and all computations made each step of the way.

Figure 21–4 is intended to show the correct method for listing items on the sheet. The measurements are taken from the plans, appropriate specification notes are included, the quantities are determined, and material costs introduced and entered into the right-hand column, which is reserved for totals.

When taking off the quantities, make it a point to break each item down into different sizes, types, and materials. This involves checking the specifications for each item you are listing. For example, in listing concrete blocks, you must consider the different sizes required, the bond pattern, the color of the unit, and the color of the mortar joint. If any of these items varies, it should be listed separately. It is important that the take off be complete in all details; do not simply write "wire mesh," but "wire mesh 6 x 6 10/10"—the size and type are very important. If the mesh is galvanized, it will increase your material cost by about 20 percent, so note it on your sheet.

Building Peterson
Location New Salem Rd - Salem Conn
Architect HY Assoc.
Subject Conc. Work - Ftg & Block

Estimate no. 735
Sheet no. 3 of
Date Nov. 15 –
Estimator 3D CH. T.D.

Description of work	No.	Dimensions L	W	H	Forms S.F.	Conc. C.Y.				
Conc. Footings: 3000 psi										
Front		~~72~~ 79	2	1	158	6				
LEFT SIDE		68	2	1	136	5				
Rear		109	2	1	218	8				
Right Side		62	2	1	124	5				
Interior		22	2	1	44	2				
					580	26 c.y.				26 c.y.
				x	2					
Forms 2X12					1160 B.F.					1160 S.F.
Conc. Pier Footings 3000 psi	No.	L	W	H	Forms S.F.	Conc. C.Y.				
Ext.	2	5	3	1	32	1.5				
Int.	4	7~~2~~	7	1	112	5				
Ext.	2	4	4	1	32	1.5				
					176	8 c.y.				8 c.y.
Forms 2x12 - 2 SIDES				x	2					
2x12					352 B.F.					352 B.F.

Figure 21-4. Workup sheet.

_____ 21.13 _____
SUMMARY SHEET

All of the costs contained on the workup sheets are condensed, totaled, and included on the summary sheet. All items of labor, equipment, material, plant, overhead, and profit must be likewise included. Only the most pertinent information is mentioned: the project name, location, architect, type of building, date, initials of the estimator and checker, number of stories, an estimate number, and sheet number are placed on each sheet. Figure 21-5 gives a typical summary sheet breakdown.

It should be noted that the work is broken down into classifications that coincide with the company's cost account numbers, which are used in connection with its accounting systems.

The summary sheet should list all of the information required, but none of the calculations and sketches that were used on the workup sheets. It should list only the essentials, yet still provide information complete enough for the person pricing the job not to have to continually look up required sizes, thicknesses, strengths, and similar types of information.

RECAPITULATION

Project _GUILD HALL Condos_   Estimate no. _754_
Location _WEST SALEM_   Sheet No. _2 of_
Architect
Engineer _MAC ASSOC._   Date _MARCH 12 -_
Summary by _Ø_   Priced by _Ø_   Checked by _D_

No.	Description	Quantity	Unit	Unit Price	Total Est. Materials	Unit Price	Total Est. Labour Cost	Total
0310	Concrete Formwork	10,300	M.F. B.M.	175		250-		
		12,500	S.F.	.78		.85		
0320	Concrete Reinforcing							
	#6 bars	5,250	lbs	.11		.06		
	#3 bars	1,450	lbs	.13		.07		
	Mesh 6x6 10/10 plain	42	rolls	2/-		7.50		
0330	Cast-in-place conc.							
	Footings - 3500 psi	112	c.y.	22-		31-		
	Slab-on-grade-2500psi	185	c.y.	20.50		27.50		
0335	FINISHES							
	Slab-machine trowel	15,800	S.F.			.03		
0340	Precast Concrete							
0343	Prestressed tees	31,600	S.F.					94800-

Figure 21-5. Summary sheet.

Chapter 22

Contracts, Bonds, Insurance, and Specifications

22.1
THE CONTRACT SYSTEM

Contracts may be awarded either by a single contract for the entire project or by separate contracts for the various phases required for completion of the project. The *single contract* comprises all of the work required for the completion of a project and is the responsibility of a single prime contractor. This centralization of responsibility provides that one of the distinctive functions of the prime contractor is to direct and coordinate all parties involved in completing the project. All of the subcontractors (including mechanical and electrical) and material suppliers involved in the project are responsible directly to the prime contractor, who in turn is responsible directly to the owner. The prime contractor is responsible for work completed in accordance with the contract documents, completion of the work on schedule, and payment of all construction costs related to the project.

Under the system of *separate contracts* the owner signs separate agreements for the construction of various portions of a project. The separate awards are often broken into the following phases:

1. General construction.
2. Plumbing.
3. Heating (ventilating, air conditioning).
4. Electrical.
5. Sewage disposal (if applicable).

In this manner the owner retains the opportunity to select the contractors for the various important phases of the project. Also, the responsibility for the installation and operation of these phases is directly between the owner and contractors rather than through the general contractor; and the owner or his agents must provide the coordination between the contractors.

There is disagreement in the construction industry itself as to which system is better. There are states that require by law the award of separate contracts when public money is involved. Most general contractor organizations favor single contracts, but in contrast, most large specialty contract groups favor separate contracts. However, in most instances the owner must decide. Under the single contract the prime contractor will include a markup on the subcontracted work for the coordination of the project. Often it is this markup that encourages the owner to use separate contracts. If the general contractor does not assume the responsibility for management and coordination of the project, someone else will have to or the result may be. utter confusion. For an added fee the architect will provide this service.

On occasion the owner himself will attempt to coordinate the project. However, unless he has a good deal of construction experience, his efforts will prove futile as he attempts to keep each contractor's work on schedule and get each contractor to coordinate all phases of the work.

No contract should ever be signed until it is checked by the lawyers of all parties. Each party's lawyer will normally give attention only to matters that pertain to his client's welfare. All contractors should employ the services of a lawyer who understands construction agreements and will take the time to review them carefully.

Types of agreements generally used are as follows:

1. Lump-sum agreement (stipulated-sum, fixed-price) (Section 22.3).
2. Cost-plus-fee agreements (Section 22.4).

22.2
TYPES OF AGREEMENTS

The *owner-contractor agreement* formalizes the construction contract. It incorporates, by reference, all of the other contract documents. The owner selects the type of agreement that will be signed; it may be a standard form of agreement such as the American Institute of Architects (AIA) or other organizations provide their members or a form prepared for a particular project.

The agreement generally includes a description of the project and contract sum. Other clauses pertaining to alternates accepted, completion date, bonus and penalty clauses, and any other items that should be amplified are included.

It is suggested that before the contractor actually signs a contract, he should make a financial check of the owner, not only for his ability to pay, but also for his reputation as a prompt payer.

If the contractor has any questions as to the owner's ability to pay during the bidding period, a preliminary check should be made at that time.

22.3
LUMP-SUM AGREEMENT
(STIPULATED-SUM)

The contractor agrees to construct the project, in accordance with the contract documents, for a set price arrived at through competitive bidding or negotiation. The contractor agrees that he will satisfactorily complete the work, regardless of the difficulties he may encounter. This type of agreement (Figure 22–1) provides the owner with advance knowledge of construction costs. The accounting process is simple and it creates centralization of responsibility in single-contract projects.

... agrees to build the project in accordance with the contract documents herein described for the lump sum of $127,750....

Figure 22–1. Lump-sum agreement.

It is also flexible with regard to alternates and changes required on the project. However, there are some disadvantages as well. The contractor must complete the work at a guaranteed price even though the costs involved may be more than he had estimated. Other disadvantages include the possibility that unscrupulous contractors may attempt to obtain maximum profit by cutting the quality of work and materials on the project, and the fact that, based on the lowest competitive bid, an incompetent contractor may be selected.

Because of the very nature of the lump-sum price, it is important that the contractor be able to understand accurately and completely the work required on the project at the time of bidding.

22.4
COST-PLUS-FEE AGREEMENTS

In *cost-plus-fee agreements* the contractor is reimbursed for the construction costs as defined in the agreement. However, the contractor is not reimbursed for all items, and complete understanding of reimbursable and nonreimbursable items is required. This agreement is often used when speed and high quality have a higher priority than the desire for the lowest possible price. It is often used when construction must begin before the drawings and specifications are completed. Extensive accounting is required, and there should be an advance understanding of what type of accounting methods shall be followed. Work out in advance all details concerning record keeping, purchasing, and the procedure that all concerned will follow.

There are a variety of types of fee arrangements, any of which may be best in a given situation. The important point is that whatever the arrangement, it must be clearly understood by all parties, not only the amount of the fee, but also how it will be paid to the contractor throughout the period of the contract.

Percentage Fee

The owner has the opportunity to profit if prices go down and changes in the work may be readily made.

The major disadvantage is that the fee increases with construction costs, so there is no desire on the contractor's part to keep costs low.

Fixed Fee

The advantages include the owner's ability to reduce construction time by beginning construction before the drawings and specifications are completed, and removal of the temptation for the contractor to increase costs or cut quality while he maintains a professional status. Also, changes in the work are readily made. Among the disadvantages are that the exact cost of the project is not known in advance, extensive accounting is required, and that keeping costs low depends on the character and integrity of the contractor.

Fixed Fee with Guaranteed Cost

Advantages are that a guaranteed maximum cost is assured to the owner; it generally provides an incentive to the contractor to keep the costs down if he will share in any savings. Again, the contractor assumes a professional status. Disadvantages include the fact that drawings and specifications must be complete enough to allow the contractor to set a realistic maximum cost. Extensive accounting is required as in all cost-plus agreements.

Sliding Scale Fee

This provides an answer to the disadvantages of the percentage fee, since as the cost of the project goes up, the percent fee of construction goes down. The contractor is motivated to provide strong leadership so that the project will be completed swiftly at low cost. Disadvantages are that the costs cannot be predetermined, extensive changes may require modifications of the scale, and that extensive accounting is required.

Fixed Fee with a Bonus and Penalty

The contractor is reimbursed the actual cost of construction plus a fee. A target cost estimate is set up and, if the cost is less than the target amount, the contractor receives a bonus of a percentage of the savings. If the cost goes over the target figure, there is a penalty (reduction of percentage).

_____ 22.5 _____
AGREEMENT PROVISIONS

While the exact type and form of agreement may vary, there are certain provisions that are included in all of them. The contractor (as well as his lawyer) must check each item carefully before signing the agreement.

Scope of the Work

The project, drawings, and specifications are identified; the architect is listed. The contractor agrees to furnish all material and perform all of the work for the project in accordance with the contract documents.

Time of Completion

Involved in this portion is the starting and completion time. Starting time should never precede the execution date of the contract. The completion date is expressed either as a number of days or as a specific date. If the number of days is used, it should be expressed in calendar days and not working days to avoid subsequent disagreements as to the completion date. Any liquidated damages or penalty and bonus clauses would usually be included here; they should be clearly written and understood by all parties concerned.

The Contract Sum

Under a lump-sum agreement the *contract sum* is the amount of the accepted bid or negotiated amount. The accepted bid amount may be adjusted by the acceptance of alternates or by minor revisions that were negotiated with the contractor after receiving the bid. In agreements that involve cost-plus there are generally articles concerning the costs that the owner reimburses the contractor for. Customarily, not all costs are paid by the contractor and reimbursed to him by the owner; reimbursable and nonreimbursable items should be listed. The contractor should be certain that all costs incurred by him are included somewhere. Also, in cost-plus-fee agreements, the exact type of compensation should be stipulated.

Progress Payments

Prime items include due dates for payments, retained percentage, work in place, and stored materials.

The *due date* for payments is any date mutually acceptable to all concerned. The maximum time the architect can hold the contractor's Application for Payment and how soon after the architect makes out the Certificate of Payment the owner must pay the contractor should be spelled out. There should also be some mention of possible contractor action if these dates are not met. Generally, the contractor has the option of stopping work. Some contracts state that if the contractor is not paid when due, the owner must also pay interest at the legal rate in force in the locale of building.

Retained Percentage

It is customary for the owner to withhold a certain percentage of the payments. This is referred to as *retainage* and is protection for the owner to assure completion of the contract. The most typical retainage is 10 percent, but other percentages are also used. On some projects this retainage is continued through the first half of the project, but not through the last half.

Schedule of Values

The contractor furnishes the architect-engineer with a statement that shows the amount allowed for each division of the work. Often, these are broken down into the same divisions as the specifications.

Work in Place and Stored Materials

The *work in place* is usually geared as a percentage of the work to be completed. The amounts allowed for each division in the schedule of values are used as the base amounts due on each item. The contractor may also receive payment for materials stored on the site or some other location agreed upon in writing. The contractor may have to present proof of purchase, bills of sale, or other assurances to receive payment for materials stored off the job site.

Acceptance and Final Payment

This sets up a time for final payment to the contractor. When the final inspection, certification of completion, acceptance of the work, and issuance of the final payment are completed the contractor will receive the *final payment,* which is the amount of retainage withheld throughout the construction period. Many agreements are set up so that if full completion is held up through no fault of the contractor, the architect can issue a certificate for part of the final retainage.

22.6
BONDS

Often referred to as *surety bonds,* bonds are written documents that describe the conditions and obligations relating to the agreement. (In law a *surety* is one who guarantees payment of another party's obligations.) The bond is not a financial loan, but acts as an endorsement of the contractor. Under the terms of the bond, the owner is compensated for actual damages sustained from any default of the contractor. The bond guarantees that the contract documents will be complied with, and all costs relative to the project will be paid. If the contractor is in breach of contract, the surety must complete the terms of the contract and pay all costs incurred up to the face of the contract.

A corporate surety specializing in contract bonds is most commonly used by contractors. The owner will reserve the right to approve the surety company and form of bond, since the bond is worth no more than the company's ability to pay. To eliminate the risk of nonpayment, the contract documents will on occasion require that the bonds be obtained from one specified company. To the contractor this may mean doing business with an unfamiliar company, and he may be required to submit financial reports, experience records, projects, in progress and completed, as well as other material, which could make for a long delay before the bonds are approved. It is up to the owner to decide whether he is willing to accept the surety obtained by the contractor or to specify what company he would like to use. In the latter case, the contractor has the option of complying with the contractor documents or not submitting a bid on the project.

While most smaller residential construction does not involve the use of bonds, larger residences, and often those in which an architect is involved, require at least a performance bond. When an architect puts a residence out for competitive bidding a bid bond may be required.

22.7
BID BOND

The *bid bond* insures that if a contractor is awarded the bid within the time specified, he will enter into the contract and provide all other specified bonds. If he fails to do so without justification, the bond shall be forfeited to the owner. The amount forfeited shall in no case exceed the amount of the bond or the difference between the original bid and the next larger bid that the owner may in good faith accept.

These bonds are usually provided by the contractor's surety for free or a small annual service charge of from $5.00 to $25.00. The usual contract requirements for bid bonds specify that they must be 5 to 10 percent of the bid price, but higher percentages are sometimes used. The contractor should inform his surety company once he decides to bid a project, especially if it is a larger amount than he usually bids or if he already has a great deal of

work. Once a surety writes a bid bond for a contractor, he is obligated to provide the other bonds required for the project. This is why surety companies may do considerable investigation of a contractor before they will write a bid bond for him, particularly if it is a company with which they have not done business before or that has never had a bid bond before.

22.8
PERFORMANCE BOND

The *performance bond* guarantees the owner that, within limits, the contractor will perform all work in accordance with the contract documents and that the owner will receive the project built in substantial agreement with the documents. It protects the owner against default on the part of the contractor up to the amount of the bond penalty. The warranty period of one year is usually covered under the bond also.

The contractor should check the documents to see if this bond is required and in what amount. He should also make his surety company aware of all requirements. Most commonly these bonds must be made out for 100 percent of the contract price.

22.9
LABOR AND MATERIAL BOND

The *labor and material bond,* also referred to as a *payment bond,* guarantees the payment of the contractor's bill for labor and materials used or supplied on the project. It acts as protection for the third parties and the owner, who is exempted from any liabilities in connection with claims against the project.

Claims must be filed in accordance with the requirements of the bond used. Most often there is the limitation included in the bond that the claimant must give written notice to any two of three parties, general contractor, owner, or surety, within 90 days after the last day the claimant performed any work on the project or supplied materials to it.

22.10
OBTAINING BONDS

The surety company will thoroughly research a contractor before it furnishes him a bid bond. The surety checks such items as financial stability, integrity, experience, equipment, and professional ability of the firm. The contractor's relations with his sources of credit will be reviewed, as will current and past financial statements. At the end of the surety company's investigations, it will establish a maximum bonding capacity for that particular contractor. The investigation often takes time to complete so the contractor should apply well in advance of the time he desires bonding—waits of two months are not uncommon.

Each time the contractor requests a bid bond for a particular job, the application must be approved. If the contractor is below his work load limit and there is nothing unusual about the project, the application will be approved quickly. But if the maximum bonding capacity is approached, or if the type of construction is new to the particular contractor or is not conventional, a considerably longer time may be required.

The surety puts the contractor through investigations before giving him a bond for a project to be sure that the contractor is not overextended. In order for the contractor to be successful, he requires equipment, working capital, and an organization; and none of these should be spread thin. The surety thus checks the contractor's availability of credit so that if he is already overextended, he will not take on a project that is too big for him. The surety will want to know if the contractor has done other work similar to that which he is about to bid. If so, the surety will want to know how large a project it was. The surety will encourage the contractor to stay with the type of work he has the most experience in. The surety may also check progress payments and amount of work to be subcontracted.

If the surety refuses the contractor a bond, the contractor must first find out why and then attempt to demonstrate to the surety that he can solve the conditions questioned. The contractor must remember that the surety is in business to make money and can only do so if the contractor is successful. The surety is not going to take any unnecessary chances in the decision to bond a project. At the same time, some surety companies are more conservative than others. If you feel the company is too conservative or not responsive enough to your needs as a contractor, shop around, talk with other sureties, and try to find one that will work with your organization.

If a contractor is approaching a surety for the first time, he should pay particular attention to what services the company provides. Some companies provide a reporting service, which includes projects being bid and low bidders. Also, when a contractor is doing public work, the company can find out when the contractor can expect to get his payment and what state it is in at a given time. Select the company that seems to be the most flexible in its approach and offers the greatest service to the contractor.

22.11
INSURANCE

The contractor must provide insurance for his business in addition to the insurance required by the contract documents. The contractor's selection of an insurance broker is of utmost importance since the broker must be familiar with construction requirements and problems. The broker also must protect the contractor against the wasteful overlapping of protection, and yet there can be no gaps in the insurance coverage that might cause the contractor serious financial loss.

Copies of the insurance requirements in the contract documents should be forwarded immediately to the insurance broker. He should be under strict instructions from the contractor that all insurance must be supplied in accordance with the contract documents. The broker will then supply the cost of the required insurance to the contractor for consideration in the bidding proposal.

Insurance is not the same as a bond. With insurance policies, a financial responsibility is assumed for a specified loss or liability.

In addition to the insurance required by the contract documents, the contractor has his own insurance requirements. By law, the contractor is required to maintain such insurance as workman's compensation, unemployment, social security, and motor vehicle insurance. Other insurances that are usually carried include fire, liability, accident, life, hospitalization, and business interruption.

No attempt will be made in this book to describe all of the various types of insurance that are available. A few of the most common types are described.

Workmen's Compensation Insurance

This policy provides the benefits required by law and covers employees killed or injured during the course of work. The rates charged for this insurance vary in different states, for different types of work, and for different contractors, depending on their own work records with regard to accidents. Workers should be classified correctly to keep rates as low as possible. Rates are charged as a percentage of payroll, and they vary considerably. The rates may range from less than 1 percent to over 30 percent, depending on the location of project and the type of work being performed. The cost of the policy is paid in full by the contractor.

Builder's Risk Fire Insurance

This insurance protects projects under construction against direct loss due to fire and lightning. The insurance should cover temporary structures, sheds, materials, and equipment stored at the site. The cost usually ranges from $0.40 to $1.05 per $100 of valuation, depending on the project location, type of construction assembly, and the company's past experience with a contractor. If desirable, the policy may be extended to all direct loss causes by windstorms, hail, explosion, riot, civil commotion, vandalism, and malicious mischief. Also available are endorsements that cover earthquakes and sprinkler leakage.

Other policies which fall under the category of "Project and Property Insurance" are:

1. Fire insurance on the contractor's buildings.
2. Equipment insurance.
3. Burglary, theft, robbery insurance.
4. Fidelity insurance—protects the contractor against loss caused by any dishonesty on the part of his employees.

22.12
SPECIFICATIONS

Specifications, as defined by the AIA (American Institute of Architects), are the written descriptions of materials, construction systems, and workmanship. The AIA further states that the quality of materials and the results to be provided by the application of construction methods are the purpose of the specifications.

Remember, the bid you, as contractor, submit is based on the drawings *and* specifications. You are responsible for everything contained in the specifications as well as what is covered on the drawings. Read them thoroughly; review them when it is necessary.

There is a tendency among estimators to simply skim over the specifications. Reading the average set of specifications is time consuming, but many important items are mentioned only in the specifications and not at all on the drawings. Since the specifications are part of the contract documents, the general contractor is responsible for all of the work and material mentioned in them.

The specifications contain items ranging from the types of bonds and insurance required to the type, quality, and color of materials used on the job. A thorough understanding of the materials contained in the specifications may make the difference between being low bidder and not.

There is no question that skimming the specifications is risky. Either your bids will be too high, since you must allow for contingencies, or too low from not allowing for all items in the required construction of the project.

Another example of sloppy specification analysis pertains to items that might be special. This is especially true if they are only slightly different from the standard product. The specifications may be printed on the last pages of the drawings or in a separate booklet.

22.13
ALTERNATES

In many projects the owner requests prices for alternate methods or materials of construction (Figure 22–2). These alternates are

---

ALTERNATE NO. ONE

The Contractor shall state in the Proposal Form the amount to be added or deducted from the base bid should he furnish and apply one stain-resistant prime coat and one acrylic stain-resistant top-coat of paint on exterior exposed plywood and eliminate the one coat of approved stain.

ALTERNATE NO. TWO

The Contractor shall state in the Proposal Form the amount to be added to the base bid should he furnish and install a two-ton central heat pump and eliminate the electric heat furnace.

Figure 22-2. Alternates.

generally spelled out on a separate listing in the specifications and they are listed on the proposal form. The alternates may be either an *add price* or a *deduct price,* which means that you either add the price to the base bid or you deduct it from the base bid. Be sure the price you include for any alternates is complete and includes all taxes, overhead, and profit. When an owner has a limited budget, the system of alternates allows him a choice of what he spends his money on.

Since lump-sum contracts are awarded on the basis of the total base bid, plus or minus any alternates accepted, there is always concern that the owner will select alternates in a way that will help a particular contractor become low bidder. This concern has become so great that some contractors will not bid projects with a large number of alternates. In order to relieve the contractor of this concern, many architects include in the contract documents the order of acceptance of the alternates. Alternates deserve the same estimating care and consideration as the rest of the project; do not rush through them or leave them until the last minute.

Chapter 23

Project Scheduling

23.1
SCHEDULING

One of the basic underlying effects of the cost of a project is how long it will take to complete it once the contracts have been awarded to a contractor. This length of time especially affects the estimator in regard to his overhead items, wages paid to supervisory and home office personnel, rental on trailers and toilet facilities. It also affects the estimate in terms of how long equipment will be required on the project. So, traditionally, the estimator has estimated an approximate length of time and that was the basis used in the estimate.

With the increased use of electronic computers it became possible to calculate the approximate length of time required for a project, to plan what the sequence of work trades would be on the project and when materials should be ordered and delivered to the project in such a way that the dates and lengths of time were tabulated electronically. When carefully planned, this procedure provides an accurate guide to the estimator of how long it will take to build the project.

But its use went far beyond just the estimator. To the owner it meant that his project was actually being planned out and that the estimated completion date meant more than the approach of the contractor stating, "Well, it should take about six months."

On many large projects, the owner requires each contractor submitting bids to also provide a schedule of time for the work required to build the building. Commonly referred to as CPM, ProMIS, CPM precedence method, CPM sequence method, PERT, these computer programs provide a valuable service from bidding time to the end of construction.

The basics of scheduling the project break down into 5 areas:

1. List all activities required for the completion of the project.

   a. Work—each item or portion of work required must be listed. For example, it is not sufficient to list "concrete"; instead each portion of the work must be listed such as footings, interior slabs, sidewalks, driveways, etc.

   b. Ordering materials—while some materials are readily available and can be ordered the day before they are required, many materials require special ordering months in advance of when they are actually required.

   c. Delivery of materials—while it is sometimes possible for material ordered in the morning to be delivered in the afternoon, it is not always so simple. For materials from out-of-state the time required for truck or train delivery

399

may run several days. In addition, the firms may be so busy that they have their delivery schedule worked out two or three weeks ahead. In this portion of the activities it is necessary to estimate how long it will take for delivery and order the material in ample time to receive it when needed.

2. Assign the time required for each of the activities listed in step 1. It is most important that *all* of the times be accurate; inaccurate times (or lengths of time) will result in inaccurate results.

3. For each activity it is then necessary to list the succeeding activity which can be completed once it is done. For example, the concrete can be poured after the reinforcing is installed. Similarly, the reinforcing can be installed after the forms are built and the reinforcing has been delivered.

4. For easy reference (and for entry into the computer) each activity is then given a code (number, letters, or a combination of numbers and letters).

5. The information may then be fed into a properly programmed computer and the time required for the project can be obtained. The computer can also:
   a. Determine when materials must be ordered.
   b. Determine when delivery must be arranged.
   c. Determine how much variation in time is available, or what is the latest date an activity can be performed and still keep the project on schedule for completion.

Continuous use of these computerized printouts may be used for:

1. Updating analysis of actual cost for each portion of work as compared with the estimate.

2. Calculating how many men and how much equipment will be required for a portion of work.

3. Determining the totals of men and equipment required at any given time. (Keep in mind that one front-end loader cannot work in two places at once. This helps coordinate how many of what type equipment and number of workers will be required at any given time.)

4. Preparing monthly job status cost reports, broken down into labor, materials, equipment, and overhead if desired.

5. Figuring unit cost of installed work.

6. Establishing what equipment is inactive and available for use.

7. Determining where each piece of equipment is being used.

8. Processing payroll and labor costs.

It is beyond the scope of this text to show the complete workings of computerized scheduling. However, the basics can be explained by the use of a small example. The small office shown in Figure 23–1 will be used.

Now, in the general order given for breaking up scheduling:

1. List work activities (Figure 23–2).

2. List ordering of materials required (Figure 23–3).

3. List delivery of materials required (Figure 23–4).

4. Assign time required for each activity.

5. Give each activity a code.

6. List succeeding activity required for each activity. (Many contractors list both preceding and succeeding activities.) (Figure 23–5).

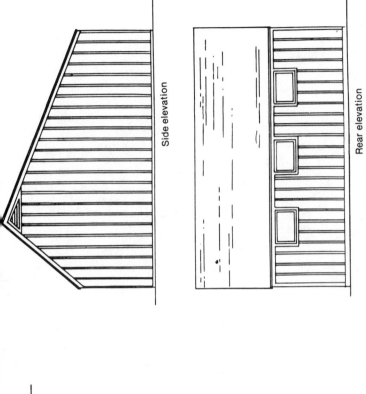

Side elevation

Rear elevation

Floor: Carpet over concrete slab

Walls: Studs, gypsum board, plywood, insulation

Roof: Trusses, plywood, insulation gypsum board

Doors: Anodized aluminum

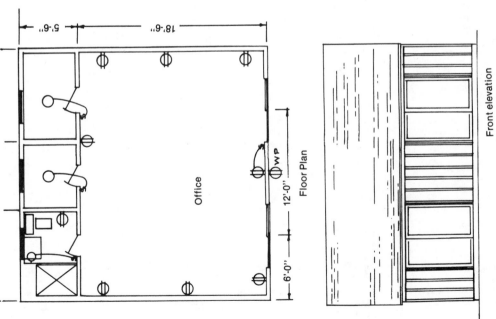

Office

Floor Plan

Front elevation

18'-6"

5'-6"

12'-0"

6'-0"

WP

Figure 23-1. Plan and three elevations.

401

Work Activities

Clear site
Scrap topsoil
Gravel fill
Plumbing, rough in
Forms, concrete slab
Pour slab and finish
Rough carpentry
Electrical, rough in
Insulation
Roofing
Plumbing, set shower
Drywall
Interior trim
Exterior trim
Telephone, rough
Plumbing, finish
Heating, rough in
Heating, finish
Painting
Stain, exterior
Carpet
Windows
Glass doors
Wood doors
Final grade
Seed
Electrical, finish

Figure 23-2. Work activities.

Ordering Activities	Duration	Event
Plumbing fixtures	15	15403A
8' anodizied aluminum doors	8	08100A
Windows	10	08600A
Heating unit	12	15802A
Carpet	18	09680A
Electrical fixtures	3	16012A

Figure 23-3. Materials required.

Delivery Activities	Duration	Event
8' anodizied aluminum doors	2	08100B
Windows	2	08600B
Gravel	1	03301B

Figure 23-4. Deliveries required.

Work Activities	Duration	Event	Succeeding Event
Clear site	1/2	02100	02200
Scrap topsoil	1/2	02200	03100
Gravel fill	1/2	03301	15401
Plumbing, rough in	2	15401	03300
Forms, concrete slab	1/2	03100	15401
Pour slab and finish	2	03300	06100
Rough carpentry	3	06100	15402, 08100, 16700, 16011, 15801, 07300
Electrical, rough in	1	16011	07200
Insulation	1-1/2	07200	09250
Roofing	1	07300	07200, 06201
Plumbing, set shower	1/2	15402	09250
Drywall	4	09250	08200
Interior trim	2	06200	09900
Exterior trim	1-1/2	06201	09901
Telephone, rough	1/2	16700	09250
Plumbing, finish	1	15403	02201
Heating, rough in	2-1/2	15801	09250
Heating, finish	1-1/2	15802	
Painting	3-1/2	09900	15802, 16012, 09680
Stain, exterior	2	09901	02201
Carpet	1/2	09680	15403
Windows	1/2	08600	
Glass doors	1/2	08100	06200
Wood doors	1/2	08200	06200
Final grade	1	02201	02800
Seed	1/2	02800	
Electrical, finish	1	16012	09900

Figure 23-5. Activity layout.

Now the work activities may be organized into the required flow. If a computer were being used, this information would be fed into it and the scheduling would be done. But a computer isn't necessary, and the principles can be shown by following these steps:

1. Cut out pieces of paper, each with one activity on it.

2. Lay out the pieces of paper in the sequence in which the work must occur. For example, the very first item is getting the permits, which leads to the surveyor coming and laying out the project, on to removing the topsoil, then to digging the footings and so on.

3. Once satisfied with the layout of activities, make a sketch of the chart of activities (Figure 23–6).

4. Next, write the duration of time for each activity (Figure 23–7).

5. By checking the lengths of time through the "circuits" of the chart, the time required for each circuit or path can be determined (Figure 23–8).

6. The circuit with the most time is the "critical path" which will govern the length of time it takes to complete the project (Figure 23–9).

Such a chart or computerized report is only as good, or as accurate, as the information used. For example, if two days are allowed for ordering and delivery of windows and it will actually take four to five weeks, it will be difficult to be accurate. Therefore, a careful, accurate list of all information is the most important requirement of scheduling.

The procedure of planning a scheduling chart clearly shows the flow of materials and activities on a project and emphasizes the need for orderly management of a project.

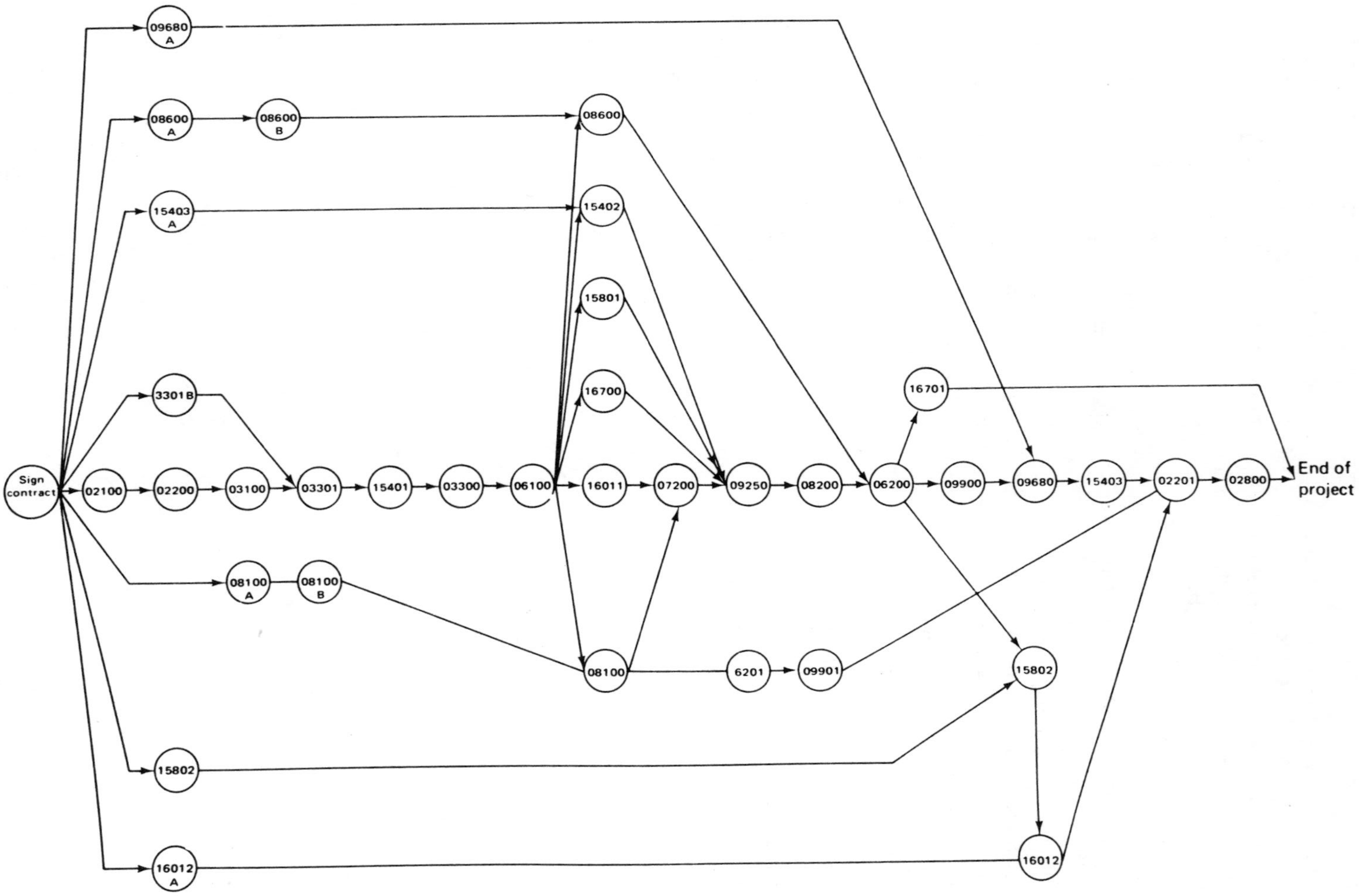

Figure 23-6. Sketch the chart of activities.

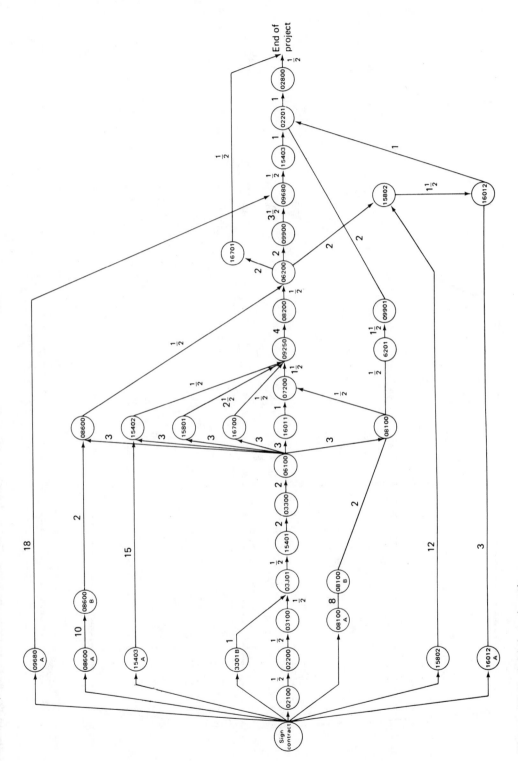

Figure 23-7. Add time duration.

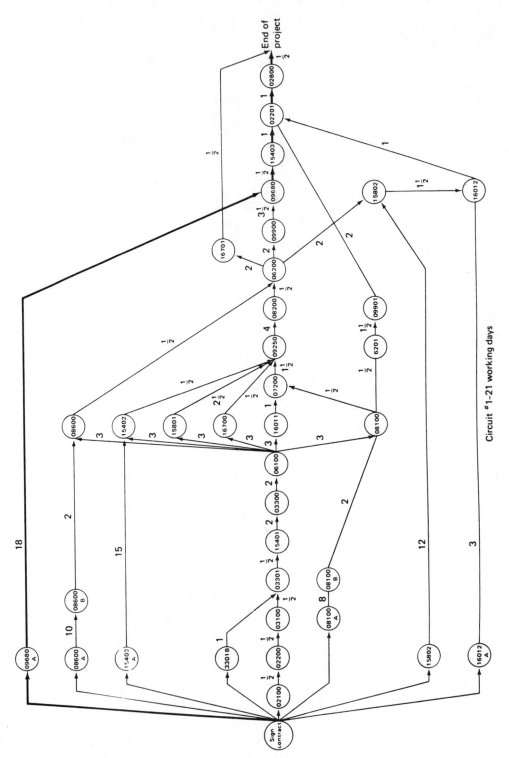

Figure 23-8. Determine the time required for each circuit.

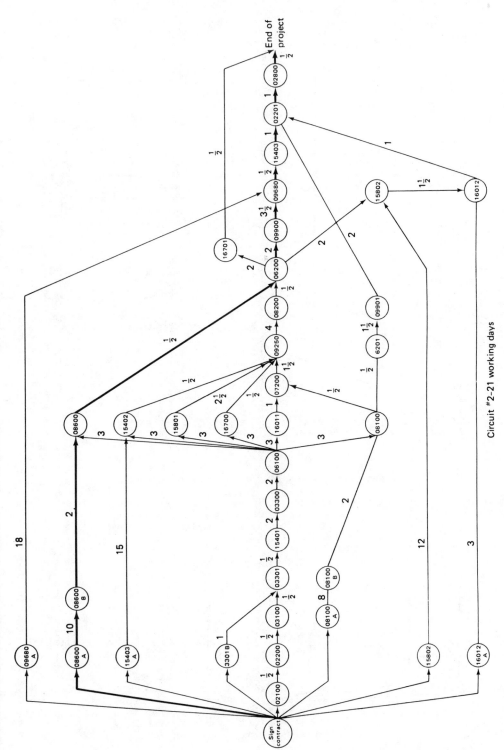

Figure 23-8. (Continued).

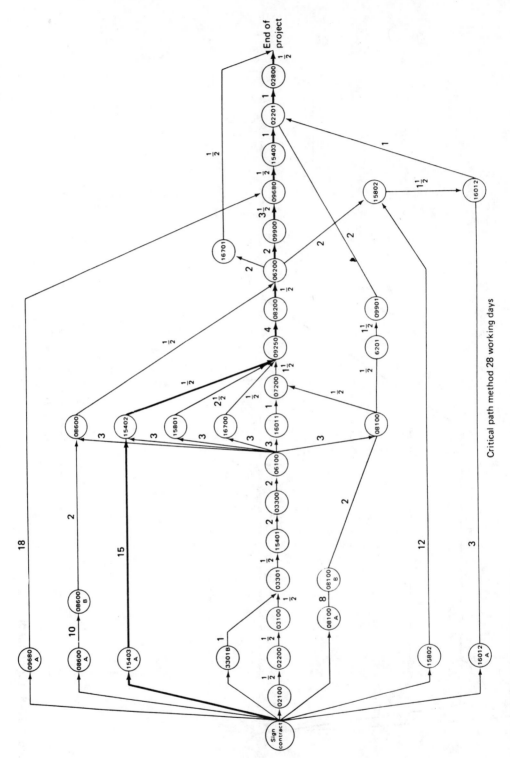

Figure 23-9. Critical path chart.

Chapter 24

Labor

GENERAL

Estimating the labor required is not the most difficult portion of the estimate to the experienced estimator, who will use his experience and job work studies that he has made, but the inexperienced estimator must tread carefully in this area.

The unit of time for estimating labor is the hour. This will eliminate the effects of varying lengths of work days. The estimator must consider that the skill and mental attitude of the workers will affect the length of time required to do a certain piece of work.

Allowances must also be made for variations in wages during the construction of the job, working conditions, availability of skilled and unskilled workers, climatic conditions, and supervision. When there are a lot of jobs for workers, there is a tendency for them to "slow down," and the time required for completion of a piece of work will increase. When jobs are scarce, workers tend to perform their tasks at a greater rate of speed. Further, the estimator must realize that a worker seldom works 60 minutes during the hour. Studies of the amount of actual working time ranges from about 30 to 50 minutes per hour. Keep in mind the time taken to "start up" in the morning, coffee breaks, trips to the toilet and for a drink of water, lunches that start a little early and end a little late, and clean-up

time all tend to shorten the work day. It is this long list of variables, as well as other possible work interruptions such as waiting for materials, that make labor the most difficult portion of the estimate.

WAGES AND RULES

The local labor situation must be surveyed carefully before making the estimate. Local union members must be hired and their work rules should be given particular attention since they may affect the contractor in a given community. There are localities in which the union rules and regulations limit the hourly output of the worker; if these rules are not followed, chances are there will be trouble with the unions—perhaps a strike. There are labor rules that can cause the increase of construction costs by as much as 35 to 80 percent. The estimator will also have to determine if the skilled men required for the construction are available, if he will have to bring in workers, or if he will be forced to use a large percentage of unskilled workers. All of these items must be considered in determining how much work will be accomplished on any particular job in one hour.

While surveying the labor market, the estimator will have to get information on the prevailing hourly wages, fringe benefits, and holidays.

24.3
FIELD REPORTS

The estimator should receive reports from the construction site. These *field reports* (Figure 24–1) should include an accurate record of the number of units of work completed; the number of workers used (broken down into trades); the number of hours spent on each portion of the work (broken down into trades); and a description of job conditions, climatic conditions, and any other circumstances that might affect the production of the labor force. Equipment utilized is also important and should be noted. These reports are made out daily.

Most field personnel dislike paper work—it seems tedious to them, so they avoid it as much as possible. When they do fill out a report, there is a tendency to "get it done" whether it is accurate or not. If they know that a particular portion of the work is running high, they often "hide" or "bury" the true cost by charging off some of the labor to another area. The estimator must take the time to explain why the reports are so important, show the field personnel how the information is used, and provide a full explanation of how each portion is utilized for future planning and estimating in the office.

The accumulation of accurate labor costs is the estimator's fund of knowledge and is one of the most important and valuable items in the contractor's office.

24.4
PRICING LABOR

To price labor, first estimate the man-hours required to do a unit of work. Be sure to include all trades required for the work—masons, carpenters, laborers, etc. Remember that all will probably not work exactly the same number of hours on the work since they may only be performing one portion of the job. Once the man-hours for each trade have been estimated, they are multiplied by the wages per hour and totaled. This will give the total labor cost per unit of work. The hourly scale should include all fringe benefits required in the agreement.

The estimator is responsible for deciding the number of man-hours required to perform a unit of work. This can be done only through experience. If the estimator has some doubt about how to handle a particular labor cost, he should check past project records and then consult with the general superintendent.

Field Report				
Proj. No. Supt.			Date Weather	
Work code	Total man hours	Skilled	Laborers	Work
06100	32	24	8	Erected roof trusses, sheathed walls and roof

Figure 24-1. Field report.

Example 24.1: Labor costs per square foot of surface area for brick are as follows: For the first 4 feet in height on the wall, it will required 15 mason-hours per 100 square feet and 18 helper-(laborer-) hours to mix the mortar, and bring mortar and brick to the masons. The mason will lay the brick and square it, "tool" the joints, and clean up at the end of the day.

The wage per hour is multiplied by the appropriate number of man-hours, and a cost per 100 square feet is calculated. Assuming that the mason earns $11.45 per hour and the laborer's wage scale is $9.75 per hour, we can compute the total cost of labor for 100 square feet to be $11.45 (15) + $9.75 (18) = $347.25. Therefore, the cost per square foot is $3.48 for labor.

Remember, as the building rises in the air, the labor costs also go up. The higher brick-work will cost more, since the laborers will have to set up scaffolds and move the brick and mortar farther. The time required for the masons to actually place the brick in the wall will remain about the same, providing they are working on a solid scaffold and do not have to wait for materials.

An obvious caution should be added to the estimator at this point: It is imperative that wages be kept up-to-date and that the level of productivity upon which the estimate is based be accurate and confirmed by experience.

_____ 24.5 _____

LABOR DIAGRAMS

Tables and diagrams are used by estimators to simplify the work, but often shortcuts may be taken to make the work conform to the diagrams. Never revise man-hours just to be able to use a table or diagram; it is better simply to work them out as shown in the example.

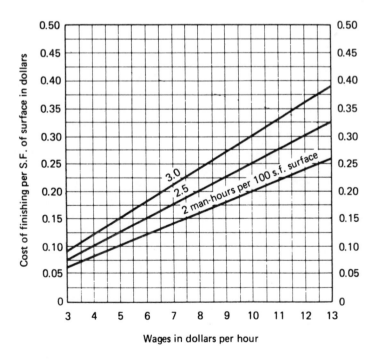

Figure 24-2. Labor diagram.

Generally, the wage scale per hour is listed on the horizontal scale and the unit costs are listed on the vertical scale. The lines representing the varying man-hours required for labor for a given unit of work are drawn on the diagram.

Figure 24–2 shows a typical labor diagram for finishing concrete. To arrive at a cost per square foot, the estimator must first decide the number of man-hours required per 100 square feet and must know the hourly wage scale. The number of man-hours will vary, depending on the type of finish required. For this example assume that a trowel finish for slab on grade is desired and that two man-hours are required for 100 square feet. With an hourly wage scale of $9.50 the cost per square foot is $9.50 times 2 divided by 100 or $0.19 per square foot. This result may also be read from the diagram. The estimator plots the lines on the diagram by designating three of the points and connecting them with a line. In this manner the estimator may make as many diagrams as he feels he needs.

Chapter 25

Estimating Excavation

25.1
GENERAL

Figuring the quantities of work that must be excavated is considered by some to be one of the most difficult portions of the estimator's task—it is loaded with problem areas. Under the title of "Excavation" in the specifications there is often a great deal of work included. The number of cubic yards to excavate is sometimes easy enough to compute, but figuring the cost for this portion of the work is difficult because of the various hidden items that may affect the cost. These include such variables as type of soil, required slope of bank in the excavated area, whether bracing or sheet-piling will be required, whether water will be encountered and pumping required, etc.

Several questions demand answers: What is the extent of work covered? What happens to the excess excavated material? Can it be left on the site or must it be removed? If the excess must be removed, how far must it be hauled? All of these things and more must be considered. Who does the clearing and grubbing?; removes trees? Must the topsoil be stockpiled for future use? Where? Who is responsible for any trenching required for the electrical and mechanical trades?

One of the first items the estimator must consider is the type of soil that will be encountered at the site. It is common practice for the estimator to investigate the soil conditions when he visits the site. Bring a long-handled shovel or a post hole digger and check the soil for yourself. Be certain that you record all that you see in your notebook.

25.2
EXCAVATION—UNIT OF MEASURE

Excavation is measured by the cubic yard for a quantity take-off (27 cubic feet = 1 cubic yard). Before excavation and in an undisturbed condition, earth weighs about 100 pounds per cubic foot; rock weighs about 150 pounds per cubic foot.

Once excavation begins, the earth and rocks are disturbed and they begin to swell. This means they expand to assume a larger volume; this expansion represents the amount of *swell* and is generally expressed as a percent gained above the original volume. When loose material is placed and compacted (as fill) on a project, it will be compressed into a smaller volume than when it was loose; this reduction in volume is referred to as *shrinkage*. Refer to Figure 25–1 for typical percentages of swell.

In surveying the dimensions are given in feet and decimals of a foot. There is usually no reason to change to units of feet and inches; however, there are times when inches must be changed to decimals. Remember that when estimating quantities, the computations need

Percentages of Swell

Material	Swell Percentages
Sand and gravel	10 to 18%
Loam	15 to 25%
Dense clay	20 to 35%
Solid rock	40 to 60%

Example: If 180 cubic yards of dense clay were

excavated with an average swell of

30% how many cubic yards would

have to be hauled away on trucks?

180 x 1.30 = 234 cubic yards to be

hauled away.

Figure 25-1. Percentages of swell.

not be worked out to an exact calculation. The following example illustrates the degree of accuracy required.

Example 25.1: In the calculation of an excavation for an area of 52.83 x 75.75 feet with an average depth of 6.33 feet, we could give an acceptable answer of 25,330 cubic feet. There is no reason to give the exact answer of 25,331.852 cubic feet.

Conversion of inches to decimals, and decimals to inches is shown in Appendix A. The estimator will have to learn the approximate values so that he will not have to continuously check for them.

_____ 25.3 _____

PERIMETERS AND AREAS

Throughout the estimate there is some basic information that is used over and over again. The *perimeter* of a building is one such basic

dimension that must be calculated. The perimeter is the distance around the building; it is the total length around the building expressed in lineal feet.

The *area* of a plan is the surface included within specified limits—in this case the building, roadway, parking lot, or plot. It is expressed in square feet.

Example 25.2: What is the perimeter and the area of the building shown in Figure 25-2?

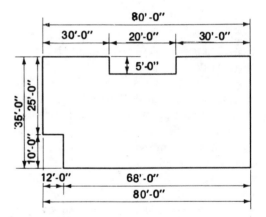

Figure 25-2. Plan-building lines.

Perimeter

Basic perimeter: 35′ + 80′ + 35′ + 80′
= 230 l.f.
Recessed walls: 5′ + 5′
=   10 l.f.
240 l.f. (lineal feet)

This could also have been figured by adding up the footage around the building, starting in the upper left corner of the building and proceeding clockwise:

30 + 5 + 20 + 5 + 30 + 35 + 68 + 10 + 12 + 25 = 240 l.f.

Area
Find the basic area of the plan and deduct the areas which are recessed and offset.

Basic area: 80'-0" × 35'-0"
= 2800 s.f. (square feet)

Recesses and offsets:
Southwest offset 12 × 10 = 120 s.f.
North recess   5 × 20 = 100 s.f.

Total deductions	− 220 s.f.
Net area of building	2580 s.f.

25.4
TOPSOIL REMOVAL

The removal of topsoil to a designated area where it is to be stockpiled for finished grading and future use is included in many specifications. Thus, the estimator must determine the depth of topsoil, where it will be stockpiled, and what equipment should be used to strip the topsoil and move it to the stockpile area. Topsoil is generally removed from all building, walk, roadway, and parking areas. Volume of topsoil is figured in cubic yards. A clearance around the entire basic plan must also be left to allow for the slope required for the general excavation—usually about 5 feet is allowed on each side of a building and 1 to 2 feet for walks, roadways, and parking areas (Figure 25–3).

Example 25.3: The unit of measure used for excavation is the cubic yard (1 cubic yard = 27 cubic feet). Therefore:

$$\frac{4050 \text{ s.f.} \times 1 \text{ ft. deep}}{27} = 150 \text{ c.y.}$$

Equipment selection will probably be limited to either a bulldozer or a front-end loader. Assume that a front-end loader with a 1-cubic yard capacity bucket was selected and that it is estimated it can scrap and stockpile an average of 24 cubic yards per hour. Mobilization time is estimated at 2.5 hours, the operating cost per hour for the equipment is estimated at $8.65, and the cost for an operator is $13.50 per hour. Estimate the number of hours required for the project and the cost for this item (Figure 25–4).

First, the total hours required to complete this phase of work must be calculated. Divide the total cubic yards

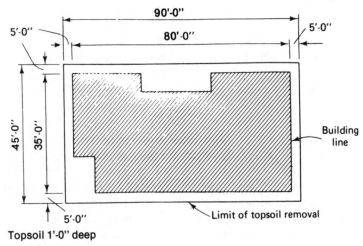

Topsoil 1'-0" deep

Figure 25-3. Topsoil quantities.

Soil	Dozer				Tractor shovel		Front end loader		Backhoe	
	50' haul		100' haul		No haul		50' haul	100' haul	No haul	
	50 hp	120 hp	50 hp	120 hp	1 c.y.	$2\frac{1}{4}$ c.y.	1 c.y.	$2\frac{1}{4}$ c.y.	$\frac{1}{2}$ c.y.	1 c.y.
Medium	40	100	30	75	40	80	24	30	25	55
Soft, sand	45	110	35	85	45	90	30	40	25	60
Heavy soil or stiff clay	15-20	40	10-15	30-35	15-20	35	10	12	10	15

Load and haul		
Truck size	Haul	c.y.
6 c.y.	1 mile	12-16
	2 miles	8-12
12 c.y.	1 mile	18-22
	2 miles	12-14

Figure 25-4. Equipment capacity.

to be excavated by the rate of work done per hour, and add the mobilization time; the answer is the total hours for the phase of work.

$$\frac{150}{24} + 2.5 = 8.75 \text{ hours}$$

The total number of hours is then multiplied by the cost of operating the equipment per hour plus the cost of crew for the period of time.

8.75 x ($8.65 + $13.50) = 8.75 hours
x $22.15 per hour = $193.82

This is the estimated cost to strip and stockpile the topsoil on this particular project.

_____ 25.5 _____
GENERAL EXCAVATION

Included under *general (mass) excavation* is the removal of all types of soil that can be handled in fairly large quantities, such as excavations required for a basement. Power equipment such as power shovels, front-end loaders, bulldozers, and graders are typically used in this type of project.

When calculating the amount of excavation to be done for a project, be certain that the dimensions used are the measurements of the outside face of the footings and not those of the outside buildings. The footings usually project at least 6 inches (and sometimes much more) beyond the wall. Also, add about 6 inches to 1 foot on all sides of the footing to allow the workman to install and remove forms (if used). The estimator must also allow for the sloping of the banks so that they will not cave in. The amount of slope required must be determined by the estimator as he considers the depth of excavation, type of soil, and possible water conditions. Some commonly used slopes (referred to as "angles of repose") are given in Figure 25-5. Allow about 1 foot all around for working space.

The actual depth of cut is the distance from the top of grade to the bottom of the fill material used under the concrete floor slab. If topsoil has been stripped, the average depth

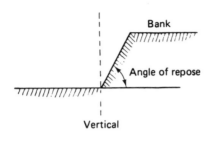

Material	Angle		
	Wet	Moist	Dry
Gravel	15–25	20–30	25–40
Clay	15–25	25–40	40–60
Sand	20–35	35–50	25–40

Figure 25-5. Earthwork slopes.

of topsoil is deducted from the depth of cut. If fill material such as gravel were not used under the concrete floor, the depth would be measured to the bottom of the floor slab. Since the footings usually extend below the fill material, a certain amount of excavation will be required to bring the excavation down to the proper elevation before footings can be placed. This would also be included under the heading of "general excavation," but kept separate from topsoil.

Before the estimator can select equipment, he will have to determine what must be done with the excess excavation—whether it can be placed elsewhere on the site or whether it must be hauled away. If it must be hauled away, he should decide how far. The answers to these questions will help determine the types and amount of equipment required for the most economical completion of this phase of the work.

To determine the amount of general excavation it is necessary to determine:

1. Size of building (building dimensions).

2. The distance that the footing projects out beyond the wall (the distance labeled "A" in Figure 25-6).

3. The amount of working space required between the edge of the footing and the beginning of excavation (the distance labeled "B" in Figure 25-6).

4. The elevation of the *existing* land, by checking the existing contour lines on the plot (site) plan.

5. The type of soil which will be encountered. This is determined by first checking the soil borings (may be on the drawings), but this *must* be checked during the site investigation (Section 21.9). Almost every specification clearly states that the soil borings are for the contractor's information *but* they are not guaranteed.

6. Whether the excavation will be sloped as shown in Figure 25-6 or shored. Slope angles (angles of repose) are given in Figure 25-5.

7. The depth that the excavation will have to be. This is done by determining the bottom elevation of the cut to be made. Then take the existing elevation, deduct any topsoil removed, and subtract the bottom elevation of the cut. This will determine the depth of the general excavation.

To practice estimating general excavation, examples are included which will show:

1. Residence with basement (Appendix B).

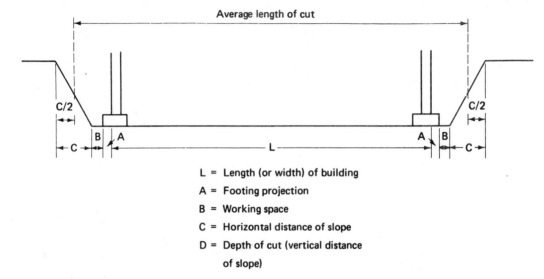

L = Length (or width) of building
A = Footing projection
B = Working space
C = Horizontal distance of slope
D = Depth of cut (vertical distance
    of slope)

Figure 25-6. Typical excavation.

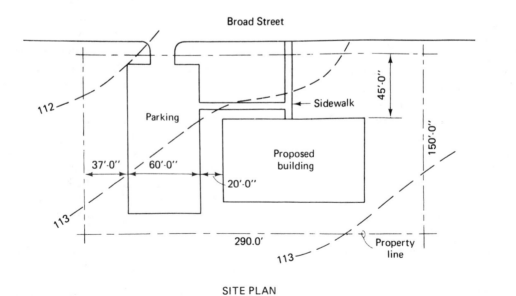

SITE PLAN

Figure 25-7. Plot plan.

418

2. Residence with slab on grade (Appendix C).

3. Residence with crawl space (Appendix D).

Example 25.4: Determine the amount of general excavation required for the building shown in Appendix B.

1. From the floor plan, the building size is 24'-0" × 50'-0".

2. From the wall section, the footing projects out 6 inches from the wall.

3. The work space between the edge of the footing and the beginning of the excavation will be 1'-0" in this example.

4. The elevation of the existing land, by checking the existing contour lines on the plot (site) plan, is about 113'-0" (Figure 25–7).

5. A check of the soil borings (assuming that this has been confirmed in the site investigation) indicates that:

    a. the soil is a gravel-clay mixture.

    b. the topsoil averages about 8 inches deep.

6. The gravel-clay soil mixture could (from experience and the table in Figure 25–5) have an angle of repose of about 65°, or a 2:3 slope (which means for every three feet of vertical depth an additional two feet of horizontal width is needed). Since the alternative is shoring or sheet-piling on this project, the sloped excavation will be used.

7. The bottom elevation of the general excavation cut will be at the bottom of the gravel. Since this elevation is rarely given it may have to be calculated. Generally, the drawings will give the elevation of the basement slab or bottom of the footing and the depth of cut is calculated from these. In Figure 25–8 the depth of cut is shown on a wall section when the basement slab elevation is given. In Figure 25–9 the depth of cut is shown when the bottom of footing is given. In this excavation the total depth is 9'-18", and once the 8 inches of topsoil is deducted, a 9'-0" cut is required.

To calculate the total area to be excavated, add the areas of the column footings and working space. The working space and the allowance for the slope of the soil are added only to the length of the column footing.

All of the information is put on a workup sheet as it is gathered. The sketch of this excavation problem is shown in Figure 25–10.

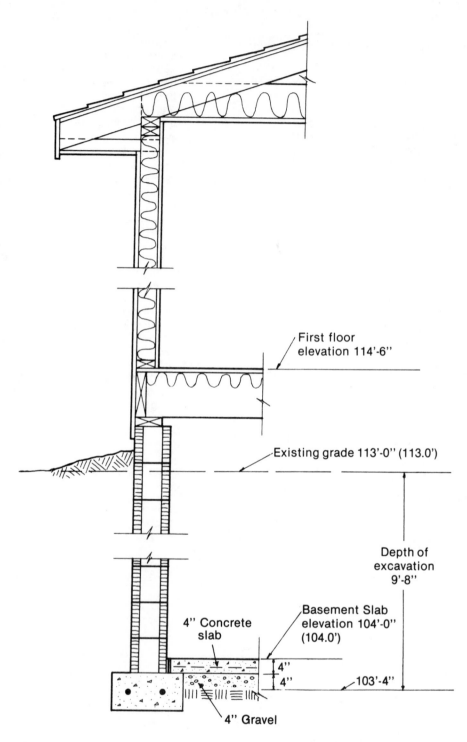

First floor
elevation 114'-6"

Existing grade 113'-0" (113.0')

Depth of
excavation
9'-8"

Basement Slab
elevation 104'-0"
(104.0')

4" Concrete
slab

4"
4"

103'-4"

4" Gravel

Figure 25-8. Depth of cut.

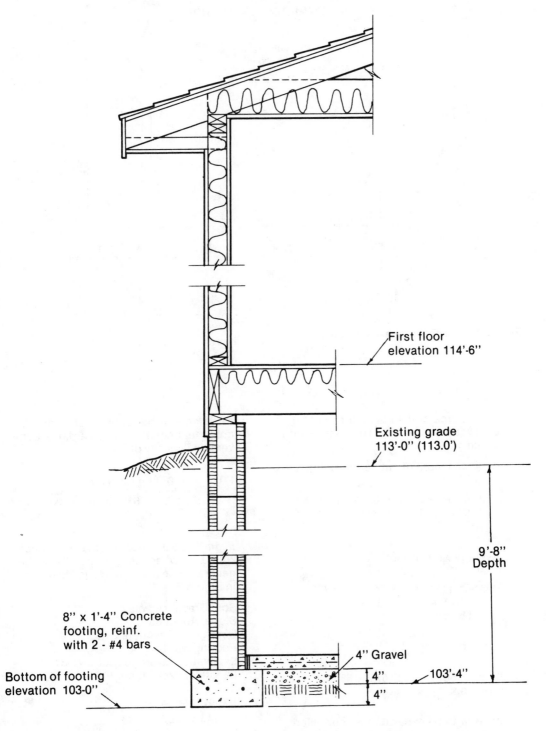

First floor
elevation 114'-6"

Existing grade
113'-0" (113.0')

9'-8"
Depth

8" x 1'-4" Concrete
footing, reinf.
with 2 - #4 bars

4" Gravel

4"
103'-4"

Bottom of footing
elevation 103-0"

4"

Figure 25–9. Depth of cut.

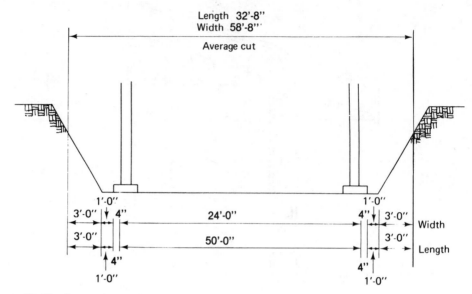

Figure 25-10. General excavation.

The area of the major portion of the excavation is the length times the width of the cut.

Average length of cut (left to right):

$3'\text{-}0'' + 1'\text{-}0'' + 0'\text{-}4'' + 24'\text{-}0''$
$+ 0'\text{-}6'' + 1'\text{-}0'' + 3'\text{-}0''$
$= 32'\text{-}8''.$

Average width of cut:

$3'\text{-}0'' + 1'\text{-}0'' + 0'\text{-}4'' + 50'\text{-}0''$
$+ 0'\text{-}4'' + 1'\text{-}0'' + 3'\text{-}0''$
$= 58'\text{-}8''.$

Area of building:

$32'\text{-}8'' \times 58'\text{-}8'' = 32.67' \times 58.67'$
$= 1917 \text{ s.f.}$

The area to be excavated is multiplied by the average depth of the cut (in feet) to determine the cubic feet. Since 27 cubic feet equal 1 cubic yard, the conversion is easily done.

$1,917 \times 9'\text{-}0'' = 17,253 \text{ c.f.}$
$$\frac{17,253}{27} = 639 \text{ c.y. to be excavated}$$

Example 25.5: For the small residence with a crawl space in Appendix D, the general excavation sketch is illustrated in Figure 25–11.

Soil: Gravel and clay
Slope: About 65°
Depth of excavation: 2'-8"
Footing width: 1'-4"
Excavation, average width: 4'-8"
Length of excavation: 148 l.f.

Excavation =
$$\frac{148 \times 4.67 \times 1.33}{27} = 34 \text{ c.y.}$$

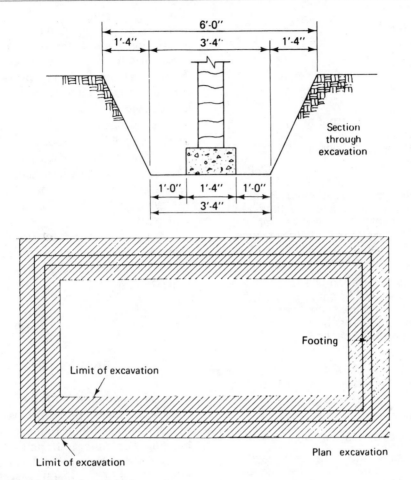

Figure 25-11. General excavation.

Labor

In these examples, the labor costs for the general excavation will be determined by using the time charts in Figure 25-4 and the labor costs as shown in the examples. Labor examples are given for each of the general excavation examples worked above.

Example 25.6: Residential Building (Appendix B)

Estimate the time and cost of general excavation for the general excavation calculated in Example 25-4. Assume a 120 h.p. dozer with a front-end loader is used and the excavated material will be spread on the side (100-foot haul, soft sand).

General excavations = 647 c.y.
Mobilization: 3 hours at $32 per hour
Rate of work: 85 c.y. per hour
Dozer: $9.85 per hour
Operator: $14. 75 per hour

Excavation time:

$$\frac{647}{85} = 7.6 \text{ hours}$$

Excavation cost:

Mobilization
      (3 hours at $32 per hour) = $96.00
Excavation
(7.6 hours at $24.60 per hour) = $\underline{186.96}$
                                      $282.96

Example: 25.7: Residential Building (Appendix D). Estimate the time and cost of general excavation for the general excavation calculated in Example 25.5. Assume a backhoe with a $^1/_2$ c.y. capacity is used.

      General excavation = 34 c.y.
      Mobilization: 1.5 hours
      Rate of work: 15 c.y. per hour
      Backhoe: $5.45 per hour
      Operator: $6.25 per hour

Excavation time:

$$\frac{34}{15} + 1.5 = 2.3 + 1.5 = 3.8 \text{ hours}$$

Excavation cost:

      3.8 hours x $11.70 per hour = $44.46

Note: Many subcontractors have a minimum charge for work done— perhaps a $50, $75, or $100 minimum to come and do any work at all.

_____ 25.6 _____
SPECIAL EXCAVATION

Usually the *special excavations* are the portions of the work that require hand excavation, but also included may be any excavation that requires special equipment; that is, equipment used for a particular excavating portion other than general (mass) excavation.

Portions of work most often included under this heading are footing holes, small trenches, and the trench-out below the general excavation for wall and column footings, if required. On a large project a backhoe may be brought in to perform this work, but a certain amount of hand labor is required on almost every project.

The various types of excavation must be kept separately on the estimate and, if there is more than one type of special or general excavation involved, each should be considered separately and then grouped together under the headings "special excavation" or "general excavation."

In calculating the special excavation, the estimator must calculate the cubic yards, select the method of excavation, and determine the cost.

_____ 25.7 _____
BACKFILLING

Once the foundation of the building has been constructed, one of the next steps in construction is the backfilling required around the building. *Backfilling* is the putting back of excess soil that was removed from around the building during the general excavation. After the topsoil, and general and special excavation have been estimated, it is customary to calculate the amount of backfill.

The material may be transported by wheelbarrows, scrapers, front-end loaders with scoops or buckets, scrapers, bulldozers, and perhaps trucks if the soil must be transported a long distance. The selection of equipment will depend on type of soil, weather conditions, and distance the material must be moved. If tamping or compaction is required, special equipment will be needed, and the rate of work per hour would be considerably lower than if no tamping or compaction were required.

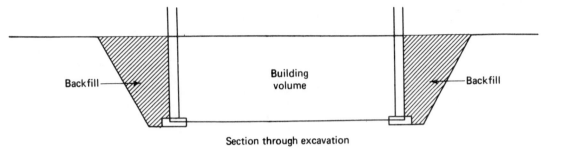

Section through excavation

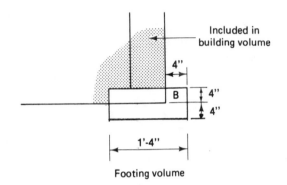

Included in
building volume

B

4"

4"

4"

1'-4"

Figure 25-12. Backfilling.

Footing volume

One method for calculating the amount of backfill to be moved is to determine the total volume of the building that will be built within the area of the excavation. This would be the total volume of the basement area, figured from the underside of fill material, and would include the volume of all footings, piers, and foundation walls. This volume is deducted from the volume of excavation that had been previously calculated. The volume of backfill required is the result of this subtraction. The figures should not include the data for top-soil, which should be calculated separately.

A second method for calculating backfill is to compute the actual volume of backfill required. The estimator usually makes a sketch of the actual backfill dimensions and finds the required amount of backfill.

The following examples illustrate how to calculate backfill quantities.

Example 25.8: Residential Building (Appendix B)

Building volume (Figure 25-12):

$$\frac{24' \times 50' \times 9'}{27} = \frac{10,800}{27} = 400 \, \text{c.y.}$$

Footing volume (Figure 25-12):

A. $\dfrac{152'\text{-}0'' \times 1'\text{-}4'' \times 4''}{27}$

$= \dfrac{152' \times 1.33' \times 0.33'}{27} = 2.5 \, \text{c.y.}$

B. $\dfrac{152'\text{-}0'' \times 4'' \times 4''}{27}$

$= \dfrac{152 \times 0.33 \times 0.33}{27} = 0.6 \, \text{c.y.}$

Total construction
volume $= -403.1 \, \text{c.y.}$

Backfill $= 235.9$

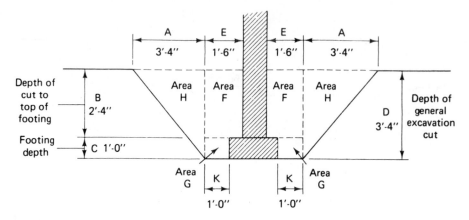

Figure 25-13. Backfill volume.

Example 25.9: Residential Building (Appendix D) (Figure 25–13)

Wall volume (8″ thick, 148 l.f., 2′ -0″ high):

$$\frac{0.67 \times 148 \times 2}{27} = 7.3 \text{ c.y.}$$

Footing volume (8″ high, 148 l.f., 1′ -4″ wide):

$$\frac{0.67 \times 148 \times 1.33}{27} = 4.9 \text{ c.y.}$$

Total volume =	12.2 c.y.
General excavation =	34 c.y.
Wall and footing volume =	− 12.2 c.y.
Backfill =	21.8 c.y.

Example 25.10: Residential Building (Appendix B)

Backfill = 235.9 c.y. (Example 25.8)
Tractor shovel (2¼ c.y. capacity):
$8.65 per hour
Operator: $13.50 per hour
Laborer: $5.50 per hour
Rate of work: 120 c.y. per hour
Mobilization: 2 hours

Backfill time:

$$\frac{235.9}{120} \text{ (allow 2 hours)} + 2 = 2 + 2 = 4 \text{ hours}$$

Backfill cost:
4 hours at $22.15 per hour = $ 88.60
4 hours at $ 5.50 per hour =   22.00

$110.60

Labor

The time required to do the backfilling is taken from the charts in Figure 25–4. The hourly wages used are based on local labor and union conditions.

_____ 25.8 _____

EXCESS AND BORROW

Once the excavation has been calculated in terms of excavation, backfill, and grading, the estimator must compare the total amounts of cut and fill required and determine if there will be an *excess* of materials that must be disposed of, or if there is a shortage of materials that must be brought in (*borrow*). Topsoil is not included in the comparison at this time; topsoil must be compared separately since it is much more expensive than the other fill that might be required, and excess topsoil is easily sold.

The specifications must be checked for what must be done with the excess material. Some specifications state that it may be placed in a particular location on the site, but many times they direct the removal of excess materials from the site. If the material is to be hauled away, the first thing the estimator must know is how many cubic yards are required and then he must find a place to haul them to. Remember that soil swells; if you calculated a haul-away of 100 cubic yards and the swell is estimated at 15 percent, then you will actually have to haul away 115 cubic yards. Finding a place that the excess material can be hauled to is not always a simple matter since it is desirable to keep the distance as short as possible from the site. The estimator should check into this when he makes his visit to the site. If material must be brought in, the estimator must first calculate the amount required and then set out to find a supply of material as close to the job site as is practical. Check the specifications for any special requirements pertaining to the type of soil that may be used. Keep in mind that the material being brought in is loose and will be compacted on the job. If it is calculated that 100 cubic yards are required, then the contractor will have to haul in at least 106 to 114 cubic yards of soil, and even more if it is clay or loam.

The next step is to select equipment for the work to be done; this will depend on the amount of material, type of soil, and the distance it must be hauled.

25.9
SUBCONTRACTORS

Contractors who specialize in excavation are available in most areas. There are certain advantages the specialized subcontractor has over many general contractors. For instance, they own a large variety of equipment, are familiar with the soil encountered throughout a given area, and know where fill can be obtained and excess cut can be hauled to. The subcontractor may bid the project as a lump sum or by the cubic yard. Either way the estimator must still prepare a complete estimate—in the first place to check the subcontractor's price since it must be checked to be certain it is neither too high or too low, and in the second place since the estimator will need the quantities to arrive at a bid price and must check all subcontractor's bids anyway.

Always discuss with the subcontractor exactly what he will include in his proposal. Put it in writing so that both parties agree on what he is bidding. It is customary that the general contractor perform all hand excavation and sometimes the trenching. Selection of a subcontractor is most often based on cost, but also to be considered are the equipment owned by the subcontractor and his reputation for speed and dependability.

25.10
EXCAVATION CHECKLIST

Clearing Site:

Removing trees and stumps

Clearing underbrush

Removing old materials from premises

Removing fences and rails

Removing boulders

Wrecking old buildings, foundations

Underpinning existing buildings

Disconnecting existing utilities

Clearing shrubbery

Excavation (including backfilling):

Basement

Footings

Foundation walls

Sheet-piling

Pumping

Manholes

Catchbasins

Backfilling

Tamping

Blasting

Grading (rough and fine)

Utility trenches

Grading and seeding lawns

Trees

Shrubbery

Topsoil removal

Topsoil—brought in

_____ 25.11 _____

ASPHALT PAVING

The asphalt paving required on the project is generally subcontracted out to a paving specialist. The general contractor's estimator will make an estimate to check the subcontractor's price.

Asphalt paving will, most commonly, be hot-mix and generally classified by traffic (heavy, medium, or light) and use (walks, driveways, courts, streets, etc.).

The estimator will be concerned with subgrade preparation, subdrains, soil steril-

ization, insulation course, subbase course, base courses, prime coats, and the asphalt paving required. Not all items are required on any given project so the estimator should determine which items will be required, material and equipment necessary for each portion of the work, and the requisite thickness and amount of compaction.

Specifications

Check the requirements for compaction, thickness of layers, total thicknesses, and materials required for each portion of the work. The drawings will also have to be checked for some of these items. The drawings will show the location of most of the work to be completed, but the specifications should also be checked. The specifications and drawings will list different requirements for the various uses. (These are called traffic requirements.)

Estimate

The number of square feet (or square yards) of surface area to be covered is determined, and the thickness (compacted) of each course and type of materials required is noted. Base courses and the asphalt paving are often taken off by the ton since this is the unit in which these materials must be bought. The type of asphalt and aggregate size required must also be noted. Two layers of asphalt paving are required on some projects; a coarse base mix may be used with a fine topping mix.

Equipment required may include a steel-wheel roller, trailers to transport equipment, dump trucks, paving machines, and various small tools.

To estimate the tons of material required per 1,000 s.f. of surface area, refer to Figure

Compacted thickness (inches)	Asphalt[1] paving	Granular[2] material	Subgrade[3] material
1"	6.5	5.25	4.6
2"	13.0	10.5	9.2
3"	19.5	15.75	13.8
4"	26.0	21.0	18.4
5"	32.5	26.25	23.0
6"	39.0	31.5	27.6
8"	52.0	42.0	36.8
10"	65.0	52.5	46.0
12"	78.0	63.0	55.2

* Per 100 S.F. of surface area, figures include 10% waste.

1. Asphalt paving, 140–150 lbs per cu. ft.

2. Granular material, 110–120 lbs per cu. ft.

3. Subgrade material, 95–105 lbs per cu. ft.

Figure 25-14. Approximate asphalt paving tonnage.

25-14. Different requirements will be listed for the various uses (walk, driveway, etc.), and the different spaces must be kept separately.

In many climates the asphalt paving has a cut-off date in cold weather, and the paving that is not placed when the mixing plants shut down will not be laid until the start-up time in the spring. The plants may be shut down for as long as four months or more, depending on the locale.

Chapter 26

Estimating Concrete

CONCRETE WORK

The concrete for a project may be either ready-mixed or mixed on the job. Most of the concrete used on residential work is ready-mixed and delivered to the job by the ready-mix company. Quality control, proper gradation, water, and design mixes are easily obtained by the ready-mix producers, which have, in many cases, fully automated and computerized their operations. When ready-mix is used, the estimator must determine the amount of concrete required, the type of cement, aggregates, and admixtures, which are discussed with the supplier, who will then give a proposal for supplying the specific material.

26.2

ESTIMATING CONCRETE

Concrete is estimated by the cubic yard, or by the cubic foot and converted into cubic yards. The cubic yard is used since it is the pricing unit of the ready-mix companies, and most tables and charts available relate to the cubic yard.

Floor slabs, slabs on grade, pavements, and sidewalks are most commonly measured and taken off in length, width, and thickness and converted to cubic feet and cubic yards (27 cubic feet = 1 cubic yard). Often, irregularly shaped projects are broken down into smaller areas for more accurate and convenient manipulation.

Footings, columns, beams, and girders are estimated by taking the lineal footage of each item times its width and depth. The cubic footage of the various items may then be tabulated and converted to cubic yards.

In estimating the footings for buildings with irregular shapes and many jogs, include the corners only once. If necessary, mark lightly on the plans which portions of the footing have been figured as you proceed. When taking measurements, keep in mind that the footing projects out from the wall, and therefore the footing length is greater than the wall length.

In completing quantities, make no deductions for holes smaller than 2 square feet or for the space that reinforcing bars or other miscellaneous accessories take up. Waste ranges from 5 percent on footings, columns, and beams to 8 percent for slabs.

The procedure which should be used to estimate the concrete on a project would be:

1. Review the specifications to determine the requirements for each area in which concrete is used separately (such as footings,

floor slabs, and walkways) and list the following:

a. type of concrete
b. strength of concrete
c. color of concrete
d. any special curing or testing.

2. Review the drawings to be certain that all concrete items shown on the drawings are covered in the specification.

3. List each of the concrete items required on the project.

4. Determine the quantities required for the working drawings. Footing sizes are checked on the wall sections and foundation plans. Watch for different size footings under different walls.

Concrete slab information will most commonly be found on wall sections and floor plans. Exterior walks and driveways will most likely be identified on the plot (site) plan and in sections and details.

An estimate of the concrete required for the residential building in Appendix B and Appendix C will be prepared.

Example 26.1: Residential Building (Appendix B)

Footings (Figure 26-1):

$$50.67 + 22 + 22 + 50.67 = 145.33 \text{ l.f.}$$

$$\frac{8'' \times 1'\text{-}4'' \times 145.33'}{27} =$$

$$\frac{0.67 \times 1.33 \times 145.33}{27} = 4.8 \text{ c.y.}$$

Concrete for footings:

$$
\begin{array}{r}
4.80 \text{ c.y.} \\
+ \text{ waste} = \underline{.20 \text{ c.y.}} \\
5 \quad \text{c.y. required}
\end{array}
$$

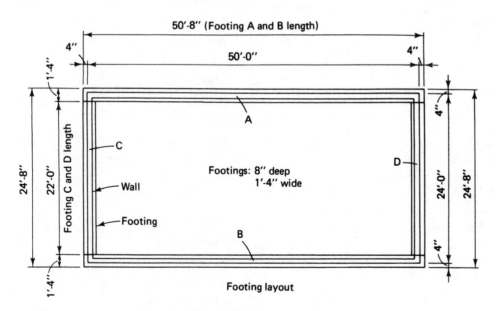

Figure 26-1. Concrete footings-residential buildings.

Example 26.2: Residential Building (Appendix C)

Slab on grade (Review Appendix C):

4″ thick
24′-0″ × 50′-0″ = 1200 s.f.

$$\frac{1200 \times 0.33}{27} = 14.7 \text{ c.y.}$$

Concrete required = 14.7 c.y.
    + Waste   .3 c.y.
                15.0 c.y.

Note: The footing detail shown in Appendix C also requires a footing. The section or detail drawing for the outside edge of the slab must always be carefully checked to determine the type of construction required.

Labor

The time required for the concrete work has been estimated from the chart in Figure 26–2. The hourly wages are based on local labor and union conditions.

Type of Placement	Man hours per c.y.
Columns	1–4
Footings (up to 8 c.y.)	1–1.5
Footings (over 8 c.y.)	2–4
Slabs (basement, 1st floor)	
up to 1,000 s.f.	1–2
1,000 to 5,000 s.f.	2–3
over 5,000 s.f.	3–4
Finishing machine	1–2
Stairs	1.5–4
Structural slabs	1.5–3
Thin walls (6 in. or less)	1.5–4
Thick walls	2–5

Figure 26–2. Man-hours for concrete work.

Example 26.3: Residential Building (from Example 26.1):

Footings = 5 c.y.
Rate of work: 2.5 c.y. per man-hour
Labor: $7.50 per hour
Time = $\frac{5}{2.5}$ = 2 man-hours
Cost = 2 hours × $7.50 = $15.00

Example 26.4: Residential Building (From Example 26.2)

Slab on grade: 15.0 c.y.
Rate of work: 2 c.y. per man-hour to place concrete
Labor: $7.50 per hour
Placement time: $\frac{15}{2}$ = 7.5 man-hours
Cost: 7.5 man-hours × $7.50 = $56.25

For example, when you are taking off the wall footings, list all of the work to be considered.

Wall Footings:

Forms

Accessories

Reinforcing

Concrete (3,000 psi)

Curing

If the portion of work being considered is a concrete floor, it may include the following items:

Concrete Floors:

Vapor barrier

Reinforcing

Expansion joints

Concrete (2,500 psi)

Finishing

26.3
REINFORCING

The reinforcing used in concrete may be either reinforcing bars, welded wire mesh, or a combination of the two. Reinforcing bars are listed (noted) by the bar number, which corresponds to the bar diameter in eighths of an inch. For example, a #7 bar (deformed) contains as much as metal as a ⅞-inch diameter bar. The #2 bar is a plain round bar, but all the rest are deformed round bars. The bar numbers, diameters, areas, and weights are given in Figure 26-3.

Reinforcing bars are taken off by lineal feet. The take-off (workup) sheet should be set up to include the numbers of the bars, number of pieces, lengths, and bends. Since reinforcing bars are usually priced by the hundredweight (100 pounds, cwt), the weight of reinforcing required must be calculated. Reinforcing bars purchased at smaller local warehouses are generally bought in 20-foot lengths, and cut and bends are made by the contractor. Refer to Figure 26-4 for manhours required for reinforcing bar bending and placing. An allowance for splicing (lapping) the bars (Figure 26-5) must be included also (lap splicing costs may range from 5 to 15 percent, depending on the size of the bar and yield strength of steel used). Waste may range from less than 1 percent for precut and preformed bars to 10 percent when the bars are cut and bent on the job site.

Bar no.	Nominal size — inches		Area sq. in.	Weight lbs/ft
2	$\frac{1}{4}''$	ϕ	0.05	0.167
3	$\frac{3}{8}''$	ϕ	0.11	0.376
4	$\frac{1}{2}''$	ϕ	0.20	0.668
5	$\frac{5}{8}''$	ϕ	0.31	1.043
6	$\frac{3}{4}''$	ϕ	0.44	1.502
7	$\frac{7}{8}''$	ϕ	0.60	2.044
8	1.0''	ϕ	0.79	2.670
9	1.13''	ϕ	1.00	3.400
10	1.27''	ϕ	1.27	4.303
11	1.41''	ϕ	1.56	5.313
14	1.		2.41	8.18
18			3.98	13.52

No. 2 bar is a plain round bar. All other bars are deformed.

Figure 26-3. Weights and areas of reinforcing bars.

Size of bar inches	Place 1000 L.F. inc. chairs, stirrups, ties and wiring	Place 1000 L.F. tie wire only	Place 100 pieces short lengths and tie wire
$\frac{1}{2}$ and less	2 − 4	1.5 − 3	1.5 − 4
$\frac{5}{8}$ and $\frac{3}{4}$	2.5 − 5	2.0 − 4	1.8 − 4.5
$\frac{7}{8}$, 1 and 1$\frac{1}{8}$	3.0 − 6	2.5 − 5	2.3 − 5
1$\frac{1}{4}$ and 1$\frac{1}{2}$	3.5 − 7	3.0 − 6	3 − 6

*Approximate

Figure 26-4. Man-hours required for placing reinforcing bars.

Bar size	Splice required when specified as number of bar diameter	
	24 d	30 d
3	N.R.	N.R.
4	1'-0"	1'-3"
5	1'-3"	1'-7"
6	1'-6"	1'-11"
7	1'-9"	2'-3"
8	2'-0"	2'-6"
9	2'-4"	2'-10"
10	2'-7"	3'-3"
11	2'10"	3'-7"

The recommended minimum length of splice is 1'-0" or 24 bar diameters whichever is greater.

N.R. − No recommended, use 1'-0".

Figure 26-5. Splice requirements.

Mesh reinforcing may be welded wire mesh or expanded metal. The former is an economical reinforcing for floor and driveways, and is commonly used as temperature reinforcing. It is usually furnished in a square or rectangular mesh arrangement of the wires and usually sold in a roll or in flat sheets (of a variety of sizes, usually specially ordered). The rolls are 5 feet wide and 150 feet long. Wire mesh is designated on the drawings by wire spacing and wire gauge in the following manner: $6 \times 6\ 10/10$. This designation shows that the longitudinal and transverse wires are spaced 6 inches on center, while both wires are No. 10 gauge. Another example, $4 \times 8\ 8/12$, means that the longitudinal wires are spaced 4 inches on center while the transverse wires are spaced 8 inches on center; the longitudinal wire is No. 8 gauge while the transverse is No. 12 gauge. The take-off must be broken up into the various sizes required and the number of square feet required of each type. It is commonly specified that wire mesh have a lap of one square and this allowance must be included; for example, a 6-inch lap requires 10 percent extra mesh while a 4-inch lap requires almost 6.7 percent extra. The mesh may be either plain or galvanized, and this information is included in the specifications. Galvanized mesh may require special ordering and delivery times of two and three weeks.

Estimating Reinforcing Bars

The lineal footage of re-bars can most often be worked up from the concrete calculations. The sections and details must be checked to determine the reinforcing requirements of the various footings. The various footing sizes can generally be taken from the concrete calculations and adapted to the reinforcing take-off.

Estimating Wire Mesh

The square footage of floor area to be covered may be taken from the slab concrete calculations. Check the sections and details for the size of the mesh required. To determine the number of rolls required, add the lap required to the area to be covered and divide by 750 (the square footage in a roll). Waste averages about 5 percent unless a good deal of cutting is required; only full rolls may be purchased in most cases.

Example 26.5: Residential Building (Appendix B)

Footing: 1'-4" wide
Re-bars: Two #4 bars
Footing length: 146 l.f. (Figure 26–1)

$$146\ \text{l.f.} \times 2 = 292\ \text{l.f.}$$
$$+\ 10\%\ \text{lap and waste} = \underline{\ \ 28\ \text{l.f.}}$$
$$320\ \text{l.f. required}$$

Weight = 320 l.f. $\times$ 0.668 lb./ft.
= 214 lb.

Labor

The time required for the reinforcing has been estimated from the charts in Figure 26–4. The hourly wages are based on local labor and union conditions.

Example 26.6: Residential Building (Appendix C)

Slab on grade: 1200 s.f.
Welded wire mesh: $6 \times 6\ 10/10$
Lapping: 6"
Wire mesh required:

$$1200\ \text{s.f.} + 10\%\ \text{lap} = 1320\ \text{s.f.}$$

$$\frac{1320\ \text{s.f.}}{750\ \text{s.f. per roll}} = 2\ \text{rolls (includes waste)}$$

Example 26.7: Residential Building (Appendix B)

Footing re-bars: 320 l.f. #4 bars
Rate of work: 4.0 man-hours per 1,000 l.f.
Labor: $5.25 per hour

Time = 0.32 × 4.0 = 1.28 man-hours
Cost = 1.28 hours × $5.25 per hour
        = $6.72

Example 26.8: Residential Building (Appendix C)

Welded wire mesh: 1320 s.f. (to be placed)
Rate of work: 3 man-hours per 100 s.f.
Labor: $7.50 per hour

Time = 1.3 × 3 = 3.9 man-hours
Cost = 3.9 man-hours × $7.50
        = $29.25

26.4
VAPOR BARRIER

The *vapor barrier* placed between the gravel and the slab poured on it is usually included in the concrete portion of the take-off. This vapor barrier most commonly consists of polyethylene films or kraft papers. The polyethylene films are designated by the required mil thickness (usually 4 or 6 mil). The material used should be lapped about 6 inches, so an allowance for this must be made depending on the widths available. Polyethylene rolls are available in widths of 3, 4, 6, 8, 10, 12, 16, 18, and 20 feet, and are 100 feet long. Careful planning can significantly cut down on waste, which should average 5 percent plus lapping. Two men can place 1,000 square feet of vapor barrier on the gravel in about one hour, including the time required to get the material from storage, place, and secure it. Large areas can be covered in proportionately less time.

Estimating Vapor Barrier

For irregular-shaped buildings the most accurate method of determining the vapor barrier is to sketch a layout of how it might be on the job. Often, several sketches must be made, trying rolls of various sizes, before the most economical arrangement is arrived at.

Example 26.9: Residential Building (Appendix C)

Vapor barrier under slab: 1200 s.f.

1 roll, 12' wide, 100' long =
1200 s.f. (no waste)

Labor

The time required for the installation of the vapor barrier has been estimated from the chart in Figure 26-6. The hourly wages are based on local labor and union conditions.

Example 26.10: Residential Building (Appendix C)

Vapor barrier: 1200 s.f.
Rate of work: 1.5 man-hours per 1,000 s.f.
Labor: $7.50 per hour

Time = 1.2 × 1.5 = 1.8 man-hours
Cost = 1.8 × $7.50 – $13.50

Sq. ft. of vapor barrier	Hours per 1,000 s.f.
Up to 1,000	1–1.5
1,000 to 5,000	0.75–1.5
Over 5,000	0.5–1.5

Figure 26-6. Man-hours required for installing vapor barriers.

26.5
ACCESSORIES

Any item cast into the concrete should be included in the concrete take-off. The list of items that might be included is extensive; the materials vary depending on the item and intended usage. The accessory items may include:

Expansion Joint Filler

Materials commonly used as fillers are asphalt, fiber, sponge rubber, cork, and asphalt-impregnated fiber. These are available in thicknesses of ¼, ⅜, ½, ¾, and 1 inch; widths of 2 to 8 inches are most common. However, sheets of filler are available and may be cut to the desired width on the job. Whenever possible, filler of the width to be used should be ordered to save labor costs and reduce waste. Lengths of filler strips may be up to 10 feet, and the filler should be taken off by lineal feet plus 5 percent waste.

Estimating Expansion Joint Filler.
Determine from the plans, details and sections exactly how many lineal feet are required. A ½-inch thick expansion joint filler is required where any concrete slab abuts a vertical surface. This would mean that filler would be re-

l.f. of expansion joint	Hours per 100 l.f.
Up to 500	1 - 1.5
500 to 1,000	0.75 1.5
Over 1,000	0.5 1.5

Figure 26–7. Man-hours required for installing expansion joint filler.

quired around the outside of the slab, between it and the wall.

Labor. The time required for the installation of the expansion joint filler has been estimated from the chart in Figure 26–7. The hourly wages are based on local labor and union conditions.

26.6
CONCRETE FINISHING

All exposed concrete surfaces require some type of *finishing*. Basically finishing consists of the patch-up work after the removal of forms and the dressing up of the surface by troweling, sandblasting, and other methods.

Patch-up work may include patching voids and stone packets, and removing fins and patching chips. Except for some floor slabs (on grade), there is always a certain amount of this type of work on exposed surfaces. It varies considerably from job to job and can be kept to a minimum with good quality concrete, with use of forms that are tight and in good repair, and with careful workmanship, especially in stripping the forms. This may be included with the form stripping costs, or it may be a separate item. As a separate item, it is much easier to get cost figures and keep a cost control on the particular item rather than "bury" it in with stripping costs. Small patches are usually made with a cement-sand grout mix of 1:2; be certain the type of cement (even the brand name) is the same used in the pour, since different cements are varying shades of gray. The man-hours required will depend on the type of surface, number of blemishes, and the quality of the patch job required. Scaffolding will be required for work above 6 feet.

The finishes required on the concrete surfaces will vary throughout the project. The finishes are included in the specifications and finish schedules; sections and details should also be checked. Finishes commonly required for floors include hand or machine troweled, carborundum rubbing (machine or hand), wood float, broom, floor hardeners, and sealers. Walls may also be troweled.

The finishing of concrete surfaces is estimated by the square foot. Since various finishes may be required throughout, keep the take-off for each one separately. The man-hours required (approximate) for the various

types of finishes are shown in Figure 26-8. Charts of this type should never take the place of experience and common sense, and are included only as a guide.

The equipment required will depend on the type of finishing work being done. Trowels (hand and machine), floats, burlap, scaffolding, and small hand tools must be included with the costs of their respective items of finishing.

Type of finishing		Man hours
Troweling, machine	– slabs on ground (100 S.F.)	1–2
hand	– slabs on ground (100 S.F.)	2–4
Broom finish (brush)	– slabs (100 S.F.)	$\frac{1}{2}$–2
	– stairs (100 S.F.)	3–6
Float finish	– slabs on grade (100 S.F.)	1–3
	– slabs suspended (+ 10%)	
	– walls (100 S.F.)	2–5
	– stairs (100 S.F.)	2–4
	– curbs (100 L.F.)	5–12
Bushhammer, machine	– green concrete (100 S.F.)	2–4
	– cured concrete (100 S.F.)	4–8
hand	– green concrete (100 S.F.)	6–12
	– cured concrete (100 S.F.)	–
Rubbed	– with burlap and grout (100 S.F.)	1–4
	– with float finish (100 S.F.)	2–6
Exposed aggregate	– retarder, apply and clean concrete	2–3
Sandblasting	– light penetration	2–4
	– heavy penetration	4–6
Patching	– tie holes and honeycombing (100 S.F.)	1–3
Acid wash	– 100 (S.F.)	1–4
Hardeners and Sealers (spray) – floors (100 S.F.)		$\frac{1}{2}$–1

Figure 26-8. Man-hours required for finishing concrete surfaces.

Estimating Concrete Finishing

Areas to be finished may be taken from other concrete calculations, either for the actual concrete required or the square footage of forms required.

Floor slabs, and slabs on grade, pavements, and sidewalk areas can most easily be taken from actual concrete required. Be careful to separate each area requiring a different finish. Footing and wall areas are most commonly found in the form calculations.

Labor

The time required for the finishing of the concrete has been estimated from the chart in Figure 26–8. The hourly wages are based on local labor and union conditions.

26.7
CURING

Estimating the cost of curing means that first a determination must be made of what type of curing will be required and what type of weather the concrete will be poured in. Many large projects have concrete poured throughout the year and thus you must consider the problems involved in cold, mild, and hot weather. Smaller projects have a tendency to fall in one or perhaps two of the seasons, but there is a general tendency to avoid the very coldest of weather.

If a temporary enclosure is required (perhaps of wood and polyethylene film), the size and shape of the enclosure must be determined, a material take-off made, and the number of man-hours determined. For simple enclosures two men can erect 100 square feet in one-half to one hour, once the materials are assembled in one place. If the wood may be reused, only a portion of the material cost is charged to this portion of the work. Keep in mind that the enclosure must be erected and possibly moved during the construction phases and taken down afterward; each step costs money.

Heating of water and aggregate raises the cost of the concrete by 3 to 10 percent, depending on how much heating is needed and whether ready-mix or job-mix concrete is being used. Most ready-mix plants already have the heating facilities and equipment with the cost spread over their entire production. *Job-mixing* means the purchase and installation of equipment on the job site as well as its operation in terms of fuel, man-hours, and upkeep.

The cost of portable heaters used to heat the space is usually based on the number of cubic yards of space to be heated. One man (not an operator) can service the heaters for 100 cubic yards in about one to two man-hours, depending on the type heater and fuel being used. Fuel and equipment costs are also based on the type used.

If approximately 100,000 Btuh are required per 300 to 400 square feet (averaging 8 feet high), estimate the number of units needed (depending on net output of the unit), determine the amount of fuel that will be consumed per hour, the number of hours the job will require, and determine the equipment and fuel costs.

The continuous spray requires purchase of equipment (which is reusable) and the employment of labor to set it up. It will require between one-half and one man-hour to run the hoses and set up the equipment for an area of 100 square feet. The equipment should be taken down and stored when it is not in use.

Moisture-retaining and watertight covers are estimated by the square footage of the surface to be covered and separated into slabs or walls and beams. The materials are estimated as to their initial cost, and this is divided by the number of uses expected of them. Covers over slabs may be placed at the rate of 1,000 square feet per one to three man-hours, and as many as five uses of material can be expected (except in the case of canvas, which lasts much longer); wall and beam covers may be placed at the rate of 1,000 square feet per two to six hours. The sealing of watertight covers takes from three-quarters to one man-hour per 100 lineal feet.

Sealing compounds are estimated by the number of square feet to be covered divided by the coverage (in square feet) per gallon to determine the number of gallons required. Note whether one or two coats are required. If the two-coat application is to be used, be certain to allow for material to do both coats. For two-coat applications the first coat coverage varies from 200 to 500 square feet, while the second coat coverage varies from 350 to 650 square feet. However, always check the manufacturer's recommendations. Equipment required will vary according to the job size. Small areas may demand the use of only a paint roller, while medium sized areas may require a pressure-type hand tank or backpack sprayer. Large, expansive areas can best be covered with special mobile equipment. Except for the roller, a great deal of reuse can be expected from the equipment, usually a life expectancy of five to eight years with reasonable care. The cost of the equipment would be spread over the time of its estimated usage. Using the pressure-type sprayer, you can estimate that from one and one-half to three hours will be required per 100 square feet, while in estimating labor for mobile equipment, you should depend more on the methods in which the equipment is mobilized than on anything else, but labor costs can usually be cut by 50 to 80 percent if the mobile equipment is used.

26.8
FORMS

This portion of the chapter is not a course in form design, but it is intended to make you aware of the factors involved in forms and the consideration of forms relative to costs. No one design or system will work for all types of formwork. In general, the formwork must be true to grade and alignment, braced against displacement, resistant to all vertical and horizontal loads, resistant to leaking through tight joints and of a surface finish that produces the desired texture. The pressure on the form is the biggest consideration in the actual design of the forms.

In the design of wall and column forms the two most important factors are the rate of placement of the concrete (feet per hour) and the temperature of the concrete in the forms. From these two variables the lateral pressure (psf) may be determined. Floor slab forms are governed primarily by the actual live and dead loads that will be incurred.

Actual field experience is a big factor in imagining exactly what is required in forming and should help in the selection of the type form to be used. The types of forms, form liners, supports, and methods are many; the preliminary selections must be made during the bidding period. This is one of the phases in which the proposed job superintendent (if you are low bidder) should be included in the discussions of the methods and types of forms being considered as well as in the consideration of what extra equipment and manpower may be required.

Engineering data relative to forms and the design of forms is available from the American Concrete Institute (ACI) and Portland Cement Association (PCA), as well as from most manufacturers. This data should be included in your reference library.

The forms for concrete footings, foundations, retaining walls, and floors are estimated by the area (in square feet) of the concrete that comes in contact with the form. The plans should be carefully studied to determine if it is possible to reuse the form lumber on the building and the number of times it may be reused. It may be possible to use the entire form on a repetitive pour item, or the form may have to be taken apart and reworked into a new form.

Many types of forms and form liners may be rented, and the renting firms often provide engineering services as well as the forms themselves. Often, the cost of the forms required for the concrete work can be reduced substantially.

Wood Forms

Most forms are built of wood. The advantages of wood are that it is readily accessible and easy to work with, and that once used, it may be taken apart and reworked into other shapes. Once he has decided upon wood, the estimator must work out the quantity of lumber required and the number of uses that can be made of it. This means the construction of the forms must be decided upon in terms of the plywood sheathing, wales, studs or joists, bracing, and ties. The estimator can easily determine all of this if he knows the height of the fresh concrete pour (for columns and walls), the temperature of the placed concrete, and the thickness of the slab (for floors). The manufacturer's brochures or the ACI formwork engineering data may be used. Complete explanations of the planning, design, materials, accessories, loads, pressure tables, design tables, and more are available in the ACI publications SP-4 "Form Work for Concrete." The address is American Con-

crete Institute, P.O. Box 19150, Detroit, Michigan 48219. This publication is a must for the estimator and should be in your library (or in the company library).

Metal Forms

Prebuilt systems of metal are used extensively on poured concrete work—not only on large work, but even for foundation walls for homes. Advantages are that these systems are reusable several times, are easily adaptable to the various required shapes, are interchangeable, and require a minimum of hardware and a minimum of wales and ties, which are easily placed. They may be purchased or rented, and several time-saving methods are employed. Curved and battered walls are easily obtained, and while the plastic-coated plywood face liner is most commonly used, other liners are available. Heavy-duty forms are available for heavy construction jobs in which a high rate of placement is desired.

Engineering data and other information pertaining to the uses of steel forms should be obtained from the metal form supplier. They can give information regarding costs (rental and purchase), tie spacing, number of forms required for the project, and labor requirements. Information concerning metal (steel) forms can be obtained from Symons Mfg. Company, 200 East Touhy Avenue, Des Plaines, Illinois 60018, as well as from many other manufacturers.

Miscellaneous Forms

Column forms are available in steel and laminated plies of fiber for round, square, and rectangular columns. Many manufacturers will design custom forms of steel, fiber, and fiberglass to meet project requirements. For columns these would include tapered, fluted, triangular, and half-rounded shapes. Fiber tubes are available to form voids in cast-in-place (or precast) concrete; various sizes are available. Most of these forms are sold by the lineal footage required of a given size. The fiber forms are not reusable, but the steel forms may be reused over and over.

Estimate

The unit of measurement used for forms is the actual contact area (in square feet) of the concrete against the forms (with the exception of moldings, cornices, sills, and copings, which are taken off by the lineal foot). The forms required throughout the project must be listed and described separately. There should be no deductions in the area for openings of less than 30 square feet.

Materials in the estimate should include everything required for the construction of the forms except stagings and bridging. Thus materials that should appear are struts, posts, bracing, bolts, wire, ties, form liners (unless they are special), and equipment for repairing, cleaning, oiling, and removing.

Items affecting the cost of concrete wall forms are the height of the wall (since the higher the wall, the more lumber that will be required per square foot of contact surface) and the shape of the building, including pilasters.

Items affecting the cost of concrete floor forms include the floor-to-floor height, reusability of the forms, length of time the forms must stay in place, type of shoring and supports used, and the number of drop beams required.

The various possibilities of renting or purchasing forms, using gang forms built on ground and lifted into place, slip forming,

and so on should be considered during this phase.

While approximate quantities of materials are given (Figure 26–9) for work forms, a complete take-off of materials should be made. The information contained in Figure 26–9 is approximate and should be used only as a check.

Estimating Wood Footing Forms

First determine if the entire footing will be poured at one time or if it will be poured in segments, which would permit the reuse of forms.

Example 26.11:   Residential Building (Appendix C)

Concrete finishing: Slab 1,200 s.f.
Machine trowel
Rate of work: 1.5 man-hours per 100 s.f.
Labor: $7.50 per hour

   Time = 12 × 1.5 = 18 man-hours

   Cost = 18 × $7.50 = $135.00

Example 26.12:   Residential Building (Appendix B)

Footing forms—single pour, use 2 × 8
l.f. of footings = 146 l.f. (actually 145.33 l.f., from Fig. 26–1)
l.f. of forms = 146 × 2 = 292 l.f.
Cutting and waste: 4%

Total l.f. of forms required = 292 × 1.04 = 303 l.f. (use 300 l.f.)

$$\frac{2 \times 8 \times 300}{12} = 400 \text{ fbm}$$

Stakes, 2 × 2 wood, 5'-0" apart
$$\frac{300}{5} = 60 \text{ stakes required}$$

Type of form	Lumber fbm	Man hours			
		Assemble	Erect	Strip and clean	Repair
Footings	200–400	2–6	2–4	2–5	1–4
Walls	200–300	6–12	3–6	1–3	2–4
Floors	170–300	2–12	2–5	1–3	2–5
Columns	170–350	3–7	2–6	2–4	2–4
Beams	250–700	3–8	3–5	2–4	2–4
Stairs	300–800	8–14	3–8	2–4	3–6
*Moldings	170–700	4–14	2–8	2–6	3–6
*Sills, coping	150–600	3–12	2–6	2–4	2–6

*Values are for 100 lineal feet

Figure 26-9. Wood forms, approximate quantity of materials and man-hours.

Labor

The time required for forms has been estimated from the table in Figure 26-9, The hourly wages are based on local labor and union conditions.

Example 26.13: Residential Building

Footing formwork: 292 l.f., 400 fbm (Example 26-12)

Rate of work—complete assembly and disassembly: 8 man hours per 300 fbm

Labor: $6.25 per hour

Time = $\frac{400 \times 8}{300}$ = 10.72 man hours

Cost = 10.72 hours × $6.25 per hour = $67

26.9 CHECKLIST

Forms for:
Footings, walls, and columns
Floors
Piers
Beams
Columns
Girders
Stairs
Platforms
Ramps
Miscellaneous
Form:
Erection

Removal

Repairs

Ties

Clamps

Braces

Cleaning

Oiling

Repairs

Liners

Concrete:

Footings, walls, and columns

Floors

Topping

Piers

Beams

Columns

Girders

Stairs

Platforms

Ramps

Curbing

Sills

Walks

Driveways

Slabs

Materials, mixes:

Cement

Aggregates (fine and coarse)

Water

Color

Air entraining

Other admixtures

Strength required

Ready-mix

Heated concrete

Cooled concrete

Finishes:

Hand trowel

Machine trowel

Bushhammer

Wood float

Cork float

Broom

Sand

Rubbed

Grouted

Removing fins

Curing:

Admixtures

Ponding

Spraying

Straw

Canvas

Vapor barrier

Heat

Reinforcing:

Bars

Wire mesh

Steel grade

Galvanizing

Bends

Hooks

Ties

Stirrups

Chairs

Cutting

Chapter 27

Estimating Masonry

27.1
CONCRETE MASONRY ESTIMATING

Concrete masonry should be estimated from the drawings by the square feet of wall required and divided into the different thickness of each wall. The total square footage of each wall, of a given thickness, is then multiplied by the number of units required per square foot (Figures 27-1 and 27-2).

While estimating the quantities of concrete masonry, use the exact dimensions shown. Corners should only be taken once, and deductions must be made for all openings, no matter what size they are. In this manner, you will arrive at the exact square footage required for the project. This area is converted to units, and to this number of units must be added an allowance for waste and breakage.

While making the take-off, the estimator should note how much cutting of masonry units will be required. Cutting of the units is an expensive item and should be anticipated.

In working up the quantity take-off, the estimator must separate masonry according to:

1. Size of the units.
2. Shape of the units.
3. Colors of the units.
4. Type of bond (pattern).
5. Shape of the mortar joints.
6. Colors of the mortar joints.
7. Any other special requirements (such as fire rating).

In this manner it is possible to make the estimate as accurate as possible. When masonry is often a large percentage of the work, accuracy on the order of 99 percent is required.

Labor

The amount of time required for a mason (with the assistance of helpers) to lay a masonry unit varies with the size, weight, and shape of the unit, bond (pattern), number of openings, whether the walls are straight or have jogs in them, distance the units must be moved (both horizontally and vertically), and the shape and color of the mortar joint.

The height of the walls becomes very important in estimating labor for masonry units. The masonry work that can be laid up without the use of scaffolding is generally the least expensive. Extra labor costs arise from the erection, moving, and dismantling of the scaffolding as the building goes up. The units and

| Materials | Materials | | | |
| | Concrete block | | | Brick |
	8 x 16	8 x 18	5 x 12	$2\frac{1}{4}''$ x $3\frac{3}{4}''$ x 8''
No. of units	112.5	100	220	617
Mortar cu. ft.				
Face shell bedding	2.3	2.2	3.6	7.2
Face shell and web bedding	3.2	3.0	5.5	7.2

Values shown are net. Waste for block and brick ranges from 5 to 10%.
Waste for mortar may range from 25% to 75% and actual job experience
should be considered on this item. It is suggested that 100% waste be
allowed by the inexperienced and actual job figures will allow a downward
revision. Figures are for a $\frac{3}{8}''$ thick joint.

Figure 27-1. Materials required for 100 s.f. of face area.

mortar have to be placed on the scaffold, which further adds to the labor costs.

The weather conditions always affect labor costs since a mason will lay more brick on a clear, warm, dry day than on a damp, cold day. Winter construction requires the building and maintenance of temporary enclosures and heating materials.

| Mix by volume | Quantities | | | |
	Masonry cement (bags)	Portland cement (bags)	Lime or putty (C.F.)	Sand (C.F.)
1:3 (masonry)	0.33	—	—	0.99
1 : 1 : 6 (portland)	—	0.16	0.16	0.97

Figure 27-2. Mortar mixes to mix one cu. ft. of mortar.

Example 27.1: Residential Building

Height of block = 2'-8" (from wall section, Appendix D)
Note: 2'-8" is 4 full block courses of 8" high block

Length of foundation wall =
50' + 22'-8" + 50' + 22'-8" = 145'-4"

Block area = 145.3 × 2.67 = 388 s.f.

Units = 3.88 × 112.5 = 437 units
+ 5% waste = <u>22</u> units
460-8" × 8" × 16" units

Labor

The time required for the concrete masonry has been estimated from the charts in Figure 27-3. The hourly wages are based on local labor and union conditions.

Example 27.2: Residential Building

Concrete block: 8" wide, 460 units (3.6 squares, Example 27.1)
Rate of work per square:

Mason, 4.5 man hours
Laborer, 4.5 man hours

Labor:

Mason, $6.25 per hour
Laborer, $3.75 per hour

Time:

Mason—3.6 × 4.5 = 16.2 man hours
Laborer—3.6 × 4.5 = 16.2 man hours

Cost:

Mason—16.2 × $6.25 = $101.25
Laborer—16.2 × $3.75 = <u>60.75</u>
$162.00

27.2
BRICK ESTIMATING

The first thing to be determined in estimating the quantity of brick is the size of the brick and the width of the mortar joint. They are both necessary to determine the required number of bricks per square foot of wall area and the quantity of mortar. Brick is sold by the thousand units, so the final estimate of materials required must be in the number of units required.

To determine the number of bricks required for a given project, the first step is to obtain the length and height of all walls to be faced with brick and then calculate the area of wall. Make deductions for all openings so that the estimate will be as accurate as possible. Check the jamb detail of the opening to determine if extra brick will be required for the revel; generally if the revel is over 4 inches deep, extra brick will be required.

Once the number of square feet has been determined, the number of bricks must be calculated. This varies depending on the size of the brick, width of the mortar joint, and style of bond required. The figures must be extremely accurate since actual quantities and costs must be determined. It is only in this manner that you will increase your chances of getting work at a profit—each figure used must be as accurate as possible.

Figure 27-4 shows the number of bricks required per square foot of wall surface for various patterns and bonds. Special bond patterns require that the estimator analyze the style of bond required and determine the number of bricks. One method of analyzing the amount of brick required is to make a drawing of several square feet of wall surface, determine the brick to be used, and divide that into the total area drawn. Sketches are often made right on the workup sheets by the estimator.

Labor costs will be affected by lengths of straight walls, number of jogs in the wall, windows, piers, pilasters, and anything else that might slow the mason down, such as the weather conditions.

Man hours per 100 s.f.

Wall thickness	4"		6"		8"		10"		12"	
Workers	Mason	Laborer	Mason	Laborer	Mason	Laborer	Mason	Laborer	Mason	Laborer
Type of work										
Simple foundation	—	—	—	—	4.5-6.0	5.0-7.5	6.0-9.0	7.0-10.5	7.0-10.0	8.0-11.5
Foundation with several corners, openings	—	—	—	—	5.0-7.5	5.0-7.5	6.5-10.0	7.5-12.0	7.5-10.0	8.5-12.0
Exterior walls, to 4'-0" high	3.5-5.5	3.5-6.0	4.0-6.0	4.0-6.5	4.5-6.0	5.0-7.5	6.0-9.0	7.0-10.5	7.0-10.0	8.0-11.5
Exterior walls, 4'-0" to 8'-0" above ground or floor	3.5-6.0	4.5-7.5	4.0-6.5	4.5-7.0	4.5-6.5	6.0-9.0	6.5-10.0	7.5-12.0	7.5-10.0	8.5-12.0
Exterior walls, more than 8'-0" above ground or floor	4.5-8.0	6.0-9.5	5.0-9.0	7.0-10.0	5.0-7.0	7.0-10.0	7.0-10.5	7.5-12.0	8.0-12.0	9.0-13.0
Interior partitions	3.0-6.0	3.5-7.0	3.5-6.5	4.5-7.5	4.5-6.0	5.0-7.5	—	—	—	—

Note:
1. The more corners and openings in the masonry wall, the more man hours it requires.
2. When lightweight units are used the man hours should be decreased by 10%.
3. Man hours include simple pointing and cleaning required.
4. Special bonds and patterns may increase man hours by 20% to 50%.

Figure 27-3. Man-hours required for concrete masonry.

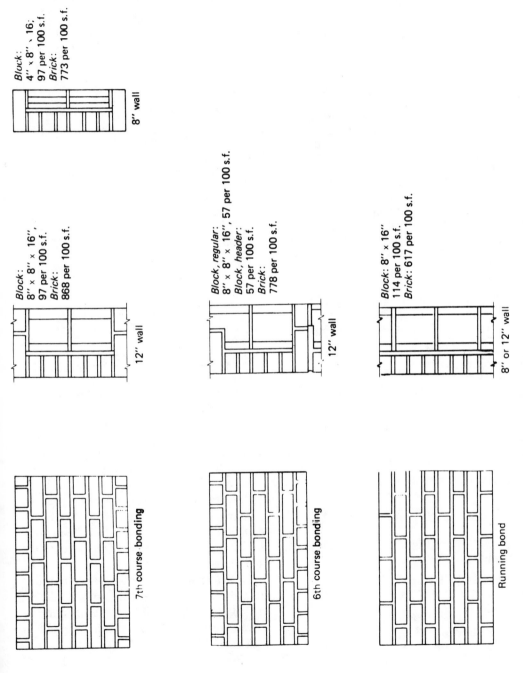

Block:
4″ × 8″ × 16″,
97 per 100 s.f.
Brick:
773 per 100 s.f.

8″ wall

Block:
8″ × 8″ × 16″,
97 per 100 s.f.
Brick:
868 per 100 s.f.

12″ wall

7th course bonding

Block, regular:
8″ × 8″ × 16″, 57 per 100 s.f.
Block, header:
57 per 100 s.f.
Brick:
778 per 100 s.f.

12″ wall

6th course bonding

Block: 8″ × 16″
114 per 100 s.f.
Brick: 617 per 100 s.f.

8″ or 12″ wall

Running bond

Figure 27-4. Bricks and mortar required per 100 s.f. of face area.

449

Also to be calculated are the amount of mortar required, and any lintels, flashing, reinforcing, and weep holes that may be specified. Note on the workup sheet any special requirements such as colored mortar, shape of joint, and type of flashing.

Example 27.3: Residential Building (Appendix B, Alternate No. 1)

Alternate for brick veneer, note that the use of this alternate would also increase the footing size requiring additional concrete.

Brick, standard size, running bond, 3/8-inch mortar joint.

Wall area, about 9'-8" high, 148 l.f. perimeter = 9.67 × 148 = 1431 s.f. gross area

Openings:
  Doors: 2 @ 3'-0" × 6'-8" =
                        40 s.f.
  Windows: 8 @ 3'-6" × 4'-4" =
                        122 s.f.
                        ‾‾‾‾‾‾‾‾‾
                        164 s.f. openings

Net wall area:
  1431 s.f. gross area
  − 164 s.f. openings
  ‾‾‾‾‾‾‾‾‾‾‾‾‾‾‾‾‾
  1267 s.f. net

From Figure 27–4, with a 3/8-inch joint, 617 bricks are required per square (100 s.f.).

1267 s.f.   = 12.67 squares

12.67 × 617   = 7818
+ 5% waste   = 392
              ‾‾‾‾‾‾‾‾‾‾
              8210 bricks required

Example 27.4: Residential Building (Appendix B. Alternate No. 1)

Brick (standard size): 8,210 bricks (8.2 thousand)
Rate of work:
  Mason, 12 man-hours per thousand (1,000) bricks
  Laborer, 13 man-hours per thousand bricks

Type bond	Mason	Laborer
Common	10.0–15.0	11.0–15.5
Running	8.0–12.0	9.0–13.0
Stack	12.0–18.0	10.0–15.0
Flemish	11.0–16.0	12.0–16.0
English	11.0–16.0	12.0–16.0

Note:
1. The more corners and openings in the wall, the more man hours it will require.

Norman brick:	+30%
Roman brick:	+25%
Jumbo brick:	+25%
Modular brick:	Same as standard
Jumbo utility brick:	+30%
Spartan brick:	+25%

Figure 27-5. Man-hours required to lay 1,000 standard size (8" × 2 3/4" × 3 3/4") bricks.

Labor:
  Mason, $10.75 per hour
  Laborer, $5.25 per hour
Time:
  Mason, 8.2 × 12 = 98.4 man-hours
  Laborer, 8.2 × 13 = 106.6 man-hours
Costs:
  Mason, 98.4 × $10.75 = $1,057.80
  Laborer, 106.6 × 5.25 =   559.65
                          ‾‾‾‾‾‾‾‾‾‾
                          $1,617.45

27.3
TILE ESTIMATING

Hollow masonry units of clay tile are estimated first by the square feet of wall area required, each thickness of wall being kept separate. Then the number of units required is determined.

To determine the amount of tile required for a given project, the first step is to take off the length, height, and thickness of units required throughout. All openings should be deducted. Check the details for any special shapes or cuts that might be specified.

When the total square footage of a given thickness has been determined, calculate the number of units. Figure the size of the unit plus the width of mortar joint; this total face dimension is divided into the square feet to determine the number of units.

The labor costs will be affected by jogs in the wall, openings, piers, pilasters, weather conditions, and the height of the building. Items that need to be included are mortar, lintels, flashing, reinforcing, weep holes, and any special shapes.

27.4
STONE ESTIMATING

Stone is usually estimated by the area in square feet with the thickness being given. In this manner, the total may be converted to cubic feet and cubic yards easily, and it still gives the estimator his basic square-foot measurement to work with. Stone trim is usually estimated by the lineal foot.

Stone is sold in various ways, sometimes by the cubic yard, often by the ton. Cut stone is often sold by the square foot; of course, the square foot price goes up as the thickness increases. Large blocks of stone are generally sold by the cubic foot. It is not unusual for the supplier to submit a lump-sum proposal whereby he will supply all of a certain type of stone required for a given amount of money. This is especially true for cut stone panels.

In calculating the quantities required, note that the length times height equals the square footage required; if the number must be in cubic feet, multiply the square footage by the thickness. Deduct all openings but usually not the corners. This will give you the volume of material required. However, the stone does not take up all the space: the volume of mortar must also be deducted. The pattern in which the stone is laid and the type of stone used will greatly affect the amount of mortar required. Cut stone may have 2 to 4 percent of the total volume as mortar, ashlar masonry 6 to 20 percent, and random rubble 15 to 25 percent.

Waste is equally hard to anticipate; cut dressed stone has virtually no waste, while ashlar patterns may have 10 to 15 percent waste.

Dressing a stone involves the labor required to provide a certain surface finish to the stone. Dressing and cutting stone require skill on the part of the mason, and this varies considerably from person to person. There is an increased tendency to have all stone dressed at the quarry or supplier's plant rather than on the job site.

The mortars used should be nonstaining mortar cement mixed in accordance with the specifications. When cut stone is used, some specifications require that the mortar joints be raked out and that a specified thickness of caulking be used. The type and quantity of caulking must then be taken off. The type, thickness, and color of the mortar joint must also be taken off.

Wall ties are often used for securing random, rubble, and ashlar masonry to the backup material. The type of wall tie specified must be noted as must the number of wall ties per square foot. Divide this into the total square footage to determine the total number of ties required.

Cleaning must be allowed for. Flashing should be taken off by the square foot and lintels by the lineal foot. In each case note the type required, the supplier, and the installer.

Stone trim is used for door and window sills, steps, copings, and moldings. This item is usually priced by the supplier by the lineal foot or as a lump sum. Some type of anchor or dowel arrangement is often required for setting the pieces. Check with the supplier as to who is supplying the anchors and dowels, and who will provide the anchor and dowel holes.

Be certain the holes are larger than the dowels being used.

ACCESSORIES

Masonry Wall Reinforcing

Steel reinforcing, in a wide variety of styles and wire gauges, is placed continuously in the mortar joints (Figure 9–22). The reinforcing is generally available in lengths up to 20′. The estimator must determine the lineal footage required. The drawings and specifications must be checked to determine the spacing required (sometimes every course, often every second or third course). The reinforcement is also used to tie the outer and inner wythes together in cavity wall construction. The reinforcing is available in plain or corrosion-resistant wire. Check the specifications to determine what is required. Then check your local supplier to determine prices and availability of the specified material.

In calculating this item, you will find that the easiest method for larger quantities (over 1,500 square feet) is to multiply the square footage by the factor given:

Reinforcement every course 1.50
every second course 0.75
every third course 0.50

For small quantities the courses involved and the length of each course must be figured. Deduct only openings in excess of 50 square feet.

Control Joints

A *control joint* is a straight vertical joint that completely cuts the masonry wall from top to bottom. The horizontal distance varies from ½ to about 2 inches. The joint must also be filled with some type of material; materials usually specified are caulking, neoprene and molded rubber, and copper and aluminum. These materials are sold by the lineal foot and in a variety of shapes. Check the specifications to find the types required and check both the drawings and the specifications for the locations of control joints. Extra labor is involved in laying the masonry, since alternate courses utilizing half-size units will be required to make a straight vertical joint.

Wall Ties

Wall ties are available in a variety of sizes and shapes including corrugated strips of metal 1¼ inches wide and 6 inches long (about 22 gauge), and wire bent to a variety of shapes. Adjustable wall ties are among the most popular since they may be used where the coursing of the inner and outer wythes is not lined up. Noncorrosive metals or galvanized steel may be used. Check the specifications for the type of ties required and their spacing. To determine the amount required, take the total square footage of masonry and divide it by the spacing. A spacing of 16 inches vertically and 24 inches horizontally requires one tie for every 2.66 square feet; a spacing of 16 inches vertically and 36 inches horizontally requires one tie for every 4.0 square feet. Often, closer spacings are required. Also allow for extra ties at control joints, wall intersections, and vertical supports as specified.

Flashing

The flashing built into the walls is generally installed by the mason. It is installed to keep moisture out and to divert any moisture that does get in back to the outside of the building. Flashing may be required under sills and copings, over openings for doors and windows, at intersections of roof and masonry wall, at floor lines, and at the bases of the buildings (a little above grade) to divert moisture out. Materials used include copper, aluminum, copper-backed paper, copper and polyethylene, plastic sheeting (elastomeric compounds), wire and paper, and copper and fabric. Check the specifications to determine the type required. The drawings and specifications must also be checked to determine all of the locations in which the flashing must be used. Flashing is generally sold by the square foot or by the roll. A great deal of labor may be required to bend metal flashings into shape. Check carefully whether the flashing is to be purchased and installed under this section of the estimate, or if it is to be purchased under the roofing section and installed under the masonry section.

Weep Holes

In conjunction with the flashing at the base of the building (above grade level), weep holes are often provided to drain out any moisture that might have got through the outer wythe. Weep holes may also be required at other locations in the construction. The maximum horizontal spacing for weep holes is about 3 feet, but specifications often require closer spacing. The holes may be formed by using short lengths of cord inserted by the mason, or they may be formed by well-oiled rubber tubing. The material used should extend up into the cavity for several inches to provide a drainage channel through the mortar droppings that accumulate in the cavity.

Lintels

A *lintel* is the horizontal member that supports the masonry above a wall opening. It spans from one side of the opening to the other. Materials used for lintels include steel angle iron, composite steel sections, lintel block (shaped like a "U") with reinforcing bars and filled with concrete, and precast concrete lintels. The lintels are usually set in place by the mason as he lays up the wall. Some specifications require that the lintel materials be supplied under this section, while other specifications require the steel angles and composite steel section to be supplied under "structural steel" or "miscellaneous accessories." Precast lintels may be supplied under "concrete"; the lintel block will probably be included under "masonry," as will the reinforcing bars and concrete used in conjunction with it.

It is not unusual for several types of lintels, in a variety of sizes, to be required on any one project. They must be broken down into the types, sizes, and lengths for each material used. Steel lintels may require extra cutting on the job so that the masonry will be able to fit around it. If the lintel is heavy, it may be necessary to use equipment (such as a lift truck or a crane) to put it in place. In determining the length, be certain to take the full masonry opening and add the required lintel bearing on each end. Lintel bearing for steel is generally a minimum of 4 inches, while lintel block and precast lintels are often required to bear 8 inches on each end. Steel is purchased by the pound, precast concrete by the lineal foot, and lintel block by the unit (note the width, height, and length).

Sills

Sills are the members at the bottoms of window or door openings. Materials used are brick, stone, tile, and precast concrete. These types of sills are installed by the mason, although the precast concrete may be supplied under a different portion of the specifications. The brick and tile sills are priced by the number of units required, while the stone and precast concrete sills are sold by the lineal foot. The estimator should check the maximum length of stone and precast concrete sill required and note it on his take-off.

Also to be checked is the type of sill required: a *slip sill,* which is slightly smaller than the width of the opening and can be installed after the masonry is complete, or a *lug sill,* which extends into the masonry at each end of the wall and must be built into the masonry as the job progresses.

Some specifications require special finishes on the sill and will have to be checked. Also, if dowels or other inserts are required, it should be noted.

Coping

The coping covers the top course of a wall to protect it from the weather. It is most often used on parapet walls. Masonry materials used include coping block, stone, tile, and precast concrete. Check the specification for the exact type required and who supplies it. The drawings will show the locations in which it is used, its shape, and how it is to be attached. The coping block and tile are sold by the unit, while the stone and precast coping are sold by the lineal foot.

Special colors, finishes, dowels, dowel holes, and inserts may be required. Check the drawing and specifications for these items and note all requirements on the workup sheet.

27.6
SUBCONTRACTORS

In most localities masonry subcontractors are available, and the estimator will have to decide whether it is advantageous to use a subcontractor on each individual project.

The decision to use a subcontractor does not mean the estimator does not have to prepare an estimate for that particular item; the subcontractor's bid must be checked to be certain it is neither too high nor too low. Even though a particular contractor does not ordinarily subcontract masonry work, it is possible the subcontractor can do the work for less money. There may be a shortage of masons, or the contractor's masonry crews may be tied up on other projects.

If the decision is made to consider the use of subcontractors, the first thing the estimator should decide is which subcontractors he wants to submit a proposal for the project. The subs should be notified as early in the bidding period as possible to allow them time to make a thorough and complete estimate. Often, the estimator will meet with the sub to discuss the project in general and go over exactly which items are to be included in the sub's proposal. Sometimes, the proposal is for materials and labor, other times for labor only. Be certain that both parties clearly understand the items that are to be included.

27.7
CHECKLIST

Masonry:

Type (concrete, brick, stone, gypsum)

Kind

Size (face size and thickness)

Load-bearing

Nonload-bearing

Bonds (patterns)

Colors

Special facings

Fire ratings

Amount of cutting

Copings

Sills

Steps

Walks

Reinforcing:

Bars

Wall reinforcings

Galvanizing (if required)

Mortar:

Cement

Lime (if required)

Fine aggregate

Water

Admixtures

Coloring

Shape of joint

Miscellaneous:

Inserts

Anchors

Bolts

Dowels

Reglets

Wall ties

Flashing

Lintels

Expansion joints

Control joints

Weep holes

Chapter 28

Estimating Wood Frame Construction

FRAME CONSTRUCTION

The wood frame construction discussed in this chapter covers the rough carpentry work, which includes framing, sheathing, subfloors, and insulation. Flooring, roofing, drywall, and wetwall construction are all included in their respective chapters and discussions of them are not repeated here.

The estimator should possess a good working knowledge of the trade, as well as a familiarity with the possible job conditions, in order to estimate quantities and costs accurately.

28.2
BOARD MEASURE

When any quantity of lumber is purchased, it is priced and sold by the thousand-feet-board measure, abbreviated to *mfbm*. The number of board feet required on the job must be calculated by the estimator.

One board foot is equal to the volume of a piece of wood 1 inch thick and 1 foot square (Figure 28-1). By the use of the following formula, the number of board feet can be quickly

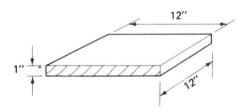

A. One foot of board measure

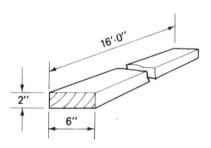

B. Sixteen feet of board measure

Figure 28-1. Board measures, examples.

determined. The nominal dimensions of the lumber are used even though the actual dimensions are smaller, once the lumber is dressed.

$$N = P \frac{T(W)}{12} L$$

N = Number of feet (board measure)
P = Number of pieces of lumber
T = Thickness of the lumber (in inches)
W = Width of the lumber (in inches)
L = Length of the pieces (in feet)

Example 28.1: Estimate the fmb of ten pieces of lumber 2 × 6 inches and 16′ -0″ long. (It would be written 10/2 × 6 – 16′-0″). (See Figure 28–1.)

$$N = 10 \frac{2(6)}{12} 16 = 160 \text{ fbm}$$

Example 28.2: Estimate the fbm in 24/2 × 8 – 16′ –0″.

$$N = 24 \frac{2(8)}{12} 16 = 512 \text{ fbm}$$

This formula may be used for any size order and any number of pieces that are required. The table (Figure 28–2) shows the fbm, per piece, for some typical lumber sizes and lengths. Multiply the fbm by the number of pieces required for the total fbm.

28.3
FLOOR FRAMING

In beginning a wood framing quantity take-off the first portion estimated would be the floor framing. As shown in Figure 28–3, the floor framing generally consists of a girder, sill, floor joist, joist headers, and subflooring.

The first step in the estimate is to determine the grade (quality) of lumber required and to check the specifications for any special requirements. Make note of this information on the workup sheets.

In the examples of wood framing the residence in Appendix B is used.

Size of lumber inches*	Length of piece in feet						
	8	10	12	14	16	18	20
2 x 4	$5\frac{1}{3}$	$6\frac{2}{3}$	8	$9\frac{1}{3}$	$10\frac{2}{3}$	12	$13\frac{1}{3}$
2 x 6	8	10	12	14	16	18	20
2 x 8	$10\frac{2}{3}$	$13\frac{1}{3}$	16	$18\frac{2}{3}$	$21\frac{1}{3}$	24	$26\frac{2}{3}$
2 x 10	$13\frac{1}{3}$	$16\frac{2}{3}$	20	$23\frac{1}{3}$	$26\frac{2}{3}$	30	$33\frac{1}{3}$
2 x 12	16	20	24	28	32	36	40
2 x 14	$18\frac{2}{3}$	$23\frac{1}{3}$	28	$32\frac{2}{3}$	$37\frac{1}{3}$	42	$46\frac{2}{3}$
2 x 16	$21\frac{1}{3}$	$26\frac{2}{3}$	32	$37\frac{1}{3}$	$42\frac{2}{3}$	48	$53\frac{1}{3}$

*nominal

Figure 28–2. Typical board feet.

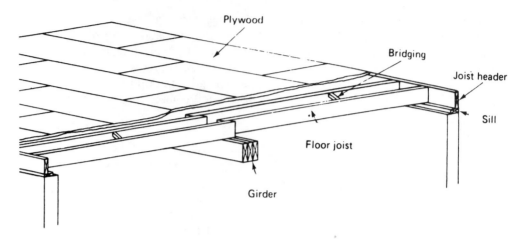

Figure 28-3. Floor framing.

Girders

When the building is of a width greater than that which the floor joists can span, a beam of some type is required. A built-up wood member, referred to as a *girder,* is often used. The sizes of the pieces used to build up the girder must be listed and the length of the girder noted.

Determine the length of the girder from the foundation plan. This length is equal to the distance between the block walls *plus* the distance that the girder rests or "bears" on each wall (Figure 28-4).

Bearing distance

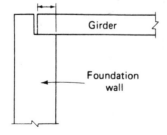

Figure 28-4. Bearing distance.

Example 28.3: Residential Building Girder (Figure 28-5)

Size	Length	L.F.	B.F.
3 – 2 × 10	49′-8″	149	248.33

$$\text{b.f.} = 149 \times \frac{2 \times 10}{12} = 248.33$$

Sills

Sill plates are most commonly 2 x 6, 2 x 8, and 2 x 10 inches. They are placed on the foundation so the length of sill plate required is the distance around the perimeter of the building. Lengths odered will depend on the particular building. Not all frame building require a sill plate so the details should be checked; but generally, where there are floor joists, there are sill plates.

The length of sill is often taken off as the distance around the building (perimeter)—in this case, 148 lineal feet (50 + 24 + 50 + 24) as shown in Figure 28-6.

An exact take-off would require that the estimator allow for the overlapping of the sill pieces at the corners. In this case there is a 5 ½ inch actual overlap at each corner. As shown

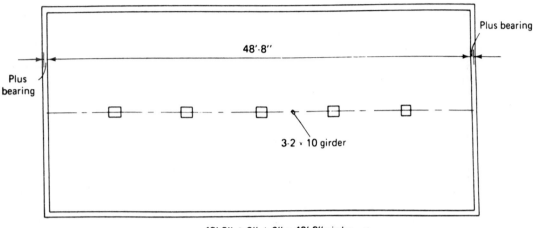

48'-8" + 6" + 6" = 49'-8" girder

Figure 28-5. Girder layout.

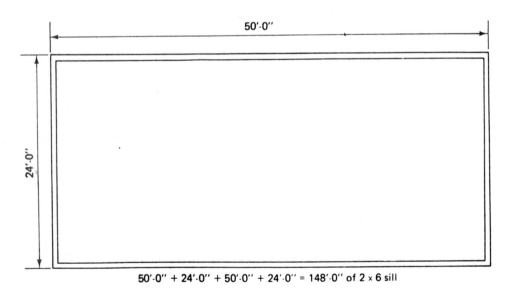

50'-0" + 24'-0" + 50'-0" + 24'-0" = 148'-0" of 2 x 6 sill

Figure 28-6. Sill layout.

in Figure 28-7, the actual length required would be 146'-2", meaning that 148 l.f. would have to be ordered. This type of ac- curacy is required only when the planning of each piece of wood is involved on a series of projects, such as mass produced housing.

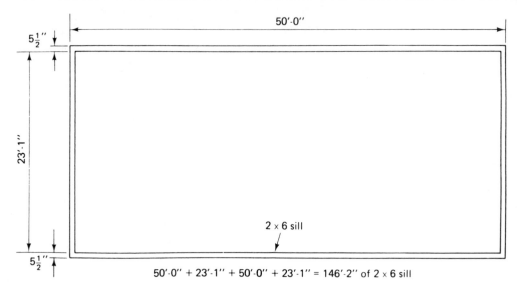

Figure 28-7. Actual sill length.

Example 28.4: Residential Building: Sill plate

Size	L.F.	B.F.
2 x 6	148	148

$$\text{b.f.} = \frac{148 \times 2 \times 6}{12} = 148$$

Wood Floor Joists

The wood joists should be taken off and separated into the various sizes and lengths required. The spacing most commonly used for joists is 16 inches on center, but spacings of 12, 20, 24, 30, and 36 inches on center are also found. The most commonly used sizes are 2 x 6, 2 x 8, 2 x 10, and 2 x 12 inches, although material 13 inches wide is sometimes used.

To determine the number of joists required for any given area, the length of the floor (in feet) is divided by the joist spacing (in feet), and then one joist is added for the extra joist

that is required at the end of the span. If the joists are to be doubled under partitions, or if headers frame into them, one extra joist should be added for each occurrence. Factors for various spacings are given in Figure 28–8.

The length of the joist is taken as the inside dimension of its span plus 4 to 6 inches at each end for bearing on the wall or bearing beam.

Example 28.5: The area to be covered is 32 feet long and 32 feet wide (Figure 28–9), with a bearing beam down the center of the building.

It will require joists 16 feet long to allow for a wall bearing on one end and a beam bearing on the other end. With a joist spacing of 16 inches on center, one joist will be required every 1⅓ feet; or 32 divided by 1⅓; or ¾ as many joists as the length of the building (the ¾ from Figure 28–8). There are 24 joists required plus 1 extra joist for the end, making a total of 25 joists times two (two sides to be framed).

O.C. spacing inches	Divide length to be framed by	OR	Multiply length to be framed by
12	1		1
16	1.33		0.75
20	1.67		0.60
24	2		0.50
30	2.5		0.40
36	3		0.33

*Answer gives number of spaces involved, add one (1) to obtain the number of framing members required. This table makes no allowance for waste, doubling of members or intersecting walls (for stud take-off).

Figure 28-8. Number of framing members required for a given spacing.

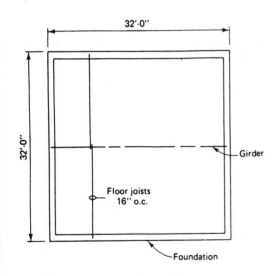

Figure 28-9. Wood floor joists.

Estimating Steps.

1. From the foundation plan and wall section (Appendix B), determine the size of the floor joists required.

2. Determine the number of floor joists required by first finding the number of spaces; then add one extra joist to enclose the last space. A sketch of the residence joist layout is shown in Figure 28-10.

3. Since there are two rows of joists, double the number of joists.

4. Add one extra for partitions which run parallel to the joists (Figure 28-11).

5. Determine the required length of the floor joists. This is done by gathering some information from the drawings:

   a. The total width of the building is given on the floor plan as 24' -0" (Figure 28-12).

Note: This dimension is from the exterior face of stud to exterior face of stud.

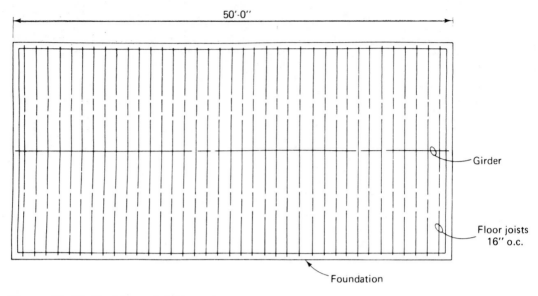

Figure 28-10. Residence joist layout.

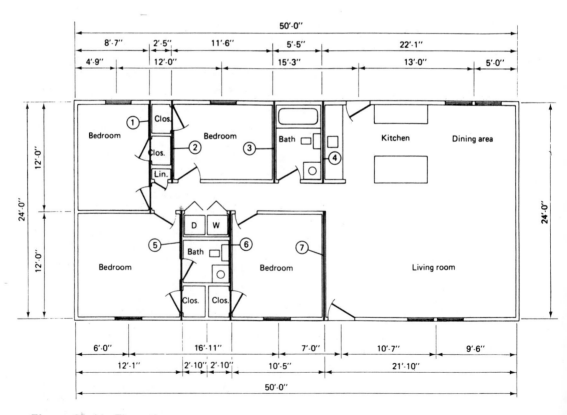

Figure 28-11. Floor plan.

462

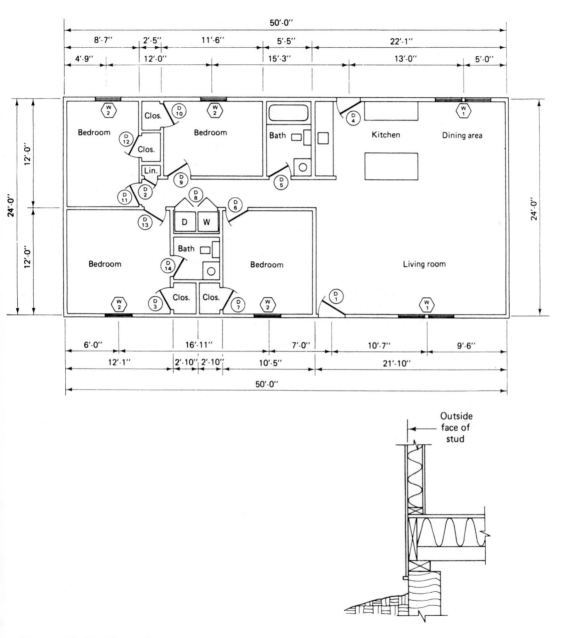

Figure 28-12. Floor plan.

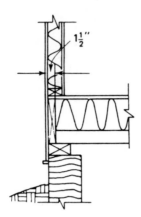

Figure 28-13. Joist header.

b. Determine the horizontal distance from the outside face of the stud to the edge of the floor joist by carefully checking the wall section. Typically this distance will be the thickness of the joist header, or 1½ inches (Figure 28-13).

c. The distance the joists must cover is found by deducting 1½ inches from each side (for each exterior wall), or a total of 3 inches from the framing

dimension of the building as shown in Figure 28-14.

$$24'\text{-}0'' - 0'\text{-}3'' = 23'\text{-}9''$$

d. Since the foundation plan requires a girder, the construction method used where the joists attach to the girder must be determined. Reviewing the typical assemblies used in Figure 28-15 shows that the detail used can easily affect the length of joist required.

If there is no detail showing the assembly required, recheck the specifications. If not in the specifications or on the drawings, it will be necessary to call the architect-engineer (or contractor, owner, etc.) and determine what method assembly is to be used.

Note: For competitive bidding it would be necessary to get this information in the form of an addenda.

One last point should be made—don't assume that the assembly will be done any one particular way. Any of the

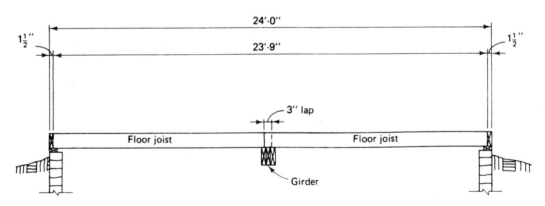

Figure 28-14. Joist length.

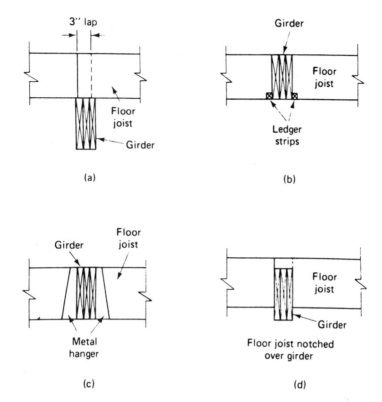

Figure 28-15. Typical joist-girder assemblies.

assemblies shown in Figure 28-15 may be required, or even some variation of the assemblies shown. Never assume that you know what is required.

In this estimate the detail shows that the joists pass over top of the girder and lap a minimum of 3 inches as shown in Figure 28-15a. This information is then sketched on the workup sheet and shown in Figure 28-16.

e. Since the girder is located in the center of the house (from the foundation plan), the distance from the edge of the joist to the center line of the girder is one-half the inside dimension found.

This is added to the sketch and shown in Figure 28-16.

f. Add one-half of the lap to the center line dimension to find the length of floor joist required (Figure 28-17).

6. List the required floor joists on the workup sheet.

Floor joists: 2 × 10, 16″ o.c., #2 KD

Four 2 × 12's 10′ long will give twelve 3′ -3″ pieces. One 2 × 12 14′ long will give two 3′ -3″ pieces, two 2′ -11″ pieces, and 1′ -8″ of waste.

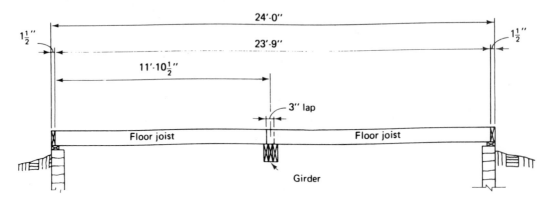

Figure 28-16. Joist sketch.

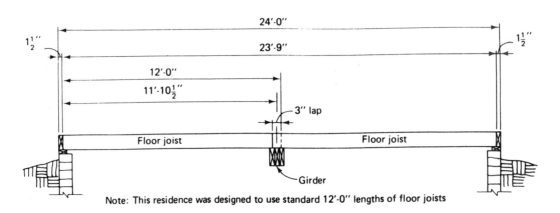

Note: This residence was designed to use standard 12'-0" lengths of floor joists

Figure 28-17. Joist length.

	No.	Length
Joists	78	12'
Extra joists under partition	7	12'

Trimmers and Headers

Openings in the floor, such as for stairs or fireplaces, are framed with trimmers running in the direction of the joists and headers which support the cut-off "tail beams" of the joist (Figure 28-18).

Unless the specifications say otherwise, when the header length (Figure 28-19) is 4 feet most codes allow single trimmers and headers to be used.

For header lengths greater than 4 feet, codes usually require double trimmers and headers (Figure 28-20) and for certain special conditions some codes even require them to be tripled.

To determine the extra material required for openings we will investigate two situations:

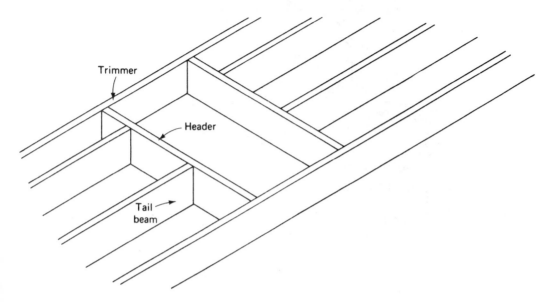

Figure 28-18. Framing for floor openings.

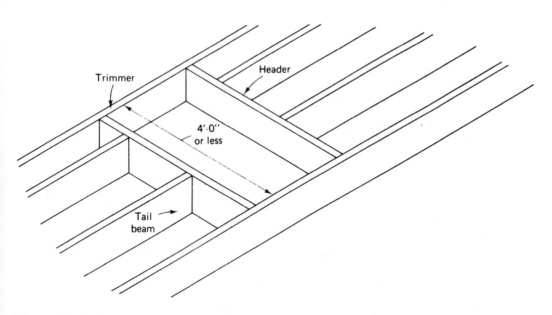

Figure 28-19. Single trimmers and headers.

467

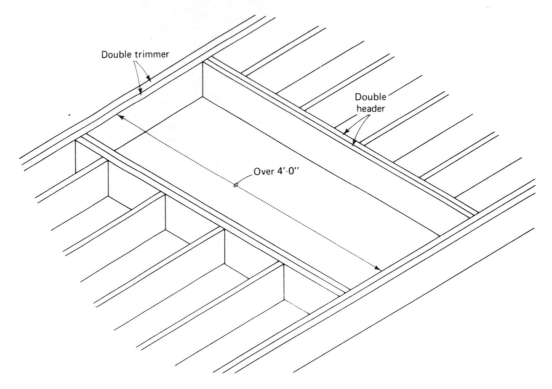

Figure 28-20. Double trimmers and headers.

1. a 3'-0" × 4'-0" opening as shown in Figure 28–21.
2. a 3'-0" × 8'-0" opening as shown in Figure 28–21.

The material is determined by:

1. Sketch the floor joists without an opening (Figure 28–22).
2. Locate the proposed opening on the floor joist sketch (Figure 28–23).
3. Sketch what trimmer and header pieces are required and how the cut pieces of the joist

may be used as headers and trimmers (Figure 28–24).

Joist headers are taken off next. The header runs along the ends of the joist to seal the exposed edges (Figure 28–25). The headers are the same size as the joists, and the length required will be two times the length of the building (Figure 28–26).

Board feet of framing material for the floor assembly is now determined and noted on the workup sheets.

Note: There are *no* floor openings in the residence.

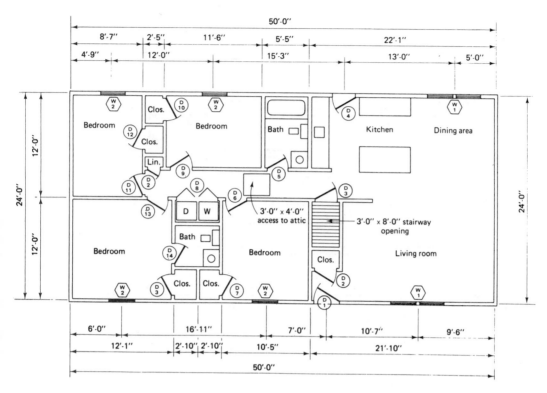

Figure 28-21. Typical floor openings.

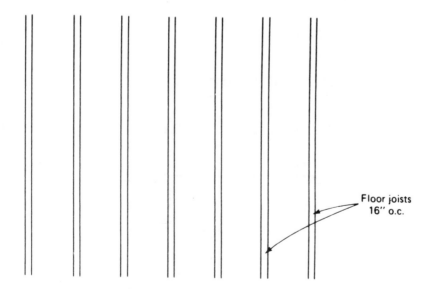

Floor joists
16" o.c.

Figure 28-22. Floor joist sketch.

469

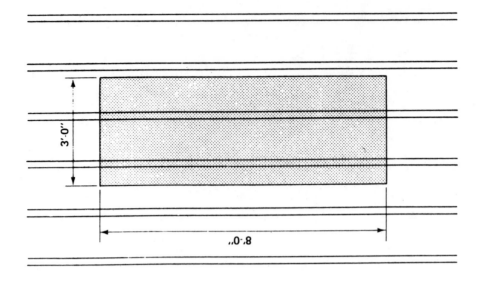

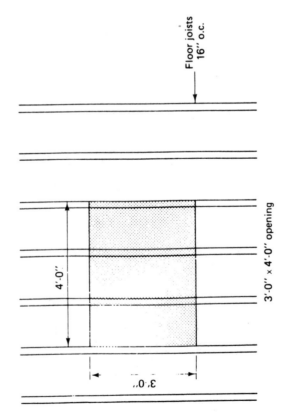

3'-0'' × 4'-0'' opening

Floor joists
16'' o.c.

Figure 28-23. Opening location.

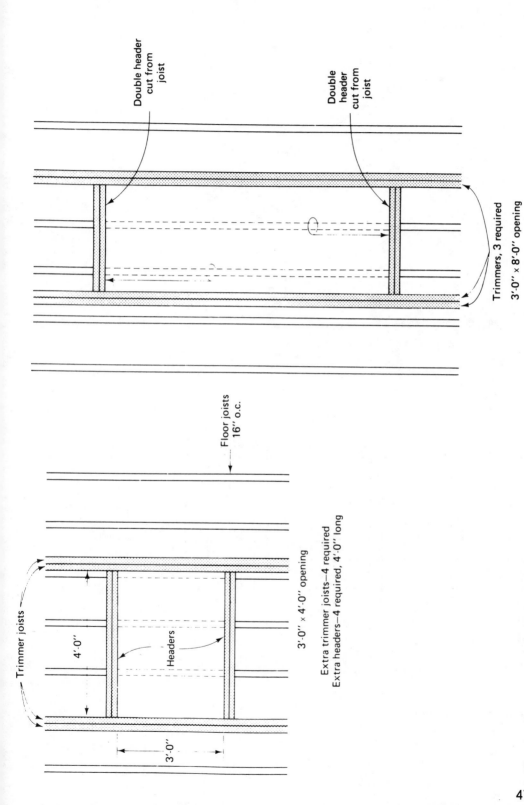

Double header cut from joist

Double header cut from joist

Trimmers, 3 required
3'-0" × 8'-0" opening
Extra trimmers — 3 required

Floor joists 16" o.c.

Trimmer joists

4'-0"

Headers

3'-0"

3'-0" × 4'-0" opening
Extra trimmer joists—4 required
Extra headers—4 required, 4'-0" long

Figure 28-24. Trimmers and headers required.

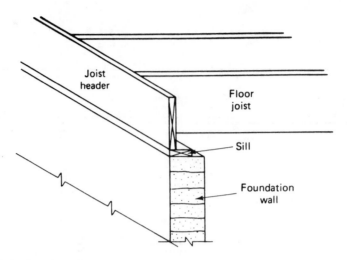

Figure 28-25. Joist header.

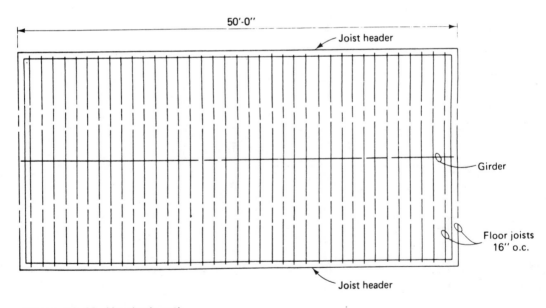

Figure 28-26. Header length.

Floor joists: $2 \times 10, 16''$ o.c., #2KD

	No.	Length	Lf.
Joists	78	12'-0"	936
Extra under partition	7	12'-0"	84
Headers	2	50'-0"	100
			1,120 l.f.

$$\text{Board feet} = 1,120 \times \frac{2 \times 10}{12} = 1,867 \text{ b.f.}$$

Bridging is customarily used with joists (except for glued nailed systems) and must be included in the costs. Codes and specifications vary on the amount of bridging required, but at least one row of cross-bridging is required between the joists. The bridging may be wood, 1×3 or 1×4 inches, or metal bridging (Figure 28–27). The wood bridging must be cut, while the metal bridging is obtained ready for installation and requires only one nail at each end, half as many nails as would be needed with the wood bridging.

Bridging for floor and flat roof joists and beams

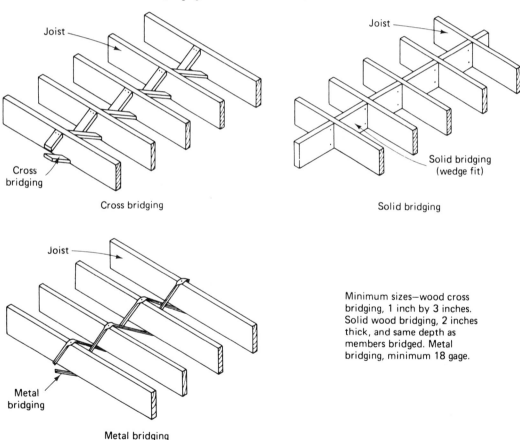

Joist

Cross bridging

Cross bridging

Joist

Solid bridging (wedge fit)

Solid bridging

Joist

Metal bridging

Metal bridging

Minimum sizes—wood cross bridging, 1 inch by 3 inches. Solid wood bridging, 2 inches thick, and same depth as members bridged. Metal bridging, minimum 18 gage.

Figure 28–27. Bridging.

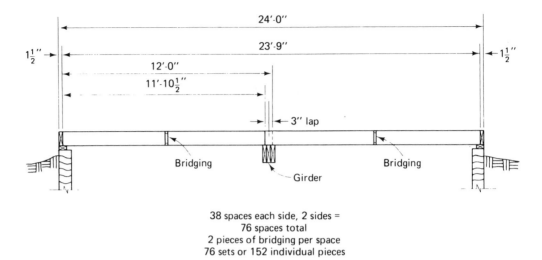

38 spaces each side, 2 sides =
76 spaces total
2 pieces of bridging per space
76 sets or 152 individual pieces

Figure 28-28. Metal bridging.

In this estimate the specifications require metal bridging (Figure 28-28) with a maximum spacing of 8 feet between bridging or bridging and bearing. A check of the joists' length shows they are about 12 feet long and that one row of bridging will be required for each row of joists.

Since each space requires two pieces of metal bridging, determine the amount required by multiplying the joist *spaces* (not the number of joists) times two (2 pieces each space) times the number of rows of bridging required.

Sheathing is taken off next. This sheathing is most commonly plywood but diagonal sheathing is still occasionally (infrequently) used. First, a careful check of the specifications should provide the plywood information required. The thickness of plywood required may be given in the specifications or in the wall sections (Figure 28-29). In addition, the specifications will spell out any special installation requirements, such as the glued-nailed system. The plywood is most accu-

rately estimated by doing a sketch of the area to be covered and planning the plywood sheet layout. Using 4 × 8 sheets a layout for the residence is shown in Figure 28-30.

Another commonly used method of estimating plywood is to determine the square footage to be covered and divide the total

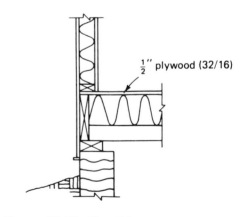

Figure 28-29. Sheathing.

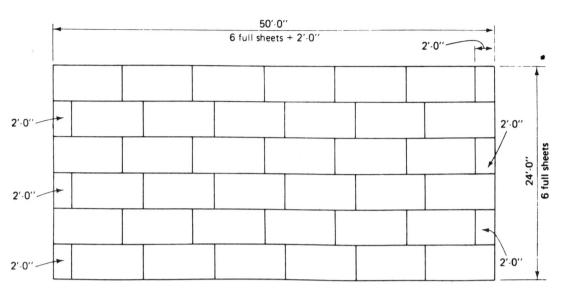

36 full sheets plus 6-2'-0" pieces = $37\frac{1}{2}$ sheets
order 38 sheets

Figure 28-30. Residence sheathing layout.

square footage to be covered by the area of one sheet of plywood (Figure 28–31).

While both methods give almost the same answer, the use of square feet alone does not allow planning of sheet layout on irregular plans (Figure 28–32) or allow properly for waste.

List the quantity of plywood required for the sheathing on the workup sheet, being certain to list all related information.

Plywood subflooring: 1/2", 32/16, 32-4 × 8 sheets, C-D.

Subflooring is sometimes used over the sheathing. Subflooring may be a pressed particle board (where the finished floor is carpet) or plywood (where the finished floor is resilient tile, ceramic tile, slate, etc.). A careful review of the specifications and draw-ings will show whether subflooring is required, the type, thickness, and location. It is taken off in sheets the same as plywood.

28.4
WALL FRAMING

In this section a quantity take-off of the framing required for exterior and interior walls is done. Since the exterior and interior walls have different finish materials on them they will be estimated separately. The exterior walls are taken off first, then the exterior.

Exterior Walls

Basically, most of the wall framing consists of sole plates, studs, double plates, headers, and finish materials (Figure 28–33).

$$\begin{array}{r} 24 \\ \times 50 \\ \hline 1200 \text{ s.f.} \end{array} \qquad \frac{1200}{32} = 37.5 \text{ sheets (order 38 sheets)}$$

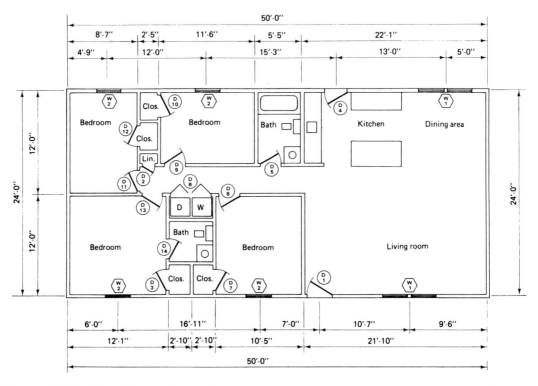

Figure 28-31. Sheathing requirements.

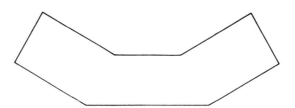

Figure 28-32. Irregular plan.

476

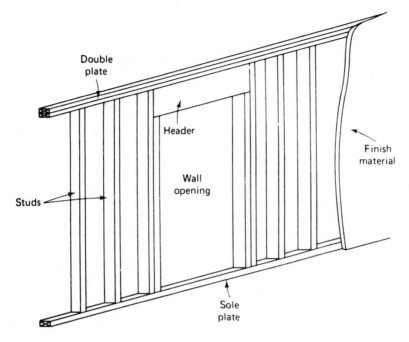

Figure 28-33. Wall framing.

Plates. The most commonly used assembly incorporates a double-top plate and a single-bottom plate, although other combinations may be used. Begin by first reviewing the specifications and drawings for the thickness of materials (commonly 2 x 4 or 2 x 6), grade of lumber to be used, and information on the number of plates required. The commonly used assembly in Figure 28-33 provides an 8'-0" ceiling height, which is most commonly used and which works economically with 4' × 8' sheets of plywood and gypsum board. When an 8'-2" ceiling height is required, a double-top and bottom plate is used (Figure 28-34).

Estimating steps.

1. Plates are required around the perimeter of the building. Since this is the same perimeter used to determine the sill

material in Section 28.3 and Example 28.4, the perimeter is already known.
2. The total lineal feet of exterior plates is determined by multiplying the lineal feet of wall times the number of plates.
3. List this information on the workup sheet and calculate the board feet required.

Wall framing:
Exterior wall—2 x 4
Plate—2 x 4

L.F.	Plates	Total L.F.	B.F.
148	3	444	296

$$\text{Board feet} = 444 \times \frac{2 \times 4}{12} = 296 \text{ b.f.}$$

Note: Interior plates will be done with the interior walls later in this portion of the take-off, but the same basic procedures shown here will be used.

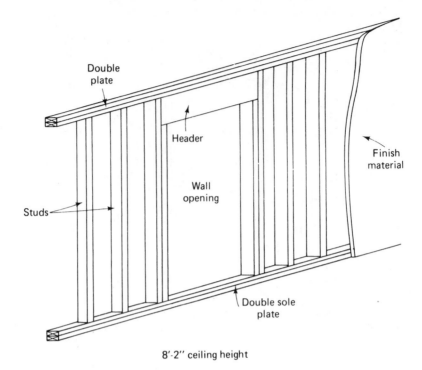

Figure 28-34. Double top and bottom plates.

Studs. The stud take-off should be separated into the various sizes and lengths required. Studs are most commonly 2 x 4, 16 inch, and 24 inch on center and 2 x 6, 24 inch on center (providing a greater width for extra wall insulation).

Estimating steps.

1. Review the specifications and drawings for the thickness and spacing of studs and lumber grade.
2. The exterior studs will be required around the perimeter of the building; the perimeter used for the sills and plates may be used.
3. Divide the perimeter (length of wall) by the spacing of the studs to determine the number of spaces. Then add one to close off the last space.

Studs: 2 × 4, 16″ o.c., 148 l.f.

$$\frac{148}{1.33} = 111.3 = 112 \text{ spaces} = 113 \text{ studs}$$

4. Add extra studs for corners, wall intersections (where two walls join), and wall openings.

   a. Corners—using 2 x 4 studs—a corner is usually made up of 3 studs (Figure 28–35). This requires 2 extra studs at each corner. Estimate the extra material required by counting up the number of corners and multiplying the number of corners by two.

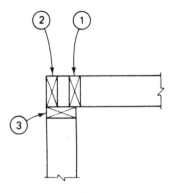

3 - Stud corner

Figure 28-35. Corners—2 × 4 studs.

Studs	No.
2 × 4, 16″ o.c., 148 l.f.	113
4 corners—4 × 2	8

b. Corners—using 2 × 6 studs—a corner usually made up of 2 studs (Figure 28-36), requiring 1 extra stud. Drywall clips are used to secure the drywall panel to the studs. Some specifications and details require that 3 studs be used for 2 × 6 corners (similar to the 2 × 4 corner).

c. Wall intersections—using 2 × 4 studs the wall intersection is made up of 3 studs (Figure 28-37). This requires 2 extra studs at each intersection. The extra material is estimated by counting up the number of wall intersections (on the exterior wall and multiplying the number of corners by two) (Figure 28-38).

d. Wall intersections—using 2 × 6 exterior studs the wall intersection may be made up of 1 or 2 studs (Figure 28-39) with drywall clips used to secure the gypsum board (drywall). Some specifications and details require that 3 studs be used for 2 × 6 corners (similar to the 2 × 4 corner).

Studs	No.
2 × 4, 16″ o.c., 148 l.f.	113
4 corners—4 × 2	8
9 intersections—9 × 2	18

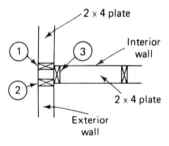

Figure 28-37. Wall intersection.

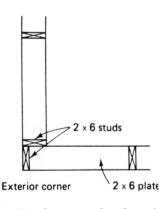

Exterior corner   2 × 6 plate

Figure 28-36. Corners—2 × 6 studs.

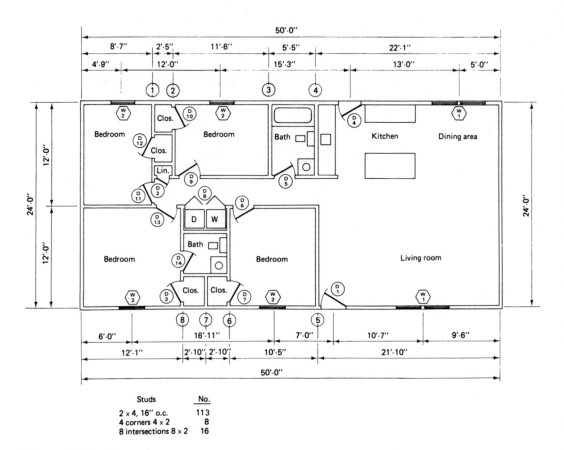

Figure 28-38. Floor plan.

Studs	No.
2 × 4, 16″ o.c.	113
4 corners 4 × 2	8
8 intersections 8 × 2	16

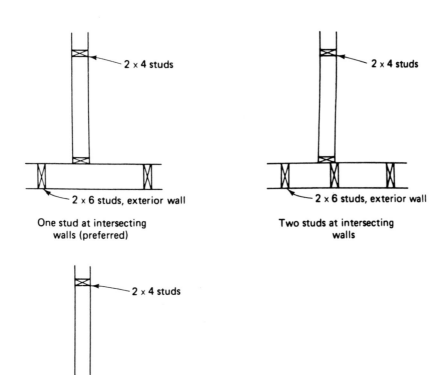

One stud at intersecting
walls (preferred)

2 x 4 studs

2 x 6 studs, exterior wall

Two studs at intersecting
walls

2 x 4 studs

2 x 6 studs, exterior wall

Three studs at
intersecting walls

2 x 4 studs

2 x 6 studs, exterior wall

Figure 28-39. Wall intersections.

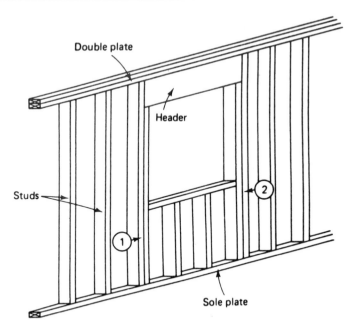

Figure 28-40. Wall opening.

e. Wall openings—typically, extra studs are required at all openings in the wall (Figure 28-40). However, when the openings are planned to fit into the framing of the building, it requires less extra material. The illustration in Figure 28-41a and 28-41b shows the material required for a window which has not been worked into the module of the framing and 28-41b shows the material required for a window which is worked into the framing module.

In this example there is no indication that openings have been worked into the framing module. For this reason, an average of 3 extra studs were included for all window and door openings.

Studs	No.
2 × 4, 16″ o.c., 148 l.f.	113
4 corners—4 × 2	8
9 intersections—9 × 2	18
8 openings—8 × 3	24

5. Additional studs are also required on the gable ends of the building (Figure 28-42) which have to be framed with studs between the double plate and the rafter (unless trusses are used, Section 28.7). The specifications and elevations should be checked to determine whether the gable end is plain or has a louver.

Estimating the studs for each gable end is accomplished by first drawing a sketch of the gable end and noting its size (Figure 28-43).

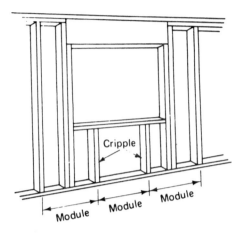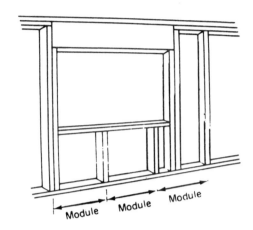

(a)   Extra stud material
4 extra studs
1 extra cripple

Window not on module

(b)   Extra stud material
3 extra studs
1 extra cripple

Window partially on module

Figure 28-41. Wall openings.

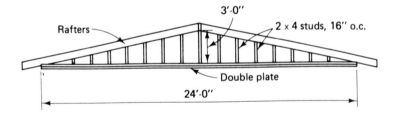

Figure 28-42. Gable end.

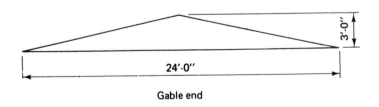

Gable end

Figure 28-43. Gable end sketch.

483

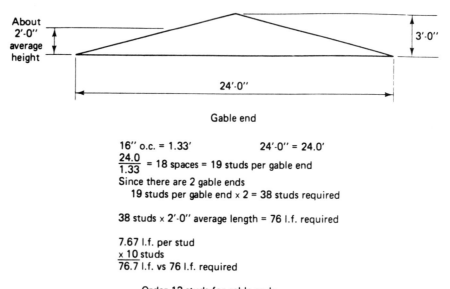

Gable end

16″ o.c. = 1.33′   24′-0″ = 24.0′

$\dfrac{24.0}{1.33}$ = 18 spaces = 19 studs per gable end

Since there are 2 gable ends

    19 studs per gable end x 2 = 38 studs required

38 studs x 2′-0″ average length = 76 l.f. required

7.67 l.f. per stud
x 10 studs
76.7 l.f. vs 76 l.f. required

Order 12 studs for gable ends

Figure 28-44. Studs required for gable end.

Find the number of studs required by dividing the length by the stud spacing. Then, from the sketch find the approximate average height of the studs required and record the information on the workup sheet. Multiply the quantity by two since there are two gable ends on this building (Figure 28-44).

6. The total stud material has now been estimated and is converted into board feet.

Studs	No.
2 × 4, 16″ o.c.	113
4 corners—4 × 2	8
9 intersections—8 × 2	18
8 openings—8 × 3	24
Gable ends	12
Total exterior studs	175

$175 \times 7'\text{-}9'' \times \dfrac{2 \times 4}{12} = 905$ b.f.

Headers. Headers are required to support the weight of the building over the openings. A check of the specifications and drawings must be made to determine if the headers required are solid wood, headers and cripples, or plywood sheathing (Figure 28-45). For ease of construction many carpenters and home builders feel that a solid header provides best results, and they use 2 x 12s as headers throughout the project, even in nonload-bearing walls. Shortages and higher costs of materials have increased the use of plywood and smaller headers.

The header length must also be considered. As shown in Figure 28-46, the header extends over the top of the studs, and it is wider than the opening. Most specifications and building codes require that headers for openings up to 6 feet wide must extend over 1 stud at each end (Figure 28-46), and headers for openings 6 feet and wider must extend over 2 studs at each end (Figure 28-46).

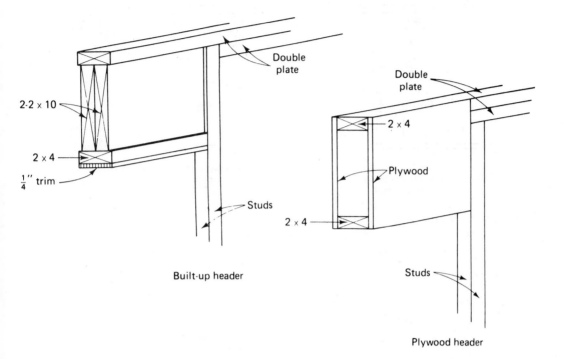

2-2 × 10

2 × 4

$\frac{1}{4}''$ trim

Double
plate

2 × 4

Plywood

2 × 4

Studs

Double
plate

Studs

Built-up header

Plywood header

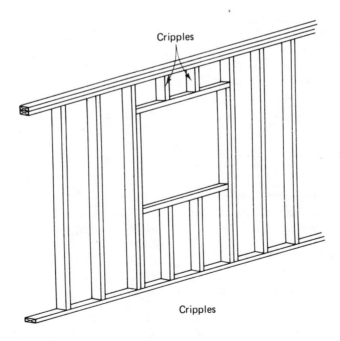

Cripples

Cripples

Figure 28-45. Headers.

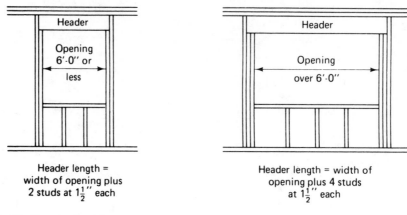

Header length =
width of opening plus
2 studs at $1\frac{1}{2}''$ each

Header length = width of
opening plus 4 studs
at $1\frac{1}{2}''$ each

Figure 28-46. Header length.

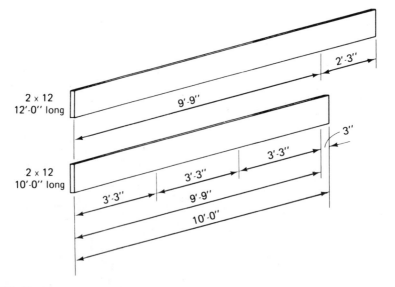

Figure 28-47. Headers.

List the number of openings, their widths, the number of headers, their lengths, and the lineal feet and board feet required.

Headers: 2 x 12 (Figure 28–47)

Four 2 × 12's 10′ long will give twelve 3′-3″ pieces. One 2 × 12 14′ long will give two 3′-3″ pieces, two 2′-11″ pieces, and 1′-8″ of waste.

	Opening	Pieces Required	Length Required	L.F.
7 openings	36″ wide	14	3′-3″	45′-6″
1 opening	2′-8″ wide	2	2′-11″	5′-10″
				50′-16″ = 51′-4″

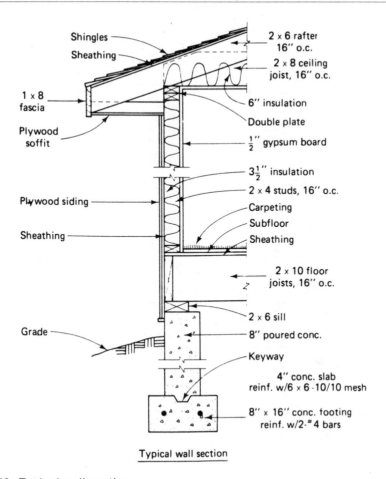

Shingles

Sheathing

2 x 6 rafter
16" o.c.

2 x 8 ceiling
joist, 16" o.c.

1 x 8
fascia

6" insulation

Double plate

Plywood
soffit

½" gypsum board

3½" insulation

2 x 4 studs, 16" o.c.

Plywood siding

Carpeting

Subfloor

Sheathing

Sheathing

2 x 10 floor
joists, 16" o.c.

2 x 6 sill

Grade

8" poured conc.

Keyway

4" conc. slab
reinf. w/6 x 6-10/10 mesh

8" x 16" conc. footing
reinf. w/2-# 4 bars

Typical wall section

Figure 28-48. Typical wall section.

$$54 \times \frac{2 \times 12}{12} = 108 \text{ b.f.}$$

Finish Material. Exterior wall sheathing may be a fiberboard material soaked with a bituminous material, insulation board (often a urethane insulation covered with an aluminum reflective coating), or plywood (Figure 28-48). Carefully check the specifications and working drawings to determine what is required (insulation requirements,

thickness). Fiberboard and insulation board sheathing must be covered by another material (such as brick, wood, or aluminum siding), while the plywood may be covered or left exposed.

All of these sheathing materials are taken off first by determining the square feet required and then determining the number of sheets required. The most accurate take-off is made by making a sketch layout of the material required (as was done with the plywood in floor framing). Check the height

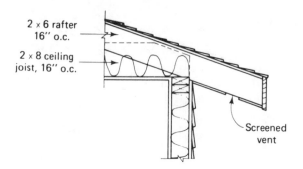

Figure 28-49. Sloped soffit.

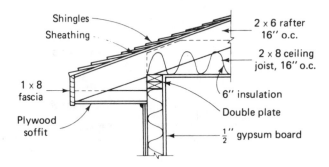

Figure 28-50. Boxed-in soffit.

of plywood carefully since a building with a sloped soffit (Figure 28–49) may require a 9-foot length, while 8 feet may be sufficient when a boxed-in soffit is used (Figure 28–50). This is especially important when plywood will be left exposed and as few joints as possible are desired.

Openings in the exterior wall are neglected unless they are large, and the plywood which would be cut out can be used elsewhere; otherwise it is considered waste.

Sheathing is also required to cover the gable ends (Figure 28–51).

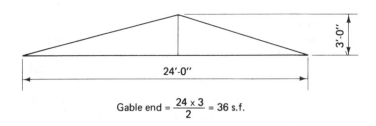

$$\text{Gable end} = \frac{24 \times 3}{2} = 36 \text{ s.f.}$$

Figure 28-51. Gable end.

Exterior plywood: ½", 32/16, A-C, 4 × 8 sheets

148 l.f. perimeter = 148 l.f. × 8 ft. high = 1,184 s.f.

$$\frac{148 \text{ l.f.}}{4 \text{ ft.}} = 37 \text{ sheets required}$$
(neglect openings)

Gable ends = 36 s.f. (Figure 28–51) × 2 ends = 72 s.f.

The six 32" × 36" pieces of plywood left from the openings can be used on part of the gable ends. These pieces amount to 6 × 2.67' × 3.0' = 48 s.f.

Total   s.f.   = 1,184 s.f.
Gable (72 - 48) = 24 s.f.
                  1,208 s.f.

Allowing for waste, order 40 sheets, 1,280 s.f.

Note: The inside of the exterior wall is covered with gypsum board (drywall). The gypsum materials will be taken off in Chapter 29.

Interior Walls

Interior walls are framed with studs, top and bottom plates, and a finish on both sides of the wall. When estimating the material for interior walls, the first step is to determine from the specifications and drawings what thickness(es) of walls is (are) required. Most commonly, 2 x 4 studs are used, but 2 x 3 studs and occasionally, 2 x 6 studs are also used. Also, the stud spacing may be different from the exterior walls.

Next, the lineal feet of interior walls must be determined. This length is taken from the plan by:

1. Using dimensions from the floor plan.
2. Scaling the lengths with a scale.

3. Using a distance measurer over the interior walls.

On large projects extreme care must be used when using a scale or distance measurer since the drawing may not be done to the exact scale shown.

Any walls which are different thicknesses (such as a 6-inch-thick wall, sometimes used where plumbing must be installed) and of special construction (such as a double or staggered wall, Figure 28–52) may require larger stud or plate sizes.

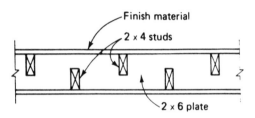

Staggered wall

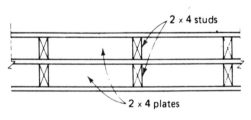

Double wall

Figure 28-52. Special wall construction.

Plates.   Refer to the discussion under "Exterior Walls" earlier in this section.

Interior walls: 2 × 4, 16" o.c., 149 l.f.
Plate: 2 x 4

L.F.	Plates	Total L.F.	B.F.
149	3	447	298

Studs. Refer to the discussion under "Exterior Walls" earlier in this section. As in exterior walls, deduct only where there are large openings and take into account all corners, wall openings and wall intersections.

L.F.	Studs
149 l.f. of plate	113
17 intersections	34
12 openings x 3	36
Interior studs	183 = 946 b.f.

Headers. Refer to the discussion under "Exterior Walls" earlier in this section.

Openings	Header Length	No. of Pieces	L.F. Required
One 1'-6" wide	1'-9"	2	3'-6"
Two 2'-4" wide	2'-7"	4	10'-4"
Eight 2'-6" wide	2'-9"	16	44'-0"
One 5'-0" wide	5'-3"	2	10'-6"
			68'-4"

One 12' length will give four 2'-9" pieces and 1'-0" of waste—you will need four 12' lengths for sixteen 2'-9" pieces.

One 8' length will give one 5'-3" piece, one 2'-7" piece and 2" of waste—you will need two 8' lengths for two 5'-3" pieces and two 2'-7" pieces.

One 10' length will give two 2'-7" pieces, two 1'-9" pieces and 1'-4" of waste. Order:

Four 2 × 12 12' long	= 48 l.f.
Two 2 × 12 8' long	= 16 l.f.
One 2 × 12 10' long	= 10 l.f.
	74 l.f.
	= 148 b.f.

At this point, the entire materials' take-off for the framing of the interior and exterior walls has been estimated. Finishes for the interior and exterior wall (drywall, interior plywood, etc.) will be covered in Chapter 29.

28.5
CEILING ASSEMBLY

In this section a quantity take-off of the framing required for the ceiling assembly of a wood frame building is done. The ceiling assembly will require a take-off of ceiling joists, headers, and trimmers, quite similar to the take-off done for the floor assembly.

First, a careful check of the specifications and drawings must be made to determine if the ceiling and roof are made up of joists and rafters (Figure 28–53), often called "stick construction" or prefabricated wood trusses (Figure 28–63) which are discussed in Section 28.11.

Ceiling joists, from the contract documents determine:

1. Size, spacing, and grade of framing required.
2. The number of ceiling joists required by dividing the spacing into the length and adding one (the same as done for floor joists).
3. The length of each ceiling joist. (Don't forget to add one-half of any required lap.)

This information is compiled on the workup sheets.

Ceiling joist: 2 × 8, 16" o.c., #2KD

	No.	Length	B.F.
Joists	78	12'	1,248

The *headers* and *trimmers* required for any openings (such as stairways, fireplace, and attic access openings) are considered next. This

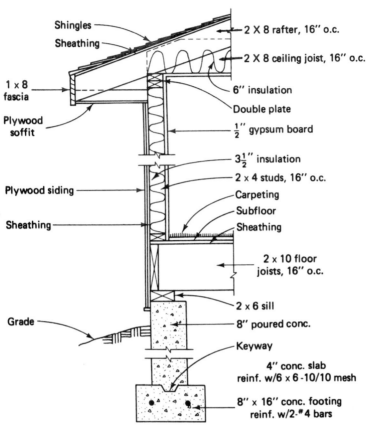

Shingles

Sheathing

2 X 8 rafter, 16" o.c.

2 X 8 ceiling joist, 16" o.c.

1 x 8 fascia

6" insulation

Double plate

Plywood soffit

$\frac{1}{2}''$ gypsum board

$3\frac{1}{2}''$ insulation

2 x 4 studs, 16" o.c.

Plywood siding

Carpeting

Subfloor

Sheathing

Sheathing

2 x 10 floor joists, 16" o.c.

2 x 6 sill

Grade

8" poured conc.

Keyway

4" conc. slab reinf. w/6 x 6 -10/10 mesh

8" x 16" conc. footing reinf. w/2-#4 bars

Typical wall section

Figure 28-53. Typical wall section.

is taken off the same as done for floor framing (Section 28.3).

The ceiling finish material (drywall or wetwall) is estimated in Chapter 29.

Note: A 3'-0" × 3'-9" attic access is required in the residence (Figure 28-54).

	No.	Length	B.F.
Joists	78	12'	1,248
Trimmers	2	12'	32
Trimmers	1	16'	22
			1,302

28.6
ROOF ASSEMBLY

In this section a quantity take-off of the framing required for the roof assembly of a wood frame building is done. The roof assembly will require a take-off of rafters, ridge, lookouts, collar ties (or supports), plywood sheathing, and the felt which covers the sheathing.

If trusses are called for, then a separate take-off for roof and ceiling assembly will not

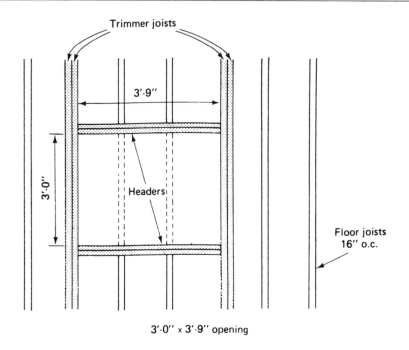

3'-0" x 3'-9" opening

Trimmer joists, 2 extra
Headers, 4 required at 3'-9" = 15'-0" (use 16'-0" length)

	No.	Joist length	B.F.
Joist	78	12'-0"	1248
Trimmers	2	12'-0"	32
Headers	1	16'-0"	22
			1302 B.F.

Figure 28-54. Headers and trimmers.

be made and it would be estimated as discussed in Section 28.11, "Trusses."

Rafters

Roof rafters should be taken off and separated into the sizes and lengths required. The spacings most commonly used for rafters are 16 and 24 inches on center, but 12, 20, 32, and 36 inches on center are also used. Rafter sizes of 2 x 6, 2 x 8, 2 x 10, and 2 x 12 are most common. The lengths of rafters should be carefully taken from the drawings or worked out by the estimator if the drawings are not to scale. Be certain to add any required overhangs to the lengths of the rafters. The number of rafters for a pitched roof can be determined in the same manner as the number of joists. The principle of pitch versus slope should be understood to reduce mistakes. Figure 28-55 shows the difference between

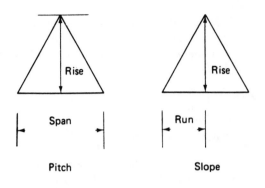

Pitch = $\dfrac{Rise}{Span}$

Slope = $\dfrac{Rise}{Run}$

Example: Span 40'-0", Rise 10'-0"

Pitch = $\dfrac{10}{40}$ = $\dfrac{1}{4}$ Pitch

Slope = $\dfrac{10}{20}$ = $\dfrac{6}{12}$ = 6 inches per foot

Figure 28-55. Pitch and slope.

Pitch of roof	Slope of roof	For length of rafter multiply length of run by
$\frac{1}{12}$	2 in 12	1.015
$\frac{1}{8}$	3 in 12	1.03
$\frac{1}{6}$	4 in 12	1.055
$\frac{5}{24}$	5 in 12	1.083
$\frac{1}{4}$	6 in 12	1.12

Note: Run measurement should include any required overhang.

Example:
For a 24'-0" span with a 2'-0" overhang on each side, $\frac{1}{8}$ pitch, what length rafter is required?

$\dfrac{\text{Span \& Overhang}}{2} = \dfrac{24 + 4}{2}$ = Run = 14'-0"

Rafter length = 14'-0" x 1.03 = 14.42' = 14'- $5\frac{1}{32}$ '

Figure 28-56. Rafter length.

pitch and slope, while Figure 28–56 shows the length of rafter required for varying pitches and slopes.

Rafters, from the contract documents, determine:

1. Size, spacing, and grade of framing required.
2. The number of rafters required (divide spacing into length and add one). If the spacing is the same as the ceiling and floor joists the same amount will be required.
3. The length of each rafter. (Don't forget to allow for slope (or pitch) and overhang).

This information is compiled on workup sheets.

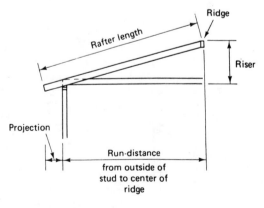

Figure 28-57. Rafters.

Roof Assembly

Rafters: 2 × 8, 16" o.c., #2KD

	No.	Length	B.F.
Rafters	78	14'	1,456

$$\text{Run} = 12'\text{-}0''$$
$$\text{Projection} = \underline{\quad 1'\text{-}6''}$$
$$\text{Horizontal distance} = 13'\text{-}6''$$

Slope: 3 inches in 12. Use 1.03 factor:

$$13'\text{-}6'' \times 1.03 = 13.5 \times 1.03 = 13.905'$$

Use 14' rafters.

Headers and Trimmers

Headers and trimmers are required for any openings (such as chimneys) just as in floor and ceiling joists. For a complete discussion of headers and trimmers refer to Section 28.3, "Floor Framing."

Collar Ties

These are used to keep the rafters from spreading (Figure 28-58), and most codes and specifications require them to be a maximum of 5 feet apart or every third rafter whichever is less. A check of the contract documents will determine:

a. Required size (usually 1 x 6 or 2 x 4).

b. Spacing.

c. Location (how high up).

Determine the number of collar ties required by dividing the total length of the building (used to determine the number of rafters) by the spacing and add one to close up the last space.

The length required can be a little harder to determine exactly. Most specifications don't spell out the exact location of the collar ties (up high near ridge, low closer to joists, halfway in between), but the typical installation has the collar ties about one-third down from the ridge with a length of 5 feet to 8 feet depending on the slope and span of the particular installation.

The collar tie take-off for the residence is compiled on the workup sheets.

Collar ties (every third rafter): 2 x 4

	No.	Length	B.F.
Collar ties	14	6'	56

78 rafters = 39 rows of rafters

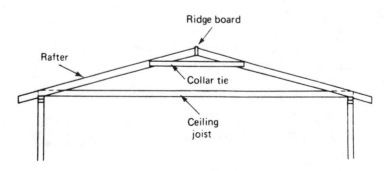

Figure 28-58. Ridge board.

$$\frac{39}{3} = 13 + 1 \text{ extra} = 14 \text{ collar ties}$$

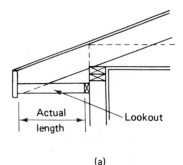

(a)

Ridge

The *ridge board* (Figure 28–58) is taken off next. The contract documents must be checked for the size ridge board required. Quite often no mention of ridge board size is made anywhere on the contract documents, and this would have to be checked with the architect-engineer (or whoever has authority for the work). Generally, the ridge is 1-inch thick and one size wider than the rafter. In this case, a 2 x 8 rafter would be used with a 1 x 10 ridge board. Other times a 2-inch thickness may be required.

The length of the ridge board required will be the length of the building plus any side overhang.

The ridge board information should be gathered and noted on the workup sheet.

Ridge Board: 1 x 10

	L.F.	B.F.
Ridge board	52	43.4

Lookouts

Lookouts are often required when a soffit is boxed-in as shown in Figure 28–50. The number of lookouts is found by dividing the total lineal feet of boxed-in soffit by the spacing and adding one extra lookout for *each* side of the house it is required on. (If required, front and back add 2 extra lookouts.) The length of the lookout is found by reviewing the sections in details. The detail in Figure

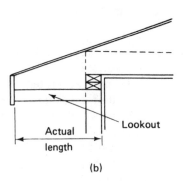

Figure 28-59. Lookouts.

28–59a indicates the lookout would be the width of the overhang minus the actual thickness of the fascia board and the supporting pieces against the wall. In Figure 28–59b the lookout is supported by being nailed to the stud wall, and its length is equal to the overhang plus stud wall minus the actual thickness of the fascia board.

The lookouts for this residence have been compiled on the workup sheet.

Lookouts: 2 × 4, 16″ o.c.

	No.	Length	B.F.
Lookouts	78	1′-4″	70

Finish Materials

Sheathing for the roof is taken off next. A review of the contract documents should provide the plywood information needed:

1. Thickness or identification index (maximum spacing of supports).
2. Veneer grades.
3. Species grade.
4. Any special installation requirements.

This information should be gathered and noted on the workup sheet.

The plywood is most accurately estimated by making a sketch of the area to be covered and planning the plywood sheet layout. For roof sheathing be certain that the rafter length that is required is used and that it takes into consideration any roof overhang and the slope (or pitch) of the roof. The roof sheathing for the residence is then compiled on the workup sheets.

Each side requires (from Figure 28–60)

19.5 sheets (for top 3 rows) × 2 = 39
7 sheets for bottom row   = 7
                                  ――
                                  46   sheets
(will take care of both sides)
    4 × 8 sheets = 46 sheets = 1,472 s.f.

Since the carpenters usually install the roofing felt (which protects the plywood until the roofers come), the roofing felt is taken off here or under roofing depending on the estimator's preference. Since felt, shingles and built-up roofing materials are estimated in Chapter 30 that is where it will be estimated for this residence.

The fascia and soffit are estimated in Section 28.8.

--------- 28.7 ---------
INSULATION

Insulation in light frame construction may be placed between the framing members (studs or joists) or nailed to the rough sheathing. It is used in the exterior walls and the ceiling and floors of most buildings.

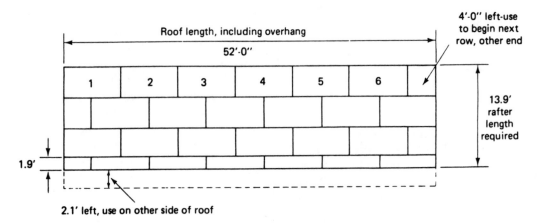

Plywood sheet layout, $\frac{1}{2}$ of roof area

Figure 28-60. Plywood sheet layout.

Insulation placed between the framing members may be pumped in or laid in rolls or in sheets, while loose insulation is sometimes placed in the ceiling. Roll insulation is available in widths of 11, 15, 19, and 23 inches to fit snugly between the spacing of the framing materials. Sheets of the same widths and shorter lengths are also available. Rolls and sheets are available unfaced, faced on one side, faced on both sides, and foil-faced. The insulating materials may be of glass fiber or mineral fiber. Nailing flanges, which project about 2 inches on each side, lap over the framing members and allow easy nailing or stapling. To determine the number of lineal feet required, the square footage of wall area to be insulated is easily determined by multiplying the distance around the exterior of the building by the height to which the insulation must be carried (often the gable ends of a building may not be insulated). Add any insulation required on interior walls to find the gross area. This gross area should be divided by the factor given in Figure 28–61. If the studs are spaced 16 inches on center, the 15-inch wide insulation plus the width of the stud equals 16 inches so a batt 1′ -0″ long will cover an area of 1.33 square feet.

Ceiling insulation may be placed in the joists or in the rafters. The area to be covered should be calculated and if the insulation is placed in the joists, the length of the building is multiplied by the width. For rafters (for gabled roofs) the methods shown in Chapter 30 (for pitched roofs) should be used; i.e., the length of the building times the rafter length from the ridge to plate times two (for both sides). Divide the area to be covered by the factor given in Figure 28–61 to calculate the lineal footage of a given width roll.

Ceiling insulation may also be poured between the joists; a material such as vermiculite is most commonly used. Such materials are available in bags and may be easily leveled to any desired thickness. The cubic feet of material required must be calculated.

Insulation board that is nailed to the sheathing or framing is estimated by the square foot. Sheets in various thicknesses and sizes are available in wood, mineral, and cane fibers. The area to be covered must be calculated and the number of sheets determined.

For all insulation add 5 percent waste when net areas are used. Net areas will give the most accurate take-off. Check the contract documents to determine the type of insulation, thickness, or R value required and any required methods of fastening.

Stud spacing inches	Insulation width	To determine L.F. of insulation required multiply S.F. of area by	OR	L.F. of insulation required per 100 S.F. of wall area
12	11	1.0		100
16	15	0.75		75
20	19	0.60		60
24	23	0.50		50

Figure 28–61. Insulation requirements.

When compiling the quantities, keep each different thickness or type of insulation separate.

Floor: 3½", R-11, 1,200 s.f. required.
1 roll 15" wide × 56' long = 75 s.f.
1,200 s.f. ÷ 75 s.f. = 16 rolls required

Ceiling: 6", R-19, 1,200 s.f. required.
1 roll 15" wide × 32' long = 43 s.f.
1,200 s.f. ÷ 43 s.f. = 27.9 rolls required

Walls: 3½", R-11
1 roll 7'-8" high, 148 l.f. = 1,135 s.f. required
15" wide × 56' long = 75 s.f.
1,135 s.f. ÷ 75 s.f. = 15.13 = 16 rolls required

—————— 28.8 ——————
TRIM

The trim may be exterior or interior. Exterior items include moldings, fascias, cornices, and corner boards. Interior trim may include base, moldings, and chair rails. Other trim items may be shown on the sections and details, or may be included in the specifications. Trim is taken off by the lineal foot required and it usually requires a finish (paint, varnish, etc.), although some types are available prefinished (particularly for use with prefinished plywood panels).

The exterior trim for the residence is limited to the fascia board and soffit.

Fascia board: 1 x 10, Fir, 160 l.f. required—order 170 l.f.

52 + 52 + 13.9 + 13.9 + 13.9 + 13.9 = 159.6 or 160 l.f.

Soffit, plywood: ½" thick, A-C, exterior.
1'-6" wide, 104 l.f. = 156 s.f. = 4.87 or 5-
4 × 8 sheets

Baseboard: 4" ranch mold, 423 l.f. required—order 440 l.f. Exterior wall = 148 l.f.; interior wall + 149 l.f. times 2 (both sides) minus doors (25 l.f.).

—————— 28.9 ——————
LABOR

Labor may be calculated at the end of each portion of the rough framing, or it may be done for all of the rough framing at once. Many estimators will use a square foot figure for the rough framing based on the cost of past work, taking into consideration the difficulty of work involved. A job such as the small residence being estimated would be considered very simple to frame and would receive the lowest square foot cost; the cost would go up as the building was more involved. Many builders use framing subcontractors for this type work and the sub may price the job by the square foot or as a lump sum. All of these methods provide the easiest approach to the estimator.

When the estimator uses his own work force and wants to estimate the time involved, he will usually use records from past jobs, depending on how organized he is. The labor would be estimated for the framing by using the appropriate portion of the table for each portion of the work to be done from Figure 28-62.

—————— 28.10 ——————
CHECKLIST

Studs
Wood joists
Rafters

Light framing	Unit	Man hours
Sills	100 l.f.	2.0 to 4.5
Joists, floor and ceiling	MBFM	16.0 to 24.0
Walls, interior, exterior (including plates)	MBFM	18.0 to 30.0
Rafters,		
gable roof	MBFM	18.5 to 30.0
hip roof	MBFM	22.0 to 35.0
Cross bridging, wood	100 sets	4.0 to 6.0
metal	100 sets	3.0 to 5.0
Plywood, floor	100 s.f.	1.0 to 2.0
wall	100 s.f.	1.2 to 2.5
roof	100 s.f.	1.4 to 2.8
Trim,		
fascia	100 l.f.	3.5 to 5.0
soffit	100 l.f.	2.0 to 3.5
baseboard	100 l.f.	1.5 to 2.5
molding	100 l.f.	2.0 to 4.0

Figure 28-62. Man-hours required for framing.

Sheathing
Insulation
Trim
Bracing
Bridging
Plates
Sills
Ridge
Collar ties
Lookouts
Headers
Trimmers
Girder

nomical, while the laminated beams and box beams are both economical and used for appearance.

Wood Trusses

Wood joists with spans in excess of 150 feet are readily available. Decking is quickly and easily nailed directly to the chords. Trusses of almost any shape (design) are possible. Ducts, piping, and conduits may be easily incorporated into the trusses. The trusses are only part of the system and must be used in conjunction with a deck of some type. Typical truss shapes are shown in Figure 28–63.

28.11
WOOD SYSTEMS

Wood is used as a component in quite a number of structural systems, among them wood trusses, laminated beams, wood decking, and box beams. Wood is used for a variety of reasons: the wood trusses are eco-

Specifications. Check the type of truss required. If any particular manufacturer is specified, note how the members are to be attached (to each other and to the building) and what stress-grade-marked lumber is required. Note also any requirements regarding the erection of the trusses and any finish requirements.

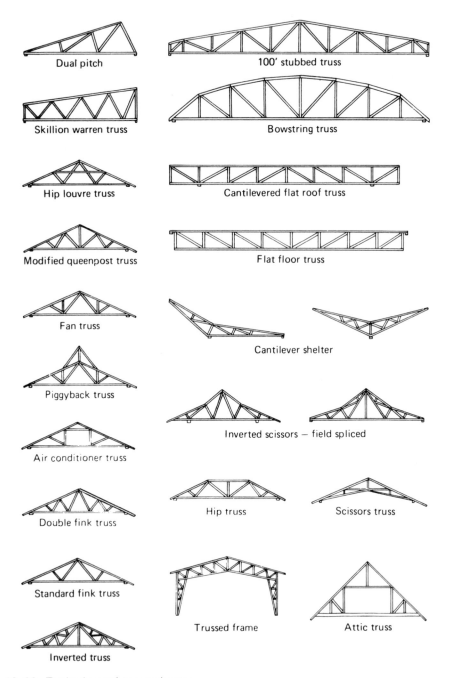

Figure 28-63. Typical wood truss shapes.

Dual pitch

Skillion warren truss

Hip louvre truss

Modified queenpost truss

Fan truss

Piggyback truss

Air conditioner truss

Double fink truss

Standard fink truss

Inverted truss

100' stubbed truss

Bowstring truss

Cantilevered flat roof truss

Flat floor truss

Cantilever shelter

Inverted scissors — field spliced

Hip truss

Scissors truss

Trussed frame

Attic truss

Estimating. Determine the number, type, and size required. If different sizes are required, keep them separate. Note any special requirements for special shapes; then get a written proposal from the manufacturer or supplier. If they are not familiar with the particular project, make arrangements for them to see the contract documents so that they will provide a complete price. Check to see how the truss is attached to the wall or column supporting it.

If the trusses are to be installed by the general contractor, an allowance for the required equipment (booms, cranes) and men must be made. The type of equipment and number of men will depend on the size and shape of the truss, and the height of erection. Trusses 100 feet long may be placed at the rate of 1,000 to 1,500 square feet of coverage per hour, using two mobile cranes mounted on trucks and seven men.

Laminated Beams and Arches

Laminated structural members are pieces of lumber glued under controlled temperature and pressure conditions. The glue used may be either interior or exterior. They may be rectangular beams or curved arches, such as parabolic, bowstring, "V," or "A." The wood used includes Douglas fir, southern pine, birch, maple, and redwood (Figure 28–64). Spans just over 300 feet have been built using laminated arches in a cross vault. They are generally available in three grades: industrial appearance, architectural appearance, and premium appearance grade. The specifications should be carefully checked so that the proper grade is used. They are also available prefinished.

Specifications. Check for quality control requirements, types of adhesives, hardware,

appearance grade, finish, protection, preservative, and erection requirements.

Estimating. The cost of materials will have to come from the manufacturer or supplier, but the take-off is made by the estimator. The take-off should list the lineal footage required of each type, and the size or style of beam or arch. Note also the type of wood, appearance grade, and finishing requirements. If the laminated shapes are not prefinished, be certain that finishing is covered somewhere in the specifications.

If the laminated beams and arches are to be installed by the general contractor, consider who will deliver the material to the site, how it will be unloaded, where it will be stored, and how much equipment and how many men will be required to erect the shapes. Erection time varies with the complexity of the project. In a simple erection job beams may be set at the rate of six to eight lengths per hour, while large, more complex jobs require four hours or more for a single arch. Check carefully the fastening details required, total length of each piece, how they will be braced during construction, and whether it is one piece or is segmental.

Wood Decking

Wood decking is available as a solid-timber decking, plank decking, and laminated decking. Solid-timber decking is available either in natural finish or prefinished; the most common sizes are 2 x 6, 3 x 6, and 4 x 6. The most commonly used woods for decking are southern pine, western red cedar, inland white fir, western white spruce, and redwood.

Plank decking is tongue-and-groove decking fabricated into panels. The most common panel size is 21 inches wide, and lengths are up to 24 feet. Installation costs are reduced by us-

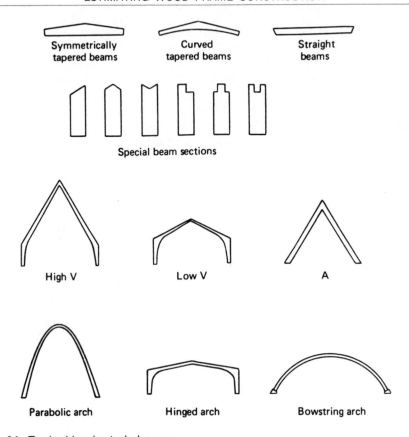

Figure 28-64. Typical laminated shapes.

ing this type of deck. Note wood species, finish, size required, and appearance grade.

Laminated decks are available in a variety of thicknesses and widths. Installation costs are reduced substantially. Note the wood species, finish, and size required. Appearance grades are also available.

Estimating.   No matter what type of wood deck is required, the take-off is the square foot area to be covered. Since the decking is sold by the board measure, the square footage is multiplied by the appropriate factor shown in Figure 28-65. Take particular notice of fastening details and size of decking required. (See board measure, Section 28.2).

Consider also if the decking price includes delivery to the job site or if it must be picked up, where the decking will be stored, how it will be moved onto the building, and the number of men required.

Solid decking requires between 12 and 16 man-hours for a mbfm (thousand board feet of measure), while plank decking may save 5 to 15 percent of installation costs on a typical project. Laminated decking saves 15 to 25 percent of the installation costs since it is attached by nails instead of with heavy spikes.

Specifications.   Check for wood species, adhesives, finishes, appearance grade, fastening, size requirements, and construction of

Block decking		Laminated decking		
nominal dimensions		actual dimensions		
3 x 6	4 x 6	$2\frac{1}{4}$ x $5\frac{1}{2}$	$3\frac{1}{16}$ x $5\frac{1}{2}$	$3\frac{13}{16}$ x $5\frac{1}{2}$
3.43	4.58	3.33	4.4	5.5

Multiply the square foot area to be covered by the factor listed to arrive at the required board measure. These factors include no waste.

Figure 28-65. Wood deck estimating factors.

the decking. Size of decking should also be noted.

Plywood Systems

The strength and versatility of plywood recommends it for use in structural systems. The systems presently in use include box beams, rigid frames, folded plates, and space planes for long-span systems. For short-span systems, stressed skin panels and curved panels are used.

The availability, size, and ease of shaping of plywood makes it an economical material to work with. Fire ratings of one hour can be obtained also.

Technical information regarding design, construction, estimating, and cost comparison may be obtained through the American Plywood Association.

Plywood structural shapes may be supplied by a local manufacturer, shipped in, or built by the contractor's work forces. The decision will depend on local conditions such as suppliers available, workmen, and space requirements.

Specifications. Note the thickness and grade of plywood, whether interior or exterior glue is required, fastening requirements, and finish that will be applied. All glue used in assembling the construction should be noted, and any special treatments such as with fire or pressure preservatives, must also be taken off.

Estimating. Take off the size and shape of each member required. Total up the lineal footage of each different size and shape. If the material is being supplied by a manufacturer, he will probably price it by the lineal foot of each size or as a lump sum. If the members are to be built by the contractor's work force, either on or off the job site, a complete take-off of lumber and sheets of plywood will be required.

_____ 28.12 _____
WOOD CHECKLIST

Wood:
Species
Finish required
Solid or hollow
Laminated
Glues
Appearance grade
Special shapes
Primers

Fastening:
Bolts
Nails

Spikes
Glue
Dowels

Erection:
Cranes
Booms
Scaffolds
Hoists

Engineering:
Fabrication
Shop drawings
Unloading, loading
Inspection
Bracing

Chapter 29

Estimating Drywall and Wetwall

29.1
DRYWALL ESTIMATING

All wallboard should be taken off by the square foot with the estimator double-checking for panel layout of the job. The first step is to determine the lineal footage of each type of wall, carefully separating any wall with different sizes or types of material, fasteners, or any other variations. Remember that there are two sides to most walls, each side requiring a finish.

After a complete listing has been made, the square footage of wallboard may be determined. Keep in mind the varieties that may be encountered and keep each separate.

Equipment required may simply be wood or metal horses, planks, platforms, and scaffolding as well as small electric tools and staplers. On projects with high ceilings, scaffolds on wheels are often used so workers can work conveniently and the scaffold may be easily moved from place to place.

Labor for drywall construction will vary depending on the type of wallboard, trim, fasteners, whether the walls are straight or jogged, height of walls or ceiling, and the presence of other construction underway at the time. Many subcontractors are available with skilled workers, specially trained for this type of work.

29.2
DRYWALL ACCESSORIES

Accessories for the application and installation of drywall construction include mechanical fasteners and adhesives, tape and compound for joints, fastener treatment, and trim to protect exposed edges and exterior corners, as well as baseplates and edge moldings.

Mechanical Fasteners

Clips and staples may be used to attach the base ply in multi-ply construction. The clip spacing may vary from 16 to 24 inches on center and may also vary depending on the support spacing. Staples should be 16-gauge, galvanized wire with a minimum of a 7/16-inch wide crown with legs having divergent points. Staples should be selected to provide a minimum of ⅝-inch penetration into the supporting structure; they are spaced about 7 inches on center for ceilings and 8 inches on center for walls.

Nails are bought by the pound. The approximate weight of the nails that would be required per thousand square feet (msf) of gypsum board varies between 5 and 7 pounds.

Screws may be used to fasten both wood and metal supporting construction and furring strips. If 8-inch on center spacing is used on the vertical joints, about 1,200 screws are required. If the boards are applied horizontally with screws spaced 12 inches on center, only about 820 screws are required.

Trim

A wide variety of trims is available in wood and metal for use on drywall construction. The trim is generally used to provide maximum protection and neat, finished edges throughout the building. The wood trim is available unfinished and prefinished in an endless selection of sizes, shapes, and costs. The metal trim is available in an almost equal amount of sizes and shapes with finishes ranging from plain steel, galvanized steel, prefinished painted to trim with permanently bonded finishes which match the wallboard; even aluminum molding, plain and anodized, is available. Most trim is sold by the lineal foot, so the take-off should also be made in lineal feet. Also to be determined is the manner in which the moldings are to be attached to the construction.

Joint Tape and Compounds

Joint tape and compounds are employed when a gypsum wallboard is used, and it is necessary to reinforce and conceal the joints between wallboard panels and to cover the fastener heads.

The amount of tape and compound required for any particular job will vary depending on the number of panels used with the least number of joints and the method of fastening specified. To finish 1,000 square feet of surface area, about 380 lineal feet of tape and 50 pounds of powder joint compound (or 5 gallons of ready-mixed compound for the average job) will be required.

Example 27.1: Residential Building (Appendix B)

Exterior walls: gypsum board, 1/2″ × 4′ × 8′
Ceiling height: 8′-0″
148 l.f. (from Section 28.3)

$$\frac{148 \times 8}{32} = 37 \text{ sheets} = 1184 \text{ s.f.}$$

Interior walls: 1/2″ gypsum board each side, 4 × 8
Ceiling height: 8′-0″
149 l.f. (from Section 28.3)

$$\frac{149 \times 8 \times 2}{32} = 75 \text{ sheets} = 2400 \text{ s.f.}$$

Ceilings: 1/2″ gypsum board, 4 × 8

24′ × 50′ = 1,200 s.f. =	38	sheets
Sheets required = +	37	(exterior walls)
+	75	(interior walls)
+	38	(ceiling)
	150	sheets
5% waste	8	
Total	158	sheets required

Metal trim:

$$\begin{aligned}
\text{Corners}-2 @ 8' &= 16 \text{ l.f.} \\
\text{At ceiling}-148 \text{ l.f.} \times 2 \text{ sides} &= \underline{296 \text{ l.f.}} \\
&\quad\ 312 \text{ l.f.}
\end{aligned}$$

Labor

Subcontractors who specialize in drywall installations generally do this type of work, and they may price it on a unit basis (per square foot) or a lump sum. The time required for the installation is shown in Figure 29-1.

The hourly wages used are based on local labor and union conditions.

29.3
WETWALL ESTIMATING

Gypsum plasters are usually packed in 100-pound sacks and priced by the ton. The estimator makes his wetwall take-off in square feet, converts it to square yards, and must then consider the number of coats, thickness

Per $\frac{1}{16}''$ thickness of plaster allow	Perforated plaster board lath — add	Metal lath Add
5.2 C.F.	4.0 C.F.	8.0 C.F.

Example: For 100 sq. yd. of wall area a $\frac{1}{4}''$ thickness over metal lath will require:

5.2(4) + 8 = 20.8 + 8 = 28.8 C.F. of plaster per 100 sq. yd.

Figure 29-2. Approximate plaster quantities.

of coats, mixes to be used, and the thickness and type of lath required. The amounts of materials required may be determined with the use of Figure 29-2. This table gives the cubic feet of plaster required per 100 square yards of surface. Depending on the type of plaster being used, the approximate quantities of materials can be determined. The mix design varies from project to project and must be carefully checked. Figure 29-3 shows some typical quantities of materials that may be required. Many projects have mixes, designed for a particular use, included in the specifications. Read them carefully and use the specified mix to determine the quantities of materials required. Once the quantities of materials have been determined, the cost for materials may be determined and the cost per square yard (or per 100 square yards) calculated.

Labor time and costs for plastering are subject to variations in materials, finishes, local customs, type of job, and heights and shapes of walls and ceilings. The ability of workmen to perform this type of work varies considerably from area to area. In many areas skilled plasterers are few. This means that the labor cost will be high and there may be a

Drywall	Man hours per 100 s.f.
Gypsum board	
nailed to studs:	
walls	1.0 to 2.2
ceiling	1.5 to 2.8
jointing	1.0 to 1.8
glued:	
walls	0.8 to 2.4
ceiling	1.5 to 3.0
jointing	0.5 to 1.2
Ceilings over 8'-0" high	add 10% to 15%
Screwed to metal studs	add 10%
Rigid insulation, glued to wall	1.2 to 3.0

Figure 29-1. Man-hours required for drywall installation.

Mix	Maximum amount of aggregate, in cu. ft. per 100 lbs of gypsum plaster	Volume obtained from mix shown (C.F.)
100:2	2	2
100:$2\frac{1}{2}$	$2\frac{1}{2}$	$2\frac{1}{2}$
100:3	3	3

Aggregate weights vary.   Sand 95–110 lb. per C.F.
Perlite 40–50 lb. per C.F.
Vermiculite 40–50 lb. per C.F.

Figure 29-3. Plaster materials.

problem with the quality of the work done. It is advisable to contact the local unions and subcontractors to determine the availability of skilled workmen. In most cases one helper will be required to work with two plasterers.

If the plastering is bid on a unit basis, be certain that you understand how the yardage will be computed since the methods of measuring vary in different localities. The yardage may be taken as the gross area, the net area, or the gross area minus the openings that are over a certain size. In addition, curved and irregular work may be charged and counted extra; it will not be done for the same costs as the flatwork.

Equipment required includes a small power mixer, planks, scaffolds, mixing tools, mixing boxes, and miscellaneous hand tools. Machine-applied plaster will require special equipment and accessories, depending on the type used.

Labor

Subcontractors who specialize in wetwall installations do this type of work and they may price it on a unit basis (per square yard) or lump sum.

Estimating Lath

The gypsum lath is sold by the sheet or 1,000 square feet, and the estimator will calculate the number of square feet required (the square yards of plaster times nine equals square feet), and divide by the number of square feet in a sheet. Note the type and thickness required. Depending on the number of jogs and openings, about 6 percent should be allowed for waste. The materials used for attachment must be estimated and a list of accessories made.

Metal lath is taken off by the square yard in the same manner as plaster. It is usually quoted at a cost per 100 square yards with the weight and finish noted. For plain surfaces add 6 to 10 percent for waste and lapping; for beams, pilasters, and columns add 12 to 18 percent. When furring is required, it is estimated separately from the lath. Determine what accessories will be required and the quantity of each.

29.4
WETWALL ACCESSORIES

The accessories available for use with wetwall construction include various types of corner beads, control and expansion joints, screeds, partition terminals, casing beads, and a variety of metal trim to provide neat edged, cased openings. Metal ceiling and floor runners are also available as are metal bases. Resilient channels may also be used. These accessories are sold by the lineal foot, so the estimator makes his take-off accordingly.

A complete selection of steel clips, nails, staples, and self-drilling screws is available to provide positive attachment of the lath. Special attachment devices are available for each particular wetwall assembly. The estimator will have to determine the number of clips or screws required on the project. The specifications will state the type of attachment required and may also give fastener spacing. The manufacturer's recommended fastener spacing may also be checked to help determine the number required.

Accessories required should be included in the subcontractor's bid, but the estimator should check the subcontractor's proposal against the contract documents to be certain they are the same size, thickness of metal, and finish.

Gypsum block
Corner beads
Accessories
Number of coats
Type of plaster
Tie wire
Molding
Stucco

Drywall:
Studs
Wallboards
Furring
Channels
Tape
Paste
Adhesives
Staples
Clips
Nails
Screws

29.5
CHECKLIST

Wetwall:
Lath, metal
Furring
Studs
Channels
Lath, gypsum

_____ Chapter 30 _____

Estimating Roofing

_____ 30.1 _____
ROOFING

Roofing is considered to include all the material that actually covers the roof deck. It includes any felt (or papers) as well as bituminous materials that may be placed over the deck, since they are installed by the roofers. If insulation is to be placed over the deck, it is most commonly installed by the roofer and included in his portion of the work. Flashing required on the roof, at wall intersections, joints around protrusions through the roof, and the like is also included in the roofing take-off. Miscellaneous items such as fiber blocking, cant strips, curbs, and expansion joints should also be included. The sections and details, as well as the specifications, must be carefully studied to determine what is required and how it must be installed.

_____ 30.2 _____
ROOF AREAS

In estimating the square footage of area to be roofed, first consider the shape of the roof (Figure 30–1). When measuring a flat roof with no overhang or with parapet walls, take the measurements of the building from the

outside of the walls. This method is used for parapet walls since it allows for the turning up of the roofing on the sides of the walls. If the roof projects beyond the building walls, the dimensions used must be the overall outside dimensions of the roof, and must include the overhang. The drawings should be carefully checked to determine the roof line around the building, as well as where and how much overhang there may be. The floor plans, wall sections, and details should be checked to determine the amount of overhang and type of finishing required at the overhang. Openings of less than 30 square feet should not be deducted from the area being roofed.

To determine the area of a gable roof, multiply the length of the ridge (A to C) by the length of the rafter (A to B) by two (for the total roof surface).

The area of a shed roof is the length of the ridge (A to C) times the length of rafter (A to B). The area of a regular hip roof is equal to the area of a gable roof that has the same span, pitch, and length. The area of a hip roof may be estimated just as the area of a gable roof is: the length of roof times the rafter times two.

The length of rafter is easily determined from the span of the roof, the overhang, and slope. Refer to Figure 28–56 for information

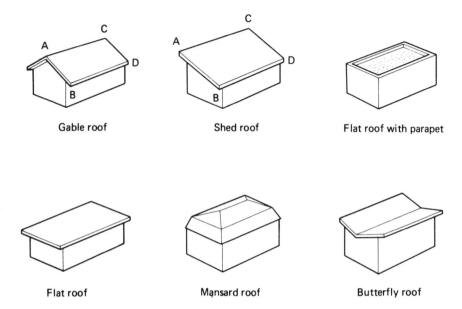

Gable roof   Shed roof   Flat roof with parapet

Flat roof   Mansard roof   Butterfly roof

Figure 30-1. Typical roof shapes.

required to determine the lengths of rafters for varying pitches and slopes.

30.3
SHINGLES

Asphalt and Asbestos Shingles

Asphalt and asbestos shingles are packed in bundles that contain enough shingles to cover 25.33 or 50 square feet of roof area. The *exposure* (amount of shingle exposed to the weather) generally is 4, 4.5, or 5 inches. Individual shingles 12 to 16 inches long are sometimes also used.

In determining the area to be covered, always allow one extra course of shingles at the eaves since the first course must always be doubled. Hips and ridges are taken off by the lineal foot and considered 1'-0" wide to determine the square footage of shingles required. Waste averages 5 to 8 percent. Galvanized, large-headed nails, ⅞ to 1¾ inches long, are used on asphalt shingles. From 1½ to 3 pounds of nails are required per square.

Asphalt shingles are generally placed over an underlayment of building paper or roofing felt. The felt is specified by the type of material and weight per square. The felt should have a minimum top lap of 2 inches and end lap of 4 inches, so to determine the square footage of felt required, multiply the roof area by a lap and waste factor of 5 to 8 percent.

Example 30.1: Residential Building

15 lb. felt paper

Roof area = 13.9' × 52' × 2 = 1,446 s.f.

Slope: 3 in 12, 2 layers felt required, lapped 19″ and exposed 17″

Roll of felt:

36″ wide × 144′ long = 432 s.f.

Roll coverage:

$$\text{s.f. in roll} \times \frac{\text{Exposure}}{\text{Width}}$$

$$= 432 \times \frac{17}{36} = 204 \text{ s.f.}$$

Rolls required:

$$\frac{1,446}{204} = 7.08 \text{ (order 8 rolls)}$$

Shingles: 235 lbs., 1,446 s.f. = 14.46 squares, plus double starter row

52 l.f. of eave × 2 eaves = 104 l.f.

Shingles: 12″ × 36″ (3′-0″)

104 l.f. ÷ 3 l.f. = 35 shingles extra for starter row (.35 squares)

  14.46
+   .35
─────────
  14.81 squares

Order 15.25 squares (this allows for waste).

Ridge: Use ridge shingles, with 5″ exposure

521 l.f. of ridge = 624 in. = 624″ ÷ 5 = 125 pieces

Labor

Subcontractors who specialize in roofing will do this work and they may price it on a unit basis or a lump sum. The time required to install shingles is shown in Figure 30–2.

Roofing materials	Man hours per square
Shingles,	
asphalt, asbestos, (strip)	0.8 to 4.0
asphalt, asbestos, (single)	1.5 to 6.0
wood (single)	3.0 to 5.0
metal (single)	3.5 to 6.0
heavyweight asphalt and asbestos	add 50%
Tile	
clay	3.0 to 5.0
metal	3.5 to 6.0
Built-up	
2 ply	0.8 to 1.4
3 ply	1.0 to 1.6
4 ply	1.2 to 2.0
aggregate surface	0.3 to 0.5

Figure 30-2. Man-hours required for shingle installation.

Wood Shingles

Available in various woods, the best of which are cypress, cedar, and redwood, wood shingles come hand-split (rough texture) or sawed, the hand-split variety commonly referred to as *shakes*. Lengths are 16, 18, and 24 to as long as 32 inches, while random widths of 4 to 12 inches are common. Wood shingles are usually sold by the square based on a 10-inch weather exposure. A double-starter course is usually required and in some installations roofing felt is also needed. Valleys will require some type of flashing, while hips and ridges require extra material to cover the joint, and thus extra long nails become requisite. Nails should be corrosion-resistant; they are 1¾ to 2 inches long and from 2 to 4 pounds per square.

─────────── 30.4 ───────────

BUILT–UP ROOFING

The take-off for built-up roofing follows the same general procedure as any other roofing:

1. The number of squares to be covered must be noted.

2. Base sheets—a base sheet or vapor barrier, when required. Note the number of plies, lap (usually 4 inches), weight per ply per square.

3. Felt—determine the number of plies, weight of felt per square, and type of bituminous impregnation (tar or asphalt).

4. Bituminous material—note the type, number of coats, and the pounds per coat per square (some specifications call for an extra heavy top coat or pour).

5. Aggregate surface—(if required) the type and size of slag, gravel, or other aggregate, and the pounds required per square should all be noted.

6. Insulation—(if required) note type, thickness, and any special requirements (refer also to Section 30.6).

7. Flashing—note type of material, thickness, width, and lineal feet required.

8. Trim—determine the type of material, size, shape and color, method of attachment, and the lineal footage required.

9. Miscellaneous materials—blocking, cant strips, curbs, roofing cements, nails and fasteners, caulking, and taping are some examples.

Bituminous materials are used to cement the layers of felt into a continuous skin over the entire roof deck. Types of bituminous materials used are coal tar pitch and asphalts. The specifications should indicate the amount of bituminous material to be used for each mopping so that each ply of felt is fully cemented to the next, and in no instance should felt be allowed to touch felt. The mopping between felts averages 25 to 30 pounds, while the top pour (poured, not mopped), which is often used, may be from 65 to 75 pounds per square. It is this last pour into which any required aggregate material will become embedded.

Felts are available in 15- and 30-pound weights, 36-inch widths, and 432- or 216-square foot rolls. With a 2-inch lap a 432-square foot roll will cover 400 square feet, while the 216-square foot roll will cover 200 square feet. In built-up roofing a starter strip 12 to 18 inches wide is applied; then over that one a strip 36 inches wide is placed. The felts that are subsequently laid overlap the preceding felts by 19, 24⅔, or 33 inches, depending on the number of plies required. Special applications sometimes require other layouts of felts. The specifications should be carefully checked to determine exactly what is required as to weight, starter courses, lapping, and plies. Waste averages about 8 to 10 percent of the required felt and this allows for the material used for starter courses.

Aggregate surfaces such as slag or gravel are often embedded in the extra-heavy top pouring on a built-up roof. This aggregate acts to protect the membrane against the elements, such as hail, sleet, snow, and driving rain. It also provides weight against wind uplift. To insure embedment in the bituminous material, the aggregate should be applied while the bituminous material is hot. The amount of aggregate used varies from 250 to 500 pounds per square. The amount required should be listed in the specifications. The amount of aggregate required is estimated by the ton.

Since various types of materials may be used as aggregates, check for the size, type, and gradation requirements. It is not unusual for aggregates such as marble chips to be specified, and this type of aggregate will result in much higher material costs than gravel or slag. Always read the specifications thoroughly.

The joints between certain types of roof

deck materials, such as precast concrete, gypsum, and wood fiber, may need to have caulking and taping of all joints. This application of flashing cement and a 6-inch-wide felt strip will minimize the deleterious effects on the roofing membrane of uneven joints, movement of the units, and moisture transmission through the joint; it also reduces the possibility of bitumen seepage into the building. The lineal footage of joints to be covered must be determined for pricing purposes.

Labor

Subcontractors who specialize in roofing will submit unit (per square) and lump-sum bids. The time required to install built-up roofing is shown in Figure 30–2.

30.5
FLASHING

Flashing is used to help keep water from getting under the roof covering and from entering the building wherever the roof surface meets a vertical wall. It usually consists of strips of metal or fabric shaped and bent to a particular form. Depending on the type of flashing required, it is estimated by the piece, lineal footage with the width noted, or square feet. Materials commonly used include copper, asphalt, tinned, painted, plastic, rubber, composition, and combinations of these materials. Bid the gauge of thickness and width specified. Expansion joints are estimated by the lineal foot and it may be necessary that curbs be built-up and the joint cover either prefabricated or job-assembled. Particular attention to the details on the drawings is required so that the installation is understood.

30.6
INSULATION

Insulation included as a part of roofing is installed on top of the deck material. This type of insulation is rigid and is in a sheet (or panel). Rigid insulation may be made of urethane, fiberboards, or perlite and is generally available in lengths of 3 to 12 feet in 1-foot increments and widths of 12, 16, 24, and 48 inches. Thicknesses of ½, ¾, 1, 1½, 2, 2½, 3, and 4 inches are available. Insulation is estimated by the square with a waste allowance of 5 percent, provided there is proper planning and utilization of the various sizes.

When including the insulation, keep in mind that its installation will require extra materials, either in the line of additional sheathing paper, moisture barriers, moppings, or a combination of these. Also, the specifications often require two layers of insulation, usually with staggered joints. This requires twice the square footage of insulation (to make up two layers), an extra mopping of bituminous material, and extra labor. Read the specifications carefully—never bid something you do not fully understand.

NOTE: Some roofing manufacturers will not bond the performance of the roof unless the insulation meets their specifications. This item should be checked and if the manufacturer, or his representative, sees any problem, the architect-engineer should be notified so that the problem may be cleared up during the bid period.

30.7
ROOFING TRIM

Trim such as gravel stops, fascia, coping, ridge strips, gutters, downspouts, and soffits, is taken off by the lineal foot. All special pieces used in conjunction with the trim (e.g., elbows, shoes, ridge ends, cutoffs, corners, and brackets) are estimated by the number of pieces required.

The usual wide variety of materials and finishes is available in trim. Not all trim is standard stock; much of it must be specially formed and fabricated. This will add considerably to the cost of the materials and to the delivery time—so check this out and never assume that the trim needed is standard stock.

installation may also be required. Built-up roofing may demand that mops, buggies, and heaters (to heat the bituminous materials) be employed; some firms have either rotary or stationary felt layers. Metal roofing requires shears, bending tools, and soldering outfits.

Equipment costs are estimated either by the square or by the job, with the cost including such items as transportation, set-up, depreciation, and replacement of miscellaneous items.

30.8
LABOR

The labor cost of roofing will depend on the hourly output and hourly wages of the workers. The output will be governed by the incline of the roof, size, irregularities in the plan, openings (skylights, etc), and the elevation of the roof above the ground, since the higher the roof, the higher the materials have to be hoisted to the work area.

Costs can be controlled by the use of work gangs that are familiar with the type of work and experienced in working together. For this reason much of the roofing done on projects is handled by a specialty contractor with equipment and trained personnel.

30.9
EQUIPMENT

Equipment requirements vary considerably, depending on the type of roofing used and the particular job being estimated. Most jobs will require hand tools for the workers, ladders, some scaffolding, and some type of hoist—regardless of the type of roofing being applied. Specialty equipment for each particular

30.10
CHECKLIST

Paper

Felt

Composition (roll)

Composition (built-up)

Tile (clay, metal, concrete)

Shingles (wood, asphalt, asbestos, slate)

Metal (copper, aluminum, corrugated, steel)

Insulation

Base

Solder

Flashing

Ridges

Valleys

Fasteners

Trim

Blockings (curbs)

Cant strips

Chapter 31

Estimating Windows and Doors

31.1
WINDOW FRAMES

Window frames may be made of wood, steel, aluminum, bronze, stainless steel, and plastic. Each material has its particular types of installation and finishes, but from the estimator's point there are two basic types of window, stock and custom-made.

Stock windows are more readily available and, to the estimator, more easily priced as to the cost per unit. The estimator can count up the number of units required and list his accessories to arrive at his material cost.

Custom-made frames cannot be accurately estimated by the estimator. Approximate figures can be worked up based on the square footage and type of window, but exact figures can be obtained only from the manufacturer. In this case the estimator will call either his local supplier or manufacturer's representative to be certain that they are bidding the job. Often, he will send them copies of the drawings and specifications, which they may use to prepare a proposal.

When checking proposals for the windows on a project, make a note whether the glass or other glazing is included, where delivery will be made, and whether installation is done by the supplier, a subcontractor, or the general contractor. Be certain the proposal includes all the accessories that may be required, including mullions, screens, sills, etc.

Example 31.1: The wood windows required for the residence have been taken off and compiled here.

4 double-hung wood windows, 42/52
(3'-6" × 4'-4")
2 sets 2 d.h. wood, 42/52

Labor

The time required to install the windows is taken from the table in Figure 31–1. The hourly wages used are based on local labor and union conditions.

Windows	Man hours per unit
Wood:	
in wood frame	2.0 to 3.5
in masonry	2.5 to 5.0
Metal:	
in wood frame	2.0 to 3.5
in masonry	2.0 to 4.0
Over 12 s.f.	add 20%
Bow and bay windows	2.5 to 4.0

Figure 31-1. Man-hours required for window installation.

Example 31.2:

Wood windows: 4 double-hung wood windows, 42/52 (3'-6″ × 4'-4″): 2 double-hung wood windows, 42/52 (3'-6″ × 4'-4″) with mullions, 2 sets.

Rate of work: 2.5 man-hours per window
Labor: Carpenter, $6.15 per hour

Time:

$$8 \times 2.5 = 20 \text{ man-hours}$$

Cost:

$$20 \times \$6.15 = \$123.00$$

Accessories

The items that may be required for a complete job include glass and glazing, screening, weatherstripping, hardware, grilles, mullions, sills, and stools. The details of the project must be checked to find out what accessories must be included, so make a list of each accessory and what restrictions there are for each.

Glass. Glass is discussed in Section 31.2. At this point note whether or not it is required, what thickness, type, and quality are re-

quired, and if it is to be a part of the unit or installed at the job site. The square footage of each type of glass must be known as well.

Screens. Screening mesh may be painted or galvanized steel, plastic-coated glass fiber, aluminum, or bronze. If specified, be certain the screens are included in the proposals received. If the screens are not bid by the supplier or subcontractor whose bid is used for frames, prices for screens must be obtained elsewhere. If so, a word of caution: Be certain that the screens and frames are compatible, and a method of attaching the screen to the frame is determined. The various sizes required will have to be noted so an accurate price for the screening can be arrived at.

Hardware. Most types of frames require hardware. It may consist only of locking and operating hardware made of the same material as the frame. Since the hardware is almost always sent unattached, it must be applied at the job site. Various types of locking devices, handles, hinges, and cylinder locks on sliding doors are often needed. Materials used may be stainless steel, zinc-plated steel, aluminum, or bronze, depending on the type of hardware and where it is being used.

Weatherstripping. Most specifications require some type of weatherstripping, and many stock frames come with weatherstripping factory installed. The only thing to be careful of in that case is that the factory-installed weatherstripping is the type specified. Some of the more common types of weatherstripping are vinyl, polypropylene-woven pile, neoprene, metal flanges and clips, polyvinyl, and adhesive-backed foam. It is usually sold by the lineal foot for the rigid type and by the roll for the flexible. Do the take-off in lineal feet, find out how it is sold, and work up a cost.

Mullions. *Mullions* are the vertical bars that connect adjoining sections of frames. The mullions may be of the same material as the frame, or a different material, color, or finish. Mullions may be small "T"-shaped sections that are barely noticeable or large elaborately designed shapes used to accent and decorate. The mullions should be taken off by the lineal footage with a note as to thicknesses, finish, and color.

Sills. The *sill* is the bottom member of the frame. The member on the exterior of the building, just below the bottom member of the frame, is another sill. This exterior sill serves to direct water away from the window itself. These exterior sills may be made of stone, brick, precast concrete, tile, metal, or wood. The details of the frame and its installation must be studied to determine the type of material, its size, and how it fits in. Sills are taken off by the lineal foot, and the take-off should be accompanied by notes and sketches that show exactly what is required.

Stools. The interior member at the bottom of the frame (sill) is called the *stool*. The stool may be made of stone, brick, precast concrete, terrazzo, tile, metal, and wood. It may be of the slip or engaged variety. The quantity should be taken off in lineal feet, with notes and sketches showing material size and installation requirements.

Flashing. Flashing may be required at the head and sill of the frame. Check the specifications and details to determine if it is required and the type required. Usually, the flashing is installed when the building is being constructed, but check the installation details to see how it is to be installed. Check also to see who buys the flashing and who is supposed to install it—don't neglect these seemingly small items since they total up to a good deal of time and expense.

31.2
GLASS

Glass is the most common material to be glazed into the frames for windows and doors. The most commonly used types are plate glass, clear window glass, wire glass, and patterned glass. Clear window glass is available in thicknesses of 0.085 to 2.30 inches thick, and the maximum size varies with the thickness and type. Generally available as single- and double-strength, heavy sheet and picture glass with various qualities is available in each classification. Clear window glass has a characteristic surface wave that is more apparent in the larger sizes.

Plate glass is available in thicknesses of $3/16$ to $1\frac{1}{4}$ inches and as polished plate glass, heavy-duty polished plate, rough, or polished plate. The more common types are regular, grey, bronze, heat-absorbing, and tempered.

Wire glass is available with patterned and polished finishes, and various designs of the wire itself. The most common thickness is $1/4$ inch and it is also available in colors. This type of glass is used where fire-retardent and safety characteristics (nonbreakage) may be required.

Patterned glass, used primarily for decoration, is available primarily in $7/32$- and $1/8$-inch thicknesses. Pattern glass provides a degree of privacy, yet allows diffused light into the space.

Other types of glass available include a structural-strength glass shaped like a channel, tempered, sound control, laminated, insulating, heat- and glare-reducing, colored, and bullet-resisting.

The frame may be single-glazed, of one sheet of glass, or double-glazed, with two sheets of glass, which provides increases sound and heat insulation. If the specifications call for double-glazed, then twice as many square feet of glass will be required.

Glass is estimated by the square foot, with sizes taken from the working drawings. Since different frames may require various types of glass throughout the project, special care must be taken to keep each type separate. Also to be carefully checked is which frames need glazing, since many windows and doors come with the glazing work already completed. The type of setting blocks and glazing compound required should be noted as well.

31.3
DOORS

Interior and exterior doors are most commonly made of wood, except for sliding glass doors and the increasingly popular metal entrance doors. Wood doors are available in two types, solid and hollow-core. Carefully check the floor plans and door schedules to determine the sizes, types, and number of doors required.

Prefinishing

Prefinishing of doors is the process of applying the desired finish on the door at the factory instead of finishing the work on the job. Doors that are premachined are often prefinished as well. Various coatings are available, including varnishes, lacquer, vinyl, and polyester films (for wood doors). Pigments and tints are sometimes added to achieve the desired visual effect. Metal doors may be prefinished with baked-on enamel or vinyl-clad finishes. Prefinishing can save considerable

job-finishing time and generally yields a better result than job-finishing. Doors which are prefinished should also be premachined so that they will not have to be "worked on" on the job. The prefinished door must be handled carefully and installed at the end of the job so that the finish will not be damaged; it is often difficult to repair a damaged finish. Care during handling and storage is also requisite.

Example 31.3: Residential Building

Doors, Interior:

1—1/6 × 6/8, hollow core, wood, prefinished
2—2/4 × 6/8, h.c., wood, prefinished
8—2/6 × 6/8, h.c., wood, prefinished
1—5/0 × 6/8, bifold, louvered, wood, prefinished

Doors, Exterior:

1—2/8 × 6/8, solid core, wood, prefinished
1—2/8 × 6/8, screen/storm, wood,
1—3/0 × 6/8, solid core, wood, prefinished

Labor

The time required to install the doors is taken from the tables in Figure 31-2. The hourly wages are based on local labor and union conditions.

Doors and frames	Man hours per unit
Residential, wood	
prehung	2.0 to 3.5
pocket	1.0 to 2.5
not prehung	3.0 to 5.0
Overhead	4.0 to 6.0
heavy duty	add 20%
Commercial	
aluminum entrance,	
per door	4.0 to 6.0
wood	3.0 to 5.0
metal, prefitted	1.0 to 2.5

Figure 31-2. Man-hours required for door installation.

Example 31.4: Residential Building

Doors:
   14 single
   1 bi-fold

Rate of work per unit:
   2.2 man-hours per unit—single
   3.5 man-hours per unit—bi-fold

Labor: Carpenter, $6.15 per hour

Time:

$$14 \times 2.2 = 30.8 \text{ man-hours}$$
$$\underline{1 \times 3.5 = 3.5 \text{ man-hours}}$$
$$34.3 \text{ man-hours}$$

Cost:

$$34.3 \times \$6.15 = \$210.95$$

─────────── 31.4 ───────────

CHECKLIST FOR DOORS AND FRAMES

1. Sizes required and the number required.

2. Frame and core types specified.

3. Face veneer specified (wood and veneer doors).

4. Prefinished or job-finished (if so, specify the finish).

5. Prehung or job-hung (if so, who will do it).

6. Special requirements:

   a. louvers

   b. windows

   c. fire rating

   d. sound control.

7. Type, size, style of frame, and the number of each required.

8. Method of attachment of the frame to the surrounding construction.

9. Finish required on the frame and who will apply it.

10. Hardware—what types are required?

11. Hardware—who will install it?

12. Accessories—types required and time to install them.

Everything takes time and costs money; someone must do it, so it must be included in the estimate.

Chapter 32

Estimating Finishes

WOOD FLOORING

Strip flooring is flooring up to 3¼ inches wide and comes in various lengths. *Plank flooring* is from 3¼ inches to 8 inches wide with various thicknesses and lengths. The most common thickness is 25/32 inch, but other thicknesses are available. It may be tongue-and-grooved, square-edged, or splined. Flooring may be installed with nails, screws, or mastic. Waste on strip flooring may range from 15 to 40 percent, depending on the size of the flooring used. This estimate is based on laying the flooring straight in a rectangular room, without any pattern involved.

Strip and plank flooring may be sold by either the square foot or board foot measure. The estimator should estimate the square footage required on the drawings, noting the size of flooring required, type of installation, and then, if required, should convert square feet to board feet (Figure 32–1). Don't forget to consider waste.

Block flooring is available as parquet (pattern) floors, which consist of individual strips of wood or larger units that may be installed in decorative geometric patterns. Block sizes range from 6 x 6 inches to 30 x 30 inches, thickness from ⁵⁄₁₆ to ¾ inch. They are available tongue-and-grooved or square edged.

Block flooring is estimated by the square foot with an allowance added for waste (2 to 5 percent). The type of flooring, pattern required, and method of installation must be noted.

The wood flooring may be unfinished or factory finished. Unfinished floors must be sanded with a sanding machine on the job and then finished with a penetrating sealer, which leaves virtually no film on the surface, or with a heavy solids type finish, which provides a high luster and protective film. The penetrating sealer will usually require a coat of wax also. The sanding of the floors will require from three to five passes with the machine. On

Size of wood Flooring (actual)	To change sq. ft. to board feet, multiply the sq. ft. amount by
$\frac{25}{32}'' \times 1\frac{1}{2}''$	1.55
$\frac{25}{32}'' \times 2\frac{1}{4}''$	1.383
$\frac{25}{32}'' \times 3\frac{1}{4}''$	1.29
$\frac{3}{8}'' \times 1\frac{1}{2}''$	1.383
$\frac{1}{2}'' \times 2''$	1.30

Values allow 5% waste.

Figure 32–1. Wood flooring, board measure.

especially fine work, hand sanding may be required. The labor required will vary, depending on the size of the space and the number of sanding operations required. The surface finish may require two or three coats to complete the finishing process. Factory-finished wood flooring requires no finishing on the job, but care must be taken during and after installation to avoid damaging the finish.

_____ 32.2 _____
RESILIENT FLOORING

Resilient flooring tiles may be made of asphalt, vinyl-asbestos, vinyl, rubber, cork, and linoleum. Resilient sheets are available in vinyl.

Estimating

Resilient tile is estimated by the square foot: the actual square footage of the surface to be covered plus an allowance for waste. The allowance will depend on the area and shape of the room. When designs and patterns are made of tile or a combination of feature strips and tile, a sketch of the floor should be drawn up and an itemized breakdown of required materials made. The cost of laying tile will vary with the size and shape of the floor, size of the tile, type of subfloor and underlayment, and the design. Allowable waste percentages are shown in Figure 32–2.

Feature strips must be taken off in lineal feet and, if they are to be used as part of the floor pattern, the square footage of feature strips must be subtracted from the floor area of tile or sheets required.

For floors that need sheet flooring, the estimator must do a rough layout of the floors to determine the widths required, where cut-

Area (sq. ft.)	Percent waste
Up to 75	10–12
75–150	7–10
150–300	6–7
300–1,000	4–6
1,000–5,000	3–4
5,000 and up	2–3

Figure 32–2. Approximate waste for resilient tile.

ting of the roll will occur, and how to keep waste to a minimum. Waste can amount to 30 and 40 percent if the flooring is not well laid out or if small amounts of different types are required.

Each area requiring different sizes, designs, patterns, types of adhesives, or anything else that may be different must be kept separately if the differences will affect the cost of material and amount of labor required for installation (including subfloor preparation).

Wall base (also referred to as *cove base*) is taken off by the lineal foot. It is available in vinyl and rubber, with heights of 2½, 4, and 6 inches and in lengths of 42 inches, 50, and 100 feet. Corners are preformed so the number of interior and exterior corners must be noted.

Adhesives are estimated by the number of gallons required to install the flooring. To determine the number of gallons, divide the total square footage of flooring by the coverage of the adhesive per gallon (in square feet). The coverage usually ranges from 150 to 200 square feet per gallon, but will vary depending on exact type used and the subfloor conditions.

Example 32.1: Residential Building

Flooring: Resilient tile, ⅛″ thick, 9 × 9 in. vinyl

    Bath 1: 8′-0″ × 5′-0″ = 40 s.f.
    Bath 2: 5′-2″ × 4′-6″ = 24 s.f.
    Kitchen: 3′-4″ × 8′-0″ = 27 s.f.
            5′-6″ × 6′-0″ = 33 s.f.

           Total   124 s.f.

Baseboard: Vinyl, ⅛ by 4 in.

    Bath 1: 13′-8″ l.f.
    Bath 2: 17′-0″ l.f.
    Kitchen: None

    Total 30′-8″ l.f.

Adhesive: coverage is 150 s.f. per gallon, 250 l.f. per gallon. Order 2 gallons of adhesive.

Labor

Subcontractors who specialize in resilient floors installation may price it on a unit basis (per square foot) or on a lump sum. The time required for resilient floor installation is shown in Figure 32-3.

32.3

CARPETING

Estimating

Carpeting is estimated by the square yard, with special attention given to the layout of the space for the most economical use of the materials. Waste and excess material may be large unless there is sufficient planning. Each space requiring different types of carpeting, cushion, or color must be figured separately. If the specifications call for the color to be selected by the architect-engineer at a later date, it may be necessary to call and try to

Tile	Man hours per 100 s.f.
Resilient squares	
9″ × 9″	1.5 to 2.5
12″ × 12″	1.0 to 2.2
Seamless sheets	0.8 to 2.4
Add for felt underlay	add 10%
Less than 500 s.f.	add 15%

Figure 32-3. Man-hours required for resilient tile installation.

determine how many different colors may be required. In this manner, a more accurate estimate of waste may be made.

Certain types of carpeting can be bought by the roll only, and it may be necessary to purchase an entire roll for a small space. In this case waste may be high since the cost of the entire roll must be charged to the project.

The cushion required is also taken off in square yards, with the type of material, design, and weight noted. Since cushions are available in a wide range of widths, it may be possible to reduce the amount of waste and excess material.

Example 32.2: Residential Building

	LENGTH OF 12′-WIDE CARPET REQUIRED
Bedroom 1	= 8′
Bedroom 2	= 8′-2″
Bedroom 3	= 11′-5″
Bedroom 4	= 10′-0″
Living-Dining room	= 21′-6″
	+ 8′-6″
Hall (to D.R.) 3′ × 30′ = 90 s.f.	+ 9′-0″
Closets	= 6′-0″
	82′-7″ of 12′-wide carpeting = 990.96 s.f.

Padding:

    82.58′ × 12′ = 990.96 s.f.

Carpet	Squares per man hour
Carpet and pad, wall to wall	8 to 20
Carpet, pad backing, wall to wall	10 to 22
Deduct for gluing to concrete slab	10%
Less than 10 squares	add 15-20%

Figure 32-4. Man-hours required for carpet installation.

Labor

Subcontractors who specialize in carpet installations may price it on a unit basis (per square yard) or lump sum. The time required for the installation is shown in Figure 32-4.

32.4
TILE

Estimating

Floor and wall tiles are estimated by the square foot. Each area must be kept separate, according to the size and type being used. It is common to have one type of tile on the floor and a different type on the walls. The different colors also vary in cost, even if the size of the tile is the same, so caution is advised. The trim pieces should be taken off by the lineal feet of each type required. There is a large variety of sizes and shapes at varying costs, so check the specifications carefully and bid what is required. If portland cement mortar is used as a base, it is installed by the tile contractor. This requires the purchase of cement, sand, and sometimes wire mesh. Tile available in sheets is much more quickly installed than individual tiles. Adhesives are sold by the gallon or sack and approximate

coverage can be obtained from the manufacturer. The amount of grout used depends on the size of the tile.

When estimating wall tile, note the size of the room, number of internal and external corners, height of wainscot, and types of trim. Small rooms require more labor than large ones. A tile setter can set more tile in a large room than in several smaller rooms in a given time period.

Accessories are also available and, if specified, should be included in the estimate. The type and style are in the specifications. They may include soap holders, tumbler holders, toothbrush holders, grab rails, paper holders, towel bars and posts, door stops, hooks, shelf supports, or combinations of these. These accessories are sold individually, so the number required of each type must be taken off. The accessories may be recessed, flush, or flanged, and this must also be noted.

32.5
PAINTING

The variables that affect the cost of painting include the material painted, the shape and location of the surface painted, the type of paint used, and the number of coats required. Each of the variables must be considered, and the take-off must list the different conditions separately.

Although painting is one of the items commonly subcontracted out, the estimator should still take off the quantities so the subcontractor's proposal can be checked. In taking off the quantities, the square feet of surface is taken off the drawings, and all surfaces that have different variables must be listed separately. With this information the amount of materials can be determined by use of the manufacturer's information on coverage per gallon.

The following methods for taking off the painting areas are suggested, with interior and exterior work listed separately.

Interior.

Walls—actual areas in square feet.

Ceiling—actual area in square feet.

Floor—actual area in square feet.

Trim—lineal footage (note width); amount of door and window trim.

Stairs—square footage.

Windows—size and number of each type, square feet.

Doors—size and number of each type, square feet.

Baseboard radiation covers—lineal feet (note height).

Columns, beams—square feet.

Exterior.

Siding—actual area in square feet.

Trim—lineal footage (note width).

Doors—square feet of each type.

Windows—square feet of each type.

Masonry—square footage (deduct openings over 50 square feet).

Interiors receive different treatment than exteriors; different material surfaces require different applications and coatings—all of this should be in the specifications. Paints may be applied by brush, roller, or spray gun, and the method to be used is also included in the specifications.

Example 32.3: Residential Building

Interior walls and ceiling: Primer plus one coat
Exterior wood: Stain one coat
Trim (doors and windows): Primer plus one coat

Interior (walls, ceiling):

1,184 + 2,384 + 1,200 = 4,768 s.f.

Exterior plywood:

148 x 8 = 1,184 + gables = 1,184 + 72
= 1,256 s.f.

Trim:

14 doors—average 17 l.f. = 238 l.f.
8 windows—42/52 = 126 l.f.
Baseboard (interior) = 148 + 149 + 149
= 446 −30 res. = 416 l.f.
Fascia = 160 l.f.
Soffit = 156 s.f.

Labor

Subcontractors who specialize in painting and staining often do this type work and they almost always price it on a lump-sum basis. The time required for painting and staining is shown in Figure 32–5.

Painting, brushes	s.f. or l.f. per hour
Interior	
primer and 1 coat	200-260 s.f.
primer and 2 coats	150-200 s.f.
stain, 2 coats	150-200 s.f.
Trim	
primer and 1 coat	120-180 l.f.
primer and 2 coats	100-160 l.f.
stain, 2 coats	100-160 l.f.
Exterior	
primer and 1 coat	160-220 s.f.
primer and 2 coats	120-180 s.f.
stain, 2 coats	200-280 s.f.

Figure 32-5. Man-hours required for painting and staining.

32.6
CHECKLIST

Floors:

Type of material

Type of fastener

Spikes

Nails

Adhesives

Screws

Finish

Thickness

Size, shape

Accent strips

Pattern

Cushion

Base

Corners

Painting:

Filler

Primer

Paint, type

Number of coats

Shellac

Varnish

Stain

Check specifications for all areas requiring paint

Chapter 33

Overhead and Contingencies

33.1
OVERHEAD

Overhead costs are generally divided into *general overhead costs* and *job overhead costs*. The general costs include items that cannot be readily charged to any one project, while the job overhead costs include all items that can be readily charged to any one project and yet are not for materials or labor.

Overhead constitutes a large percentage of costs on the job. Failure to allow sufficient amounts for overhead has forced many firms to go out of business. Consider overhead carefully, make a complete list of all required items, and the cost of each item.

Depending on where the various costs are included, the cost of overhead can vary from 12 to 30 percent of the sum of materials, labor, and equipment. No one figure is correct, since the costs of running each office must be analyzed carefully to determine its overhead. Don't be tempted to use "average" figures for this item.

33.2
GENERAL OVERHEAD (INDIRECT)

General overhead costs are not readily chargeable to one particular project—they represent the cost of doing business and fixed expenses that must be paid by the contractor. These expenses must be shared proportionally among the projects undertaken; usually the items are estimated based on a year and reduced to a percentage of the total annual business. The following are items that should be included:

Office

Rent (if owned—the cost plus a return on investment), electricity, heat, office supplies, postage, insurances (fire, theft, liability), taxes (property, water), telephone, and office machines and furnishings.

Salaries

Office employees such as executives, accountants, estimators, purchasing agent, bookkeeper, stenographer, and anyone else working in office.

Miscellaneous

Advertising, literature (magazines, books for library), legal fees (not applicable to one particular project), professional services (architects, engineers, CPA), donations, travel (including company vehicles), and club and association dues.

Depreciation

A separate account should be kept for expenditures on office equipment, calculators, typewriters, and any other equipment. A certain percentage of the cost is written off as depreciation each year and is part of the general overhead expense of running a business.

The list of expenses shown in Figure 33–1 is a typical general overhead list. Obviously, for smaller contractors the list would contain considerably fewer items and for large contractors it could fill pages; but the idea is the same. From this, it should be obvious that the more work that can be handled by a given setup, the smaller the amount that must be allowed for general overhead and the better the chance of being low bidder. As a contractor's operation grows and develops, he should pay particular attention to his office rental expense versus that involved in purchasing a building or moving to larger quarters and the addition of staff. The original staff may be able to handle the extra work if the work load is laid out carefully.

Some contractors do not allow for "General Overhead Expense" separately in their estimates; instead they figure a larger percentage for profit, or group overhead and profit together. This, in effect, "buries" or hides part of the expenses. From the estimator's point of view, it is desirable that all expenses be listed separately so they may be analyzed periodically. In this manner, the amount allowed for profit is actually figured as profit, the amount left after *all* expenses are figured.

33.3
JOB OVERHEAD (GENERAL CONDITIONS, DIRECT OVERHEAD)

Also referred to as *general conditions expense, job overhead* comprises all costs that can be readily charged to a specific project. It includes the items of job expense that cannot be charged directly to a particular branch of work, but are required to construct the project. The list of job overhead items is placed on the first of the general estimate sheets under the heading of "Job Overhead" or "General Conditions."

Itemizing each cost gives the estimator a basis for determining the amount of that expense and also provides for comparison of projects. A percentage should not be added to the cost simply to cover this item. It is important that each portion of the estimate can be analyzed for accuracy to determine whether the estimator should bid an item higher or lower next time.

Salaries

This will include the salaries paid to the project superintendent, assistant superintendent, timekeeper, material clerk, all foremen required, and a watchman if one is needed. The cost must include all traveling expenses of these people.

The salaries of the various employees required are estimated per week or per month, and that amount is multiplied by the estimated time it is expected that each will be required on the project. The estimator must be neither overly optimistic nor pessimistic in regard to the time each person will be required to spend on the job.

Temporary Office

The cost of providing a temporary job office for use by the contractor and architect during the construction of the project should also include office expenses such as light, heat, water, telephone, and office equipment. Check the specification for special require-

General Overhead Expenses For One Year

Office Rental	$ 3,600.00
Heat and Electricity	720.00
Telephone	350.00
Office Supplies (inc. stationery)	400.00
Postage	80.00
Insurance (fire and office liability)	350.00
Bookkeeper-Secretary	6,200.00
Estimator-Purchaser	14,500.00
Executive Salaries	17,000.00
Legal Services	500.00
Travel	2,500.00
Advertising, Donations	300.00
Literature	150.00
Club and Association Dues	500.00
Depreciation (office furniture and equipment)	1,000.00
Miscellaneous Expenses	1,500.00
	$49,450.00

Annual business approximately $750,000.00

General Overhead expense = 7% of total annual business

Assuming the project being bid had an estimated cost of about $150,000 the amount

of general overhead chargeable to that project would be $\frac{150,000}{750,000}$ x 49, 450 = $9,890.

Figure 33-1. General overhead expenses for one year.

ments pertaining to the office. A particular size may be required, the architect may require a temporary office for his own use or other requirements may be included.

If the contractor owns the temporary office, a charge is still made against the project for depreciation and return on investment. If the temporary office is rented, the rental cost is charged to the project. Since the rental charges are generally based on a monthly fee, estimate the number of months required carefully. At $250.00 per month, three extra months amount to $750.00 out of the profit of the project. Check whether or not the monthly fee includes setup and return of the office. If not, this must also be included.

Temporary Buildings, Barricades, Enclosures

The cost of temporary buildings includes all toolsheds, workshops, and material storage spaces. The cost of building and maintaining barricades and providing signal lights in conjunction with the barricades must also be included. Necessary enclosures include fences, temporary doors and windows, ramps, platforms, and protection over equipment.

Temporary Utilities

The costs of temporary water, light, power, and heat must also be included. For each of these items the specifications must be carefully read to determine which contractor must arrange for the installation of the temporary utilities and who will pay for the actual amounts of each item used (power, fuel, water).

Water may have to be supplied to all contractors by the general contractor. This information is included in the specifications and must be checked for each project. The contractor may be able to tie into existing water mains. In this case a plumber will have to be hired to make the connection. Sometimes the person who owns the adjoining property will allow use of his water supply, usually for a fee. Water may be drawn from a nearby creek with a pump, or perhaps a water truck will have to be used. No matter where the water comes from, temporary water will be required on many projects and the contractor must include the cost. This is also one of the items the estimator should investigate at the site. Re-

member that many water departments charge for the water used and the estimator will have to estimate the volume of water that will be required.

Electrical requirements may include light and power for the project. This item is sometimes covered in the electrical specifications, and the estimator must review it to be certain that all of the requirements of the project will be met. In small projects it may be sufficient to tap existing power lines run to a meter and string out extension lines through the project, from which lights will be added and power taken off. On large complexes it may be necessary to install poles, transformers, and extensive wiring so that all power equipment being used to construct the project will be supplied. If the estimator finds that the temporary electricity required will not meet his needs, he should first call the architect-engineer and discuss it with him. The architect may decide to issue an addendum revising the specifications. If not, the estimator will have to include the cost in the general construction estimate. All power consumed must be paid for, and how this is handled should also be in the specifications; often the cost is split on a percentage basis among the contractors, and it is necessary to make an allowance for this item also.

Heat is required if the project will be under construction in the winter. Much of the construction process requires maintenance of temperatures at a certain point. The required heat may be supplied by a portable heater using kerosene and LP or natural gas for small projects— the total cost includes the costs of equipment, fuel, and required labor (remember the kerosene type must be filled twice a day). On large projects the heating system is sometimes put into use before the rest of the project is complete; however, the labor costs run quite high since it may require around-the-clock tending. Since most of the heaters use electricity for ignition and to run the fan, the cost of power must also be included and a place to "plug in" provided.

Sanitary Facilities

All projects must provide toilets for the workers. The most common type in use are the portable toilets, which are most often rented. Large projects will require several portable toilets throughout the job.

Drinking Water

The cost of providing drinking water in the temporary office and throughout the project must be included. It is customary to provide cold or ice water throughout the summer. Keep in mind that the estimated cost must include the containers, cups, and someone to service them.

Photographs

Many project specifications require photographs at various stages of the construction and, even if they are not required, it is strongly suggested a camera, film, and flashbulbs be kept at each job site. The superintendent should make use of them at all important phases of the project to record the progress. In addition to the cost of the above item, the cost of developing and any required enlarging of the film should be considered.

Surveys

If a specification requests a survey of the project location on the property, the estimator must include the cost for the survey in his estimate. Check the specifications to see if a licensed surveyor is required, and then ask several local surveyors to submit a proposal.

Cleanup

Throughout the construction's progress, rubbish will have to be removed from the project site. Estimate how many trips will be required and a cost per trip; this means the estimator will have to determine where the rubbish can be dumped.

Before acceptance of the project by the owner, the contractor will have to clean the floors, clean up the job, and in some cases even wash the windows.

Winter Construction

When construction will run into or through the winter, several items of extra cost must be considered, including the cost of temporary enclosures, that of heating the enclosure, that of heated concrete and materials, and the cost of protecting equipment from the elements.

Miscellaneous items that should be contemplated include the possibility of damaging adjacent buildings, such as breaking windows, and the possible undermining of foundations or damages by workmen or equipment. All sidewalks and paved areas that are torn up or damaged during construction must be repaired. Many items of new construction require protection in order that they will not be damaged during construction; among these are wood floors, carpeting, finished hardware, and wall finishes in heavily traveled areas. New work that is damaged will have to be repaired or replaced. Often, no one will admit to damaging an item, and so the contractor must absorb the cost of replacing the item.

The supplementary general conditions should be carefully checked for other requirements that will add to the job overhead expense. Examples of these may be job signs, billboards, building permits, testing the soil and concrete, and written progress reports.

33.4
CHECKLIST

General Overhead (Indirect)

Salaries:

Executives

Secretaries

Estimators

Purchasing agents

Cost-keeping and bookkeeping

Draftsmen

Office personnel

Office:

Rent (or cost of owning)

Electricity

Heat

Water

Office supplies

Postage

Telephone

Furniture and fixture depreciation

Office machine depreciation

Advertising

Literature

Miscellaneous

Club and association dues

Professional services:

Lawyers

Accountants

Architects

Engineers

Vehicle:

Depreciation on company cars

Depreciation on general use vehicles

Insurance: (aside from those required on a particular project)

Fire and theft

Public liability

Automobile and vehicle

Property damage

Workmen's compensation

Social security

Unemployment

Job Overhead (Direct, General Conditions)

Salaries

Superintendent

Assistant superintendent

Timekeepers

Material clerks

Foremen

Watchmen

Traveling expenses

Temporary office:

Number of offices required

Rent (or cost of owning)

Setup and removal

Light

Heat

Water

Telephone

Office supplies

Office equipment depreciation

Bonds:

Bid (security)

Performance

Lien

Labor and material

Insurance:

Fire

Property damage

Windstorm

Automobile and vehicle liability

Public liability

Workmen's compensation

Unemployment

Social security

Boiler

Flood

Theft

Elevator

Utilities:

Temporary heat

Temporary light

Temporary power

Water

Sanitary facilities

Miscellaneous:

Temporary buildings and enclosures

Barricades and signal lights

Photographs

Engineering services

Clean-up

Repair of streets and pavements

Damage to adjoining buildings

Cutting and patching for other trades

Removal and replacement of public utilities

Permits

Licenses

Tools and equipment

Job signs

Pumping and shoring

Dust control

Chapter 34

Profit

34.1
PROFIT

Profit is not included in a chapter with other topics since it is the last thing considered and must be taken separately. Keep in mind that construction is among the tops in the percentage of business failures. Would you believe that some people even forget to add profit?

First, let us understand that by *profit* we mean the amount of money added to the total estimated cost of the project; this amount of money should be clear profit. All costs relating to the project, including project and office overhead and salaries, are included in the estimated cost of the building.

There are probably more different approaches to determining how much profit should be included than could be listed. Each contractor and his estimators seems to have a different approach. A few typical approaches are listed below:

1. Add a percentage of profit to each item as it is estimated, allowing varying amounts for the different items; for example, 8 to 15 percent for concrete work, but only 3 to 5 percent for work subcontracted out.

2. Add a percentage of profit to the total price tabulated for materials, labor, over-head, and equipment. The percent would vary from small jobs to larger jobs (perhaps 20 to 25 percent on a small job and 5 to 10 percent on a larger one), taking into account the accuracy of the take-off and pricing procedures used in the estimate.

3. Various methods of selecting a figure are employed that will make a bid low while not being too low, by trying to analyze all of the variables and other contractors who are bidding.

4. There are "strategies of bidding" that some contractors (and estimators) apply to bidding. Most of the strategies require bidding experience to be accumulated and competitive patterns from past biddings to be used as patterns for future biddings. This will also lend itself nicely to computer operations. Interested in such a strategy? If so, refer to *Construction Contracting* by Richard H. Clough (Wiley-Interscience), 2nd Ed.; 1969, Appendix L.

5. Superstition sometimes plays a part. And why not, since superstition is prevalent throughout our lives? Many contractors and estimators will use only certain numbers to end their bids; for example, some always end with a 7 or take a million dollar bid all the way down to 50 cents.

My own method is to, first, include all of the costs of the project before profit is considered. Then a review of the documents is made to find if the drawings and specifications were clear, if you understood what you were bidding on, and how accurate a take-off was made (it should always be as accurate as possible). The other factors to be considered are the architect-engineer, his reputation, and how he handles his work.

After reviewing the factors, the contractor must decide how much money (profit, over and above salary) he wants to make on this project. This is the amount that should be added to the cost of construction to give the amount of the bid (after it is adjusted slightly to take into account superstitions and strategy types of bids). Exactly what is done at this point, slightly up or down, is an individual matter, but you should definitely know your competitors, keep track of their past bidding practices, and use those against them whenever possible.

Since profit is added at the end of the estimate, the estimator has a pretty good idea of the risks and problems that may be encountered. Discuss these risks thoroughly with other members of the firm. It is far better to bid what you feel is high enough to cover the risks than to neglect the risks, bid low, and lose money. There is sometimes a tendency to "need" or "want" a job so badly that risks are completely ignored. Try to avoid this sort of foolishness—it only invites disaster. If a project entails substantial risk and it is questionable that a profit can be made, consider not even bidding it and let someone else have the heartaches and the loss. Always remember: Construction is a business in which you are supposed to make a fair, reasonable profit.

Conversions

USEFUL CONVERSION FACTORS

Cement

$$1 \text{ sack} = 1 \text{ c.f.} = 94 \text{ pounds}$$
$$1 \text{ barrel} = 4 \text{ sacks} = 376 \text{ pounds}$$

Water

$$1 \text{ U.S. gallon} = 231 \text{ c.i.} = 0.1337 \text{ c.f.} = 8.35 \text{ pounds}$$
$$1 \text{ c.f.} = 7.5 \text{ U.S. gallons} = 62.4 \text{ pounds}$$

Aggregate

$$1 \text{ ton} = 2,000 \text{ pounds} = 19 \text{ c.f.} = 0.70 \text{ c.y.}$$
$$1 \text{ c.y.} = 27 \text{ c.f.} = 2,800 \text{ pounds} = 1.40 \text{ tons (approximate)}$$

Ready-Mix Concrete

$$1 \text{ c.y.} = 27 \text{ c.f.} = 2 \text{ tons} = 4,000 \text{ pounds (approximate)}$$
$$1 \text{ ton} = 2,000 \text{ pounds} = 0.50 \text{ c.y.} = 13.5 \text{ c.f.}$$

Concrete: weights (approximate)

$$1 \text{ c.f.} = 145\# \text{ (heavyweight)}$$
$$1 \text{ c.f.} = 115\# \text{ (lightweight)}$$

CUBE

$V \text{ (Volume)} = l \times w \times h$

CONE

$$V = \frac{\text{Area of } a \times b}{3}$$

CIRCLE

$A = \pi r^2 \text{ or } 0.785 \, d^2$
$\pi = 3.14159$

circumference $= \pi d$

use $\pi = 3.14$

SQUARE

$A = s^2 \text{ or } s \times s \text{ (side squared)}$

CYLINDER

$V = \text{Area of } a \times b$

RECTANGLE

$A \text{ (area)} = b \times h$

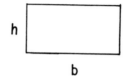

RIGHT TRIANGLE

$$A = \frac{bh}{2}$$

hypotenuse $= c = \sqrt{b^2 + h^2}$

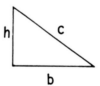

PARALLELOGRAM

$A = bh$

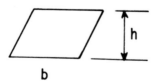

ANY TRIANGLE   TRAPEZOID

$$A = \frac{bh}{2}$$ $$A = \left(\frac{a+b}{2}\right)h$$

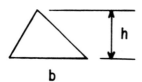   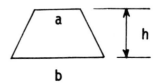

LENGTHS

12 inches = 1 foot
3 feet = 1 yard

AREAS

144 s.i. = 1 s.f.
9 s.f. = 1 s.y.
100 s.f. = 1 square

VOLUME

1,728 c.i. = 1 c.f.
27 c.f. = 1 c.y.

1 c.f. = 7.4850 gallons
1 gallon = 231 c.i.
1 gallon = 8.33 pounds
1 c.f. = 62.3 pounds

1 millimeter (mm) = 0.0394 inches
1 centimeter (cm) = 0.3937 inches
1 decimeter (dm) = 3.937 inches
1 meter (m)   = 39.37 inches
1 meter (m)   = 1.1 yards
1 kilometer (km) = 0.621 mile

1 square cm = 0.155 s.f.
1 square m = 10.764 s.f. = 1.196 s.y.

1 cubic cm = 0.06 c.i.
1 cubic dm = 61.02 c.i.
1 cubic m = 1.308 c.y.

INCHES REDUCED TO DECIMALS

INCHES	DECIMAL
½	0.041
1	0.083
1½	0.125
2	0.167
2½	0.209
3	0.250
3½	0.292
4	0.333
4½	0.375
5	0.417
5½	0.458
6	0.500
6½	0.542
7	0.583
7½	0.625
8	0.667
8½	0.708
9	0.750
9½	0.792
10	0.833
10½	0.875
11	0.917
11½	0.958
12	1.00

COMMON FRACTIONS STATED AS DECIMALS

FRACTION	DECIMAL
1/16	0.0625
1/8	0.125
3/16	0.1875
1/4	0.250
5/16	0.3125
3/8	0.375
7/16	0.4375
1/2	0.500
9/16	0.5625
5/8	0.625
11/16	0.6875
3/4	0.750
13/16	0.8125
7/8	0.875
15/16	0.9375
16/16	1.00

FRACTIONS OF *INCHES* REDUCED TO DECIMALS OF A *FOOT*

FRACTION	DECIMAL
1/16	0.00520
1/8	0.0104
3/16	0.0156
1/4	0.0208
5/16	0.0260
3/8	0.0312
7/16	0.0364
1/2	0.0416
9/16	0.0468
5/8	0.0520
11/16	0.0572
3/4	0.0624
13/16	0.0676
7/8	0.0728
15/16	0.0780
16/16	0.0833

Residence (with Basement)

OUTLINE SPECIFICATIONS

Section 1—Excavation

Topsoil excavation shall be carried 5 feet beyond all walls.

Excavation shall be carried to the depth shown on the drawings. All excess earth, not needed for fill, shall be removed from the premises.

Section 2—Concrete

Concrete for footings shall develop a compressive strength of 2,500 psi at 28 days.

Reinforcing bars shall be A-15 Intermediate Grade.

Section 3—Masonry

Concrete masonry units shall be 16 inches in length.

Mortar joints shall be gray mortar, 3/8-inch thick.

Section 4—Framing

All wood framing shall be sized in accordance with the drawings and building code requirements.

All wood framing shall be kiln dried.

Section 5—Door frames, doors, windows

Door frames shall be ponderosa pine.

Interior doors shall be 1 3/8 inches thick, hollow core, birch.

Exterior door—allow $150 for exterior doors.

Doors shall be of the sizes indicated on the drawings.

Windows shall be wood, of the size and type indicated on the drawings.

Hardware—allow $150.

Section 6—Roofing

Roofing felt shall be 15 lb., 2 layers required.

Roofing shingles shall be asphalt, 235 lbs. per square, class C wind resistant, as manufactured by Johns-Manville or approved equal.

Section 7—Finishes

All exposed exterior wood finish shall receive one coat of approved stain.

All interior drywall finishes shall receive two coats of approved alkyd paint.

Floors shall receive 1/8-inch thick, 12 inches × 2 inches from Royal Stoneglow

Base shall be 1/8-inch vinyl, 2 1/2 inches high.

Section 8—Drywall

Furnish and install in accordance with the drawings.

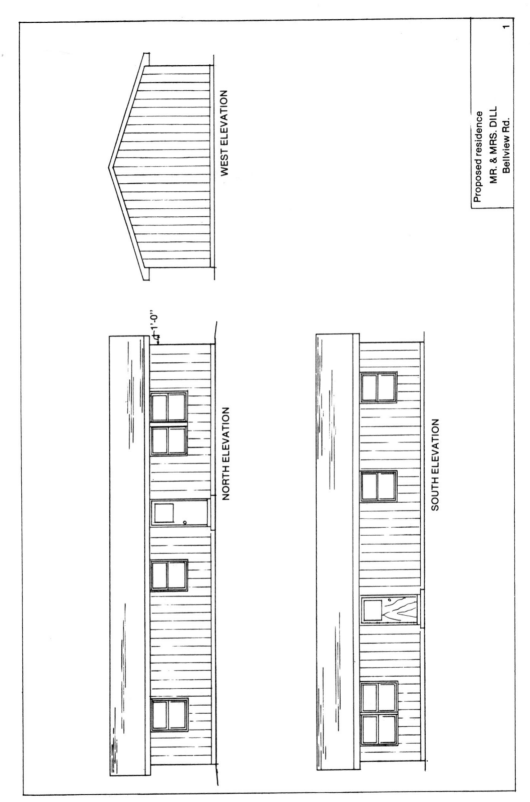

WEST ELEVATION

↕1'-0"

NORTH ELEVATION

SOUTH ELEVATION

Proposed residence
MR. & MRS. DILL
Bellview Rd.

1

542

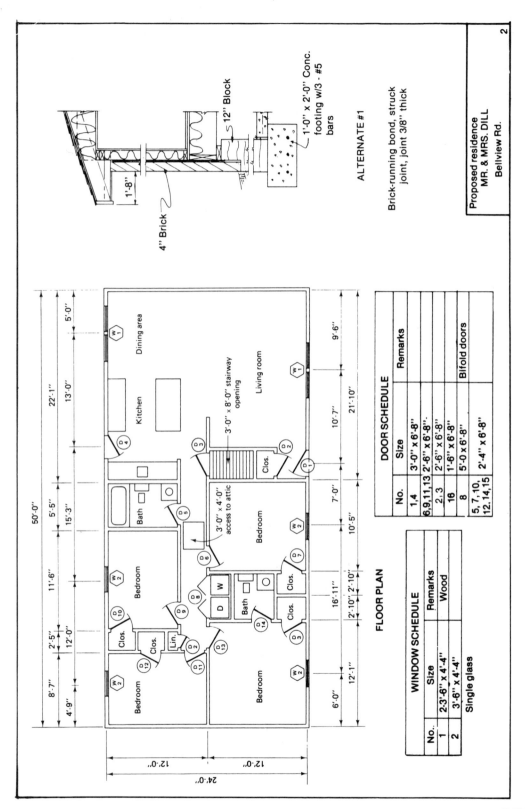

FLOOR PLAN

ALTERNATE #1

Brick-running bond, struck joint, joint 3/8" thick

4" Brick

12" Block

1'-0" x 2'-0" Conc. footing w/3 - #5 bars

3'-0" x 8'-0" stairway opening

3'-0" x 4'-0" access to attic

DOOR SCHEDULE

No.	Size	Remarks
1,4	3'-0" x 6'-8"	
6,9,11,13	2'-6" x 6'-8"	
2,3	2'-6" x 6'-8"	
16	1'-6" x 6'-8"	
8	5'-0 x 6'-8"	Bifold doors
5,7,10, 12,14,15	2'-4" x 6'-8"	

WINDOW SCHEDULE

No.	Size	Remarks
1	2'3-6" x 4'-4"	Wood
2	3'-6" x 4'-4"	

Single glass

Proposed residence
MR. & MRS. DILL
Bellview Rd.

543

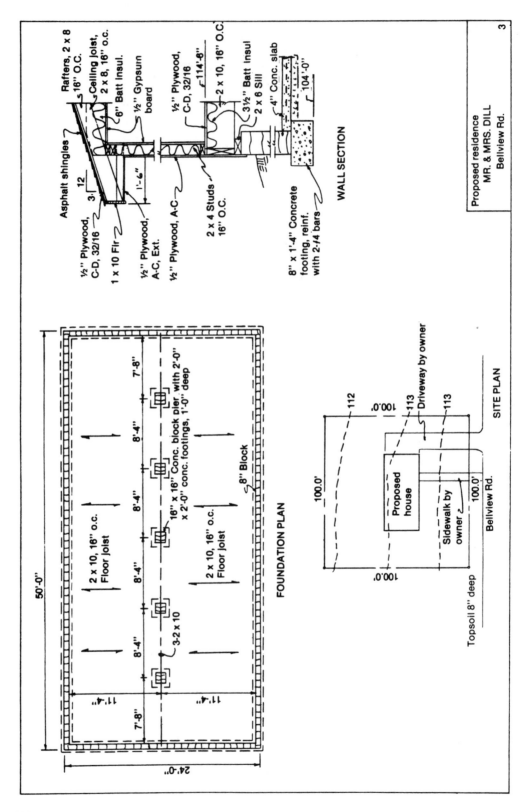

WALL SECTION

Asphalt shingles

Rafters, 2 x 8 16" O.C.

Ceiling Joist, 2 x 8, 16" o.c.

6" Batt Insul.

½" Gypsum board

½" Plywood, C-D, 32/16

½" Plywood, C-D, 32/16

1 x 10 Fir

½" Plywood, A-C, Ext.

½" Plywood, A-C

2 x 4 Studs 16" O.C.

2 x 10, 16" O.C.

3½" Batt Insul

2 x 6 Sill

4" Conc. slab

114'-6"

104'-0"

8" x 1'-4" Concrete footing, reinf. with 2-#4 bars

FOUNDATION PLAN

50'-0"

24'-0"

7'-8"

8'-4"

8'-4"

8'-4"

8'-4"

7'-8"

11'-4"

11'-4"

2 x 10, 16" o.c. Floor Joist

2 x 10, 16" o.c. Floor Joist

16" x 16" Conc. block pier, with 2'-0" x 2'-0" conc. footings, 1'-0" deep

8" Block

3-2 x 10

SITE PLAN

112

100.0'

100.0'

100.0'

100.0'

113

113

Driveway by owner

Proposed house

Sidewalk by owner

Bellview Rd.

Topsoil 8" deep

Proposed residence
MR. & MRS. DILL
Bellview Rd.

3

Residence (Slab on Grade)

OUTLINE SPECIFICATIONS

Section 1—Excavation

Topsoil excavation shall be carried 5 feet beyond all walls.

Excavation shall be carried to the depth shown on the drawings. All excess earth, not needed for fill, shall be removed from the premises.

Section 2—Concrete

Concrete for footings shall develop a compressive strength of 2,500 psi at 28 days.

Reinforcing bars shall be A-15 Intermediate Grade.

Section 3—Masonry

Concrete masonry units shall be 16 inches in length.

Mortar joints shall be gray mortar, 3/8-inch thick.

Section 4—Framing

All wood framing shall be sized in accordance with the drawings and building code requirements.

All wood framing shall be kiln dried.

Section 5—Door frames, doors, windows

Door frames shall be ponderosa pine.

Interior doors shall be 1 3/8 inches thick, hollow core, birch.

Exterior door—allow $150 for exterior doors. Doors shall be of the sizes indicated on the drawings.

Windows shall be wood, of the size and type indicated on the drawings.

Hardware—allow $150.

Section 6—Roofing

Roofing felt shall be 15 lb., 2 layers required. Roofing shingles shall be asphalt, 235 lbs. per square, class C wind resistant, as manufactured by Johns-Manville or approved equal.

Section 7—Finishes

All exposed exterior wood finish shall receive one coat of approved stain.

All interior drywall finishes shall receive two coats of approved alkyd paint.

Floors shall receive 1/8-inch thick, 12 inches × 2 inches from Royal Stoneglow

Base shall be 1/8-inch vinyl, 2 1/2 inches high.

Section 8—Drywall

Furnish and install in accordance with the drawings.

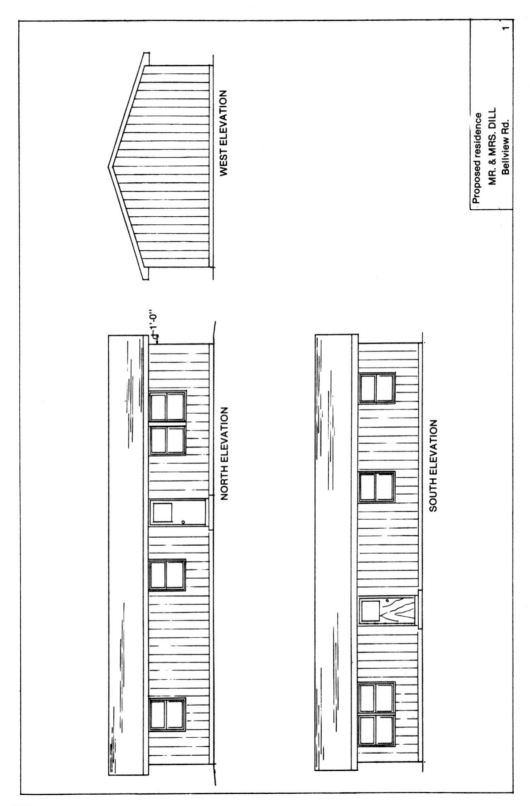

WEST ELEVATION

NORTH ELEVATION

SOUTH ELEVATION

1'-0"

Proposed residence
MR. & MRS. DILL
Bellview Rd.

1

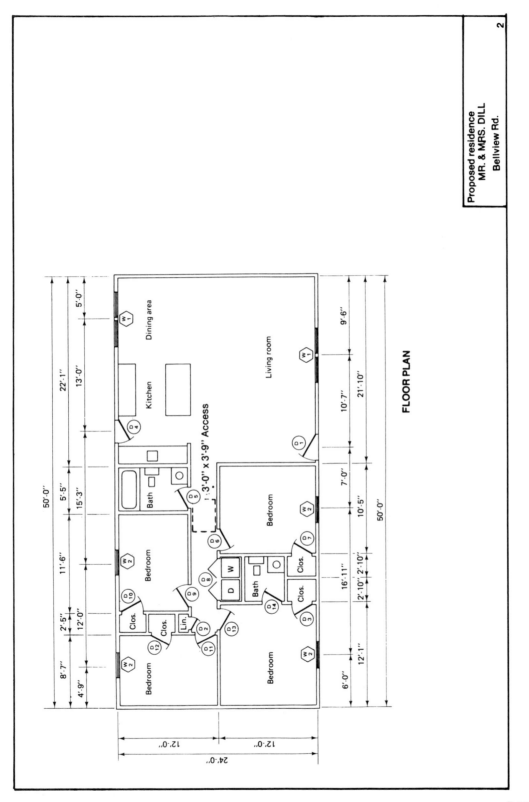

FLOOR PLAN

Proposed residence
MR. & MRS. DILL
Bellview Rd.

2

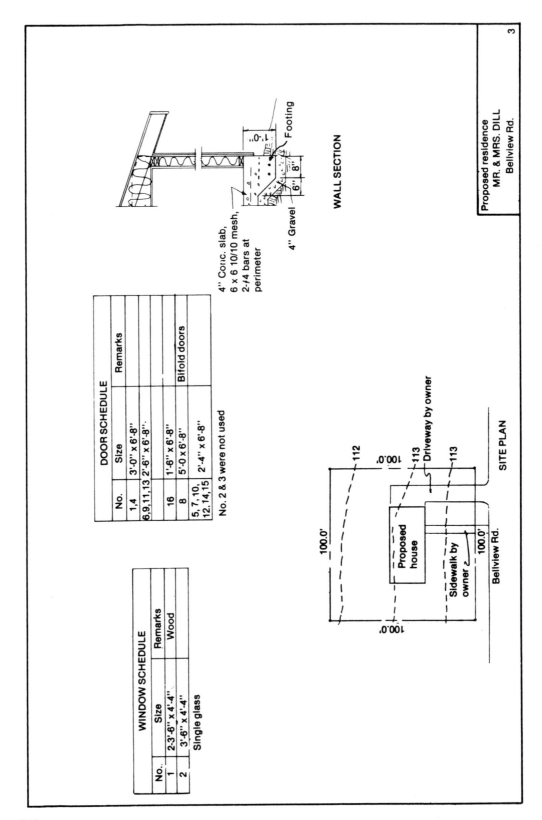

WALL SECTION

Footing

4" Conc. slab,
6 x 6 10/10 mesh,
2-#4 bars at
perimeter

4" Gravel

DOOR SCHEDULE

No.	Size	Remarks
1,4	3'-0" x 6'-8"	
6,9,11,13	2'-6" x 6'-8"	
16	1'-6" x 6'-8"	
8	5'-0 x 6'-8"	Bifold doors
5, 7, 10, 12, 14, 15	2'-4" x 6'-8"	

No. 2 & 3 were not used

WINDOW SCHEDULE

No.	Size	Remarks
1	2-3'-6" x 4'-4"	Wood
2	3'-6" x 4'-4"	

Single glass

SITE PLAN

112

113

Driveway by owner

113

100.0'

100.0'

100.0'

Proposed house

Sidewalk by owner

Bellview Rd.

100.0'

Proposed residence
MR. & MRS. DILL
Bellview Rd.

3

548

Appendix D

Residence (Crawl Space)

OUTLINE SPECIFICATIONS

Section 1—Excavation

Topsoil excavation shall be carried 5 feet beyond all walls.

Excavation shall be carried to the depth shown on the drawings. All excess earth, not needed for fill, shall be removed from the premises.

Section 2—Concrete

Concrete for footings shall develop a compressive strength of 2,500 psi at 28 days.

Reinforcing bars shall be A-15 Intermediate Grade.

Section 3—Masonry

Concrete masonry units shall be 16 inches in length.

Mortar joints shall be gray mortar, 3/8-inch thick.

Section 4—Framing

All wood framing shall be sized in accordance with the drawings and building code requirements.

All wood framing shall be kiln dried.

Section 5—Door frames, doors, windows

Door frames shall be ponderosa pine.

Interior doors shall be 1 3/8 inches thick, hollow core, birch.

Exterior door—allow $150 for exterior doors. Doors shall be of the sizes indicated on the drawings.

Windows shall be wood, of the size and type indicated on the drawings.

Hardware—allow $150.

Section 6—Roofing

Roofing felt shall be 15 lb., 2 layers required.

Roofing shingles shall be asphalt, 235 lbs. per square, class C wind resistant, as manufactured by Johns-Manville or approved equal.

Section 7—Finishes

All exposed exterior wood finish shall receive one coat of approved stain.

All interior drywall finishes shall receive two coats of approved alkyd paint.

Floors shall receive 1/8-inch thick, 12 inches × 2 inches from Royal Stoneglow

Base shall be 1/8-inch vinyl, 2 1/2 inches high.

Section 8—Drywall

Furnish and install in accordance with the drawings.

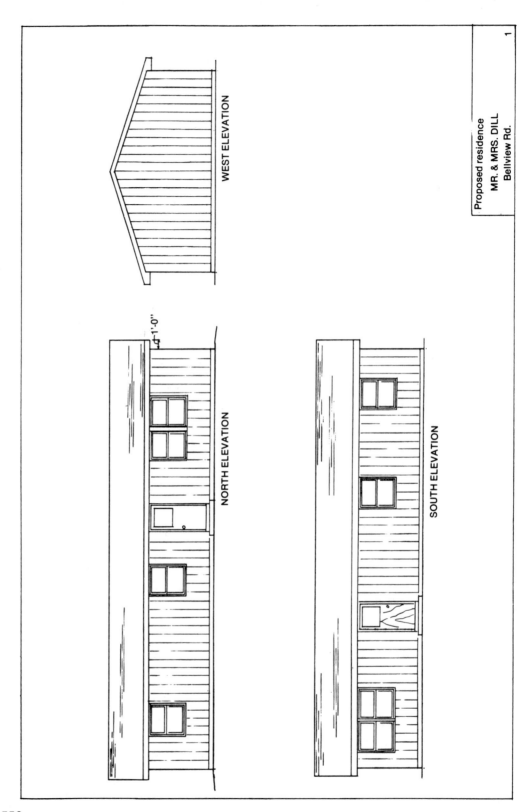

WEST ELEVATION

NORTH ELEVATION

SOUTH ELEVATION

1'-0"

Proposed residence
MR. & MRS. DILL
Bellview Rd.

1

550

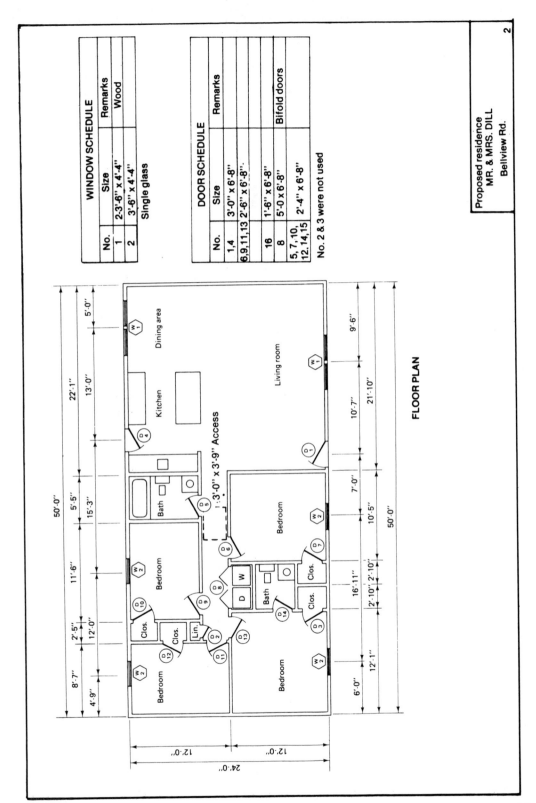

WINDOW SCHEDULE

No.	Size	Remarks
1	2'-3'-6" x 4'-4"	Wood
2	3'-6" x 4'-4"	

Single glass

DOOR SCHEDULE

No.	Size	Remarks
1,4	3'-0" x 6'-8"	
6,9,11,13	2'-6" x 6'-8"	
16	1'-6" x 6'-8"	
8	5'-0 x 6'-8"	
5,7,10, 12,14,15	2'-4" x 6'-8"	Bifold doors

No. 2 & 3 were not used

FLOOR PLAN

Proposed residence
MR. & MRS. DILL
Bellview Rd.

2

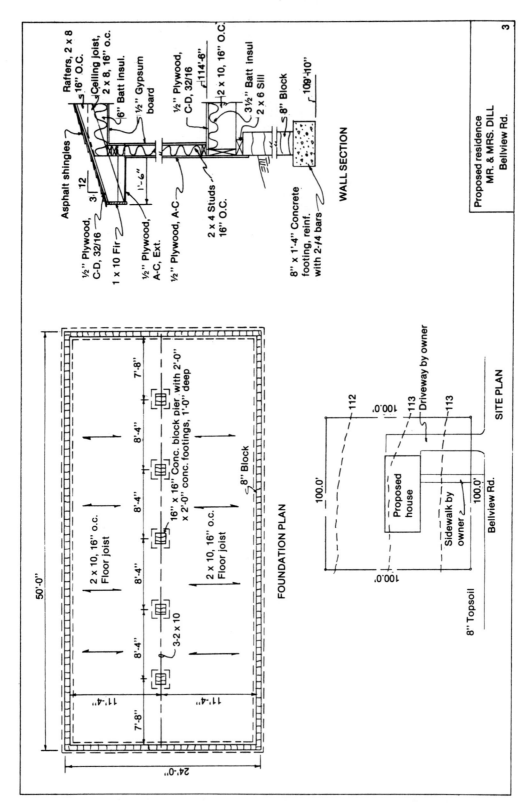

WALL SECTION

Asphalt shingles

Rafters, 2 x 8 16" O.C.

Ceiling joist, 2 x 8, 16" o.c.

6" Batt insul.

½" Gypsum board

½" Plywood, C-D, 32/16

2 x 10, 16" O.C.

3½" Batt Insul

2 x 6 Sill

8" Block

10'-'10"

½" Plywood, C-D, 32/16

1 x 10 Fir

½" Plywood, A-C, Ext.

½" Plywood, A-C

2 x 4 Studs 16" O.C.

8" x 1'-4" Concrete footing, reinf. with 2-/4 bars

12

3

1'-6"

14'-6"

FOUNDATION PLAN

50'-0"

24'-0"

7'-8"   8'-4"   8'-4"   8'-4"   8'-4"   7'-8"

16" x 16" Conc. block pier, with 2'-0" x 2'-0" conc. footings, 1'-0" deep

8" Block

2 x 10, 16" o.c. Floor joist

2 x 10, 16" o.c. Floor joist

3-2 x 10

11'-4"

11'-4"

SITE PLAN

112

100.0'

113

Driveway by owner

113

100.0'

100.0'

100.0'

Proposed house

Sidewalk by owner

Bellview Rd.

8" Topsoil

Proposed residence
MR. & MRS. DILL
Bellview Rd.

3

Common Terms Used in the Building Industry

ADDENDA: Statements or drawings that modify the basic contract documents after the latter have been issued to the bidders, but prior to the taking of bids.

ALTERNATES: Proposals required of bidders reflecting amounts to be added to or subtracted from the basic proposal in the event that specific changes in the work are ordered.

ANCHOR BOLTS: Bolts used to anchor structural members to concrete or the foundation.

APPROVED EQUAL OR: The terms used to indicate that material or product finally supplied or installed must be equal to that specified and as approved by the architect (or engineer).

AS-BUILT DRAWINGS: Drawings made during or after construction, illustrating how various elements of the project were actually installed.

ASTRAGAL: A closure between the two leaves of a double-swing or double-slide door to close the joint. This can also be a piece of molding.

AXIAL: Anything situated around, in the direction of, or along an axis.

BASEPLATE: A plate attached to the base of a column which rests on a concrete or masonry footing.

BAY: The space between column center lines or primary supporting members, lengthwise in a building. Usually the crosswise dimension is considered the *span* or *width module,* and the lengthwise dimension is considered the *bay spacing.*

BEAM: A structural member that is normally subjected to bending loads and is usually a horizontal member carrying vertical loads. (An exception to this is a purlin.) There are three types of beams:

1. *CONTINUOUS BEAM:* A beam that has more than two points of support.

2. *CANTILEVERED BEAM:* A beam that is supported at only one end and is restrained against excessive rotation.

3. *SIMPLE BEAM:* A beam that is freely supported at both ends, theoretically with no restraint.

BEAM AND COLUMN: A primary structural system consisting of a series of beams and columns; usually arranged as a continuous beam supported on several columns with or without continuity that is subjected to both bending and axial forces.

BEAM-BEARING PLATE: Steel plate with attached anchors that is set in top of a masonry wall so that a purlin or a beam can rest on it.

BEARING: The condition that exists whenever one member or component transmits load or stress to another by direct contact in compression.

BENCH MARK: A fixed point used for construction purposes as a reference point in determining the various levels of floor, grade, etc.

BID: Proposal prepared by prospective contractor specifying the charges to be made for doing the work in accordance to the contract documents.

BID BOND: A surety bond guaranteeing that a bidder will sign a contract, if offered, in accordance with his proposal.

BID SECURITY: A bid bond, certified check, or other forfeitable security guaranteeing that a bidder will sign a contract, if offered, in accordance with his proposal.

BILL OF MATERIALS: A list of items or components used for fabrication, shipping, receiving, and accounting purposes.

BIRD SCREEN: Wire mesh used to prevent birds from entering the building through ventilators or louvers.

BOND: Masonry units interlocked in the face of a wall by overlapping the units in such a manner as to break the continuity of vertical joints.

BONDED ROOF: A roof that carries a printed or written warranty, usually with respect to weathertightness, including repair and/or replacement on a prorated cost basis for a stipulated number of years.

BONUS AND PENALTY CLAUSE: A provision in the proposal form for payment of a bonus for each day the project is completed prior to the time stated, and for a charge against the contractor for each day the project remains uncompleted after the time stipulated.

BRACE RODS: Rods used in roofs and walls to transfer wind loads and/or seismic forces to the foundation (often used to plumb building but not designed to replace erection cables when required).

BRIDGING: The structural member used to give lateral support to the weak plane of a truss, joist, or purlin; provides sufficient stability to support the design loads, sag channels, or sag rods.

BUILT-UP ROOFING: Roofing consisting of layers of rag felt or jute saturated with coal tar pitch, with each layer set in a mopping of hot tar or asphalt; ply designation as to the number of layers.

CAMBER: A permanent curvature designed into a structural member in a direction opposite to the deflection anticipated when loads are applied.

CANOPY: Any overhanging or projecting structure with the extreme end unsupported. It may also be supported at the outer end.

CANTILEVER: A projecting beam supported and restrained only at one end.

CAP PLATE: A horizontal plate located at the top of a column.

CASH ALLOWANCES: Sums that the contractor is required to include in his bid and contract sum for specific purposes.

CAULK: To seal and make weathertight the joints, seams, or voids by filling with a waterproofing compound or material.

CERTIFICATE OF OCCUPANCY: Statement issued by the governing authority granting permission to occupy a project for a specific use.

CERTIFICATE OF PAYMENT: Statement by an architect informing the owner of the amount due to a contractor on account of work accomplished and / or materials suitably stored.

CHANGE ORDER: A work order, usually prepared by the architect and signed by the owner or his agent, authorizing a change in the scope of the work and a change in the cost of the project.

CHANNEL: A steel member whose formation is similar to that of a "C" section without return lips; may be used singularly or back-to-back.

CLIP: A plate or angle used to fasten two or more members together.

CLIP ANGLE: An angle used for fastening various members together.

COLLATERAL LOADS: A load, in addition to normal live, wind, or dead loads, intended to cover loads that are either unknown or uncertain (sprinklers, lighting, etc.).

COLUMN: A main structural member used in a vertical position on a building to transfer loads from main roof beams, trusses, or rafters to the foundation.

CONTRACT DOCUMENTS: Working Drawings, Specifications, General Conditions, Supplementary General Conditions, the Owner-Contractor Agreement and all Addenda (if issued).

CURB: A raised edge on a concrete floor slab, roof, or other flat surface.

CURTAIN WALL: Perimeter walls that carry only their own weight and wind load.

DATUM: Any level surface to which elevations are referred (see *Bench Mark*).

DEAD LOAD: The weight of the structure itself, such as floor, roof, framing, and covering members, plus any permanent loads.

DEFLECTION: The displacement of a loaded structural member or system in any direction, measured from its no-load position, after loads have been applied.

DESIGN LOADS: Those loads specified by building codes, state, or city agencies, or owner's or architect's specifications to be used in the design of the structural frame of a building. They are suited to local conditions and building use.

DOOR GUIDE: An angle or channel guide used to stabilize and keep plumb a sliding or rolling door during its operation.

DOWNSPOUT: A hollow section such as a pipe used to carry water from the roof or gutter of a building to the ground or sewer connection.

DRAIN: Any pipe, channel, or trench for which waste water or other liquids are carried off, i.e., to a sewer pipe.

EAVE: The line along the sidewall formed by the intersection of the inside faces of the roof and wall panels; the projecting lower edges of a roof, overhanging the walls of a building.

EQUAL, OR: (See *Approved Equal.*)

ERECTION: The assembly of components to form the completed portion of a job.

EXPANSION JOINT: A connection used to allow for temperature-induced expansion and contraction of material.

FABRICATION: The manufacturing process performed in the plant to convert raw material into finished metal building components. The main operations are cold forming, cutting, punching, welding, cleaning, and painting.

FASCIA: A flat, broad trim projecting from the face of a wall, which may be part of the rake or the eave of the building.

FIELD: The job site or building site.

FIELD FABRICATION: Fabrication performed by the erection crew or others in the field.

FIELD WELDING: Welding performed at the job site, usually with gasoline-powered machines.

FILLER STRIP: Preformed neoprene material, resilient rubber, or plastic used to close the ribs or corrugations of a panel.

FINAL ACCEPTANCE: The owner's acceptance of a completed project from a contractor.

FIXED JOINT: A connection between two members that causes them to act as a single continuous member, providing for transmission of forces from one member to the other without any movement in the connection itself.

FLANGE: That portion of a structural member normally projecting from the edges of the web of a member.

FLASHING: A sheet-metal closure that functions primarily to provide weathertightness in a structure and secondarily to enhance appearance; the metalwork that prevents leakage over windows, door, etc., around chimneys, and at other roof details.

FOOTING: That bottom portion at the base of a wall or column used to distribute the load into the supporting soil.

FOUNDATION: The substructure that supports a building or other structure.

FRAMING: The structural steel members (columns, rafters, girts, purlins, brace rods, etc.) that go together to make up the skeleton of a structure ready for covering to be applied.

FURRING: Leveling up or building out of a part of wall or ceiling by wood, metal, or strips.

GLAZE (GLAZING): The process of installing glass in window and door frames.

GRADE: The term used when referring to the ground elevation around a building or other structure.

GROUT: A mixture of cement, sand, and water used to solidly fill cracks and cavities; generally used under setting plates to obtain solid, uniform, full-bearing surface.

GUTTER: A channel member installed at the eave of the roof for the purpose of carrying water from the roof to the drains or downspouts.

HEAD: The top of a door, window, or frame.

IMPACT LOAD: The assumed load resulting from the motion of machinery, elevators, cranes, vehicles, and other similar moving equipment.

INSTRUCTIONS TO BIDDERS: A document stating the procedures to be followed by bidders.

INSULATION: Any material used in building construction for the protection from heat or cold.

INVITATION TO BID: An invitation to a selected list of contractors furnishing information on the submission of bids on a subject.

JAMB: The side of a door, window, or frame.

JOIST: Closely spaced beams supporting a floor or ceiling. They may be wood, steel, or concrete.

KIP: A unit of weight, force, or load equal to 1,000 pounds.

LAVATORY: A bathroom-type sink.

LIENS: Legal claims against an owner for amounts due those engaged in or supplying materials for the construction of the building.

LINTEL: The horizontal member placed over an opening to support the loads (weight) above it.

LIQUIDATED DAMAGES: An agreed-to sum chargeable against the contractor as reimbursement for damages suffered by the owner because of contractor's failure to fulfill his contractual obligations.

LIVE LOAD: The load exerted on a member or structure due to all imposed loads except dead, wind, and seismic loads. Examples include snow, people, movable equipment, etc. This type of load is movable and does not necessarily exist on a given member of structure.

LOADS: Anything that causes an external force to be exerted on a structural member. Examples of different types are:

1. *DEAD LOAD:* In a building, the weight of all permanent constructions, such as floor, roof, framing, and covering members.

2. *IMPACT LOAD:* The assumed load resulting from the motion of machinery, elevators, craneways, vehicles, and other similar kinetic forces.

3. *ROOF LIVE LOAD:* All loads exerted on a roof (except dead, wind, and lateral loads) and applied to the horizontal projection of the building.

4. *SEISMIC LOAD:* The assumed lateral load due to the action of earthquakes and acting in any horizontal direction on the structural frame.

5. *WIND LOAD:* The load caused by wind blowing from any horizontal direction.

LOUVER: An opening provided with one or more slanted, fixed, or movable fins to allow flow of air, but to exclude rain and sun or to provide privacy.

MULLION: The large vertical piece between windows. (It holds the window in place along the edge with which it makes contact.)

NONBEARING PARTITION: A partition which supports no weight except its own.

PARAPET: That portion of the vertical wall of a building that extends above the roof line at the intersection of the wall and roof.

PARTITION: A material or combination of materials used to divide a space into smaller spaces.

PERFORMANCE BOND: A bond that guarantees to the owner, within specified limits, that the contractor will perform the work in accordance with the contract documents.

PIER: A structure of masonry (concrete) used to support the bases of columns and bents. It carries the vertical load to a footing at the desired load-bearing soil.

PILASTER: A flat rectangular column attached to or built into a wall masonry or pier; structurally, a pier, but treated architecturally as a column with a capital, shaft, and base. It is used to provide strength for roof loads or support for the wall against lateral forces.

PRECAST CONCRETE: Concrete that is poured and cast in some position other than the one it will finally occupy; cast either on the job site and then put into place or away from the site to be transported to the site and erected.

PRESTRESSED CONCRETE: Concrete in which the reinforcing cables, wires, or rods are tensioned before there is load on the member.

PROGRESS PAYMENTS: Payments made during progress of the work, on account, for work completed and/or suitably stored.

PROGRESS SCHEDULE: A diagram showing proposed and actual times of starting and completing the various operations in the project.

PUNCH LIST: A list prepared by the architect or engineer of the contractor's uncompleted or work to be corrected.

PURLIN: Secondary horizontal structural members located on the roof extending between rafters, used as (light) beams for supporting the roof covering.

RAFTER: A primary roof support beam usually in an inclined position, running from the tops of the structural columns at the eave to the ridge or highest portion of the roof. It is used to support the purlins.

RECESS: A notch or cutout, usually referring to the blockout formed at the outside edge of a foundation, and providing support and serving as a closure at the bottom edge of wall panels.

REINFORCING STEEL: The steel placed in concrete to carry the tension, compression, and shear stresses.

RETAINAGE: A sum withheld from each payment to the contractor in accordance with the terms of the Owner-Contractor Agreement.

ROLLING DOORS: Doors that are supported on wheels that run on a track.

ROOF OVERHANG: A roof extention beyond the end or the side walls of a building.

ROOF PITCH: The angle or degree of slope of a roof from the eave to the ridge. The pitch can be found by dividing the height, or rise, by the span; for example, if the height is 8 feet and the span is 16 feet, the pitch is 8/16 or ½, and the angle of pitch is 45°. (See *Roof Slope*.)

ROOF SLOPE: The angle that a roof surface makes with the horizontal. Usually expressed as a certain rise in 12 inches of run.

SANDWICH PANEL: An integrated structural covering and insulating component consisting of a core material with inner and outer metal or wood skins.

SCHEDULE OF VALUES: A statement furnished to the architect by the contractor reflecting the amounts to be allotted for the principal divisions of the work. It is to serve as a guide for reviewing the contractor's periodic application for payment.

SEALANT: Any material that is used to close up cracks or joints.

SEPARATE CONTRACT: A contract between the owner and a contractor other than the general contractor for the construction of a portion of a project.

SHEATHING: Rough boarding (usually plywood) on outside of a wall or roof over which is placed siding or shingles.

SHIM: A piece of steel used to level or square beams or column baseplates.

SHIPPING LIST: A list that enumerates by part, number, or description each piece of material to be shipped.

SHOP DRAWINGS: Drawings that illustrate how specific portions of the work shall be fabricated and/or installed.

SILL: The lowest member beneath an opening such as a window or door; also, the horizontal framing members at floor level, such as sill girts or sill angles; the member at the bottom of a door or window opening.

SILL, LUG: A sill that projects into the masonry at each end of the sill. It must be installed as the building is being erected.

SILL, SLIP: A sill that is the same width as the opening—it will slip into place.

SKYLIGHT: An opening in a roof or ceiling for admitting daylight; also, the reinforced plastic panel or window fitted into such an opening.

SNOW LOAD: In locations subject to snow loads, as indicated by the average snow depth in the reports of the United States Weather Bureau, the design loads shall be modified accordingly.

SOFFIT: The underside of any subordinate member of a building, such as the undersurface of a roof overhang or canopy.

SOIL BORINGS: A boring made on the site in the general location of the proposed building to determine soil type, depth of the various types of soils, and water table level.

SOIL PRESSURE: The allowable soil pressure is the load per unit area a structure can safely exert on the substructure (soil) without exceeding reasonable values of footing settlements.

SPALL: A chip or fragment of concrete that has chipped, weathered, or otherwise broken from the main mass of concrete.

SPAN: The clear distance between supports of beams, girders, or trusses.

SPANDREL BEAM: A beam from column to column carrying an exterior wall and/or the outermost edge of an upper floor.

SPECIFICATIONS: A statement of particulars of a given job as to size of building, quality and performance of men and materials to be used, and the terms of the contract. A set of specifications generally indicates the design loads and design criteria.

SQUARE: One hundred square feet.

STOCK: A unit that is standard to its manufacturer. It is not custom-made.

STOOL: A shelf across the inside bottom of a window.

STUD: A vertical wall member to which exterior or interior covering or collateral material may be attached. Load-bearing studs are those which carry a portion of the loads from the floor, roof, or ceiling above as well as the collateral material on one or both sides. Nonload-bearing studs are used to support only the attached collateral materials and carry no load from the floor, roof, or ceiling above.

SUBCONTRACTOR: A separate contractor for a portion of the work (hired by the general contractor).

SUBSTANTIAL COMPLETION: For a project or specified area of a project, the date when the construction is sufficiently completed in accordance with the contract documents, as modified by any change orders agreed to by the parties, so that the owner can occupy the project or specified area of the project for the use for which it was intended.

SUPPLEMENTARY GENERAL CONDITIONS: One of the contract documents, prepared by the architect, that may modify provisions of the General Conditions of the contract.

TEMPERATURE REINFORCING: Lightweight deformed steel rods or wire mesh placed in concrete to resist possible cracks from expansion or contraction due to temperature changes.

TIME OF COMPLETION: The number of days (calendar or working) or the actual date by which completion of the work is required.

TRUSS: A structure made up of three or more members, with each member designed to carry basically a tension or a compression force. The entire structure in turn acts as a beam.

VENEER: A thin covering of valuable material over a less expensive body; for example, brick on a wood frame building.

WAINSCOT: Protective or decorative covering applied or built into the lower portion of a wall.

WALL BEARING: In cases where a floor, roof, or ceiling rests on a wall, the wall is designed to carry the load exerted. These types of walls are also referred to as load-bearing walls.

WALL COVERING: The exterior wall skin consisting of panels or sheets and including their attachment, trim, fascia, and weather sealants.

WALL NONBEARING: Wall not relied upon to support a structural system.

WATER CLOSET: More commonly known as a toilet.

WORKING DRAWING: The actual plans (drawings and illustrations) from which the building will be built. They show how the building is to be built and are included in the contract documents.

Index